MOLECULAR BIOLOGY OF HEMOPOIESIS

ADVANCES IN EXPERIMENTAL MEDICINE AND BIOLOGY

Recent Volumes in this Series

Volume 234
BIOLOGY OF GROWTH FACTORS: Molecular Biology, Oncogenes, Signal Transduction, and Clinical Implications
Edited by Jeffrey E. Kudlow, David H. MacLennan, Alan Bernstein, and Avrum I. Gotlieb

Volume 235
CENTRAL D_1 DOPAMINE RECEPTORS
Edited by Menek Goldstein, Kjell Fuxe, and Irving Tabachnick

Volume 236
NEURORECEPTORS AND SIGNAL TRANSDUCTION
Edited by Shozo Kito, Tomio Segawa, Kinya Kuriyama, Masaya Tohyama, and Richard W. Olsen

Volume 237
HISTOPHYSIOLOGY OF THE IMMUNE SYSTEM: The Life History, Organization, and Interactions of Its Cell Populations
Edited by Sigbjørn Fossum and Bent Rolstad

Volume 238
BIOTECHNOLOGICAL APPLICATIONS OF LIPID MICROSTRUCTURES
Edited by Bruce Paul Gaber, Joel M. Schnur, and Dennis Chapman

Volume 239
HOST DEFENSES AND IMMUNOMODULATION TO INTRACELLULAR PATHOGENS
Edited by Toby K. Eisenstein, Ward E. Bullock, and Nabil Hanna

Volume 240
PROTEASES: Potential Role in Health and Disease II
Edited by Walter H. Hörl and August Heidland

Volume 241
MOLECULAR BIOLOGY OF HEMOPOIESIS
Edited by Mehdi Tavassoli, Esmail D. Zanjani, Joao L. Ascensao, Nader G. Abraham, and Alan S. Levine

---

MOLECULAR BIOLOGY OF HEMOPOIESIS

Edited by

Mehdi Tavassoli
Veterans Administration Medical Center
University of Mississippi
Jackson, Mississippi

Esmail D. Zanjani
Veterans Administration Medical Center
University of Nevada
Reno, Nevada

Joao L. Ascensao and Nader G. Abraham
New York Medical College
Valhalla, New York

and

Alan S. Levine
National Heart, Lung, and Blood Institute
Division of Blood Diseases and Resources
Bethesda, Maryland

PLENUM PRESS • NEW YORK AND LONDON

Library of Congress Cataloging in Publication Data

Symposium on Molecular Biology of Hemopoiesis (3rd: 1987: Rye Brook, N.Y.)
Molecular biology of hemopoiesis / edited by Mehdi Tavassoli . . . [et al.].
p. cm.—(Advances in experimental medicine and biology; v. 241)
"Proceedings of the Third Annual Symposium on Molecular Biology of Hemopoiesis, held November 6–7, 1987, in Rye Brook, New York"—T.p. verso.
Dedicated to the centennial of the National Institutes of Health.
Includes bibliographies and index.
ISBN 978-1-4684-5573-1 ISBN 978-1-4684-5571-7 (eBook)
DOI 10.1007/978-1-4684-5571-7
1. Hematopoiesis—Congresses. 2. Molecular biology—Congresses. 3. National Institutes of Health (U.S.)—Congresses. I. Title. II. Series.
[DNLM: 1. Hematopoiesis—congresses. 2. Hematopoietic Stem Cells—congresses. 3. Molecular biology—congresses. W1 AD559 v. 241 / WH 140 S9894m 1987]
QP92.S95 1987
612′.41—dc19
DNLM/DLC 88-28836
for Library of Congress CIP

Proceedings of the Third Annual Symposium on Molecular Biology of Hemopoiesis, held November 6–7, 1987, in Rye Brook, New York

Softcover reprint of the hardcover 1st edition 1988
A Division of Plenum Publishing Corporation
233 Spring Street, New York, N.Y. 10013

PREFACE

Perhaps no scientific field in recent years has gained in techniques and applications as much as molecular biology, and it is certainly no exaggeration to say that among all the applications of molecular biology, hematology in general, and hemopoiesis in particular, have benefited most. Owing to the applications of molecular biology, we now live in a golden era of hemopoiesis. Our understanding of the intricate regulatory system in hemopoiesis has vastly expanded. The potential for future exploration is ever expanding, and finally, the possibility of gene manipulation, has provided the promise of fundamental treatment and "cure" of many genetic disorders involving hemopoietic cells.

In the ambiance of this rapidly moving scientific era, the necessity for review of what is being accomplished and where the technical potential is taking us needs no argument. This volume presents the proceedings of the third annual symposium on the Molecular Biology of Hemopoiesis, held in the Rye Town Hilton, New York, November 6 and 7, 1987, under the auspices of New York Medical College, Valhalla, New York. The fact that this was the third regular symposium covering this area in itself testifies to a need for exchange of the rapidly developing knowledge in this area. But this third symposium also coincided with the centennial of the National Institutes of Health and consequently the symposium was dedicated to this festive occasion. Biomedical scientists are an argumentative lot; but no one can argue that without generous help and direction from NIH, the impressive advances of the past century would not have been possible in biomedical sciences. Hence, it is only appropriate that in the present volume Dr. David Badman of NIH provides us with a historical role of NIH in biomedical research and Dr. Alan Levine of NIH concludes the conference with some remarks on the future perspective of biomedical research and the role of NIH.

The organizing committee are particularly indebted to Dr. Richard Levere, Chairman of the Department of Medicine at New York Medical College, without whose support this meeting would have not been possible. The organizers are also grateful to Leukemia Society of America (Westchester Chapter); Toyobo Co., Ltd.; Merck, Sharp and Dohme; Coulter Electronics; and Upjohn Laboratories for their financial support.

The Editors

CONTENTS

GROWTH FACTORS

CLOSING REMARKS

President Franklin D. Roosevelt is dedicating the present campus of NIH on October 31, 1940.

HISTORICAL ROLE OF THE NATIONAL INSTITUTES OF HEALTH IN BIOMEDICAL RESEARCH

David G. Badman

Hematology Program Director
Division of Kidney, Urologic, and Hematologic Diseases
National Institutes of Diabetes and Digestive and Kidney Diseases
National Institutes of Health
Bethesda, Maryland 20892

INTRODUCTION

First let me express my sincere appreciation for the opportunity to open this conference. I am pleased to be able to participate, and to represent the National Institutes of Health in its centennial year.

Today I would like to remind you of the origins of NIH in 1887, and of the very specific and circumscribed mission at that time. I want to remark upon its evolution and growth, and to talk about NIH today.

BEGINNINGS OF THE NIH

You may know that what now is the National Institute of Health traces its origin to the one-room Laboratory of Hygiene on Staten Island in New York Harbor. The Laboratory of Hygiene was a part of the Marine Hospital Services, which had the responsibility to provide relief of sick and disabled seamen. The Laboratory of Hygiene was established in 1887 because of the danger of communicable diseases, such as yellow fever, cholera, and tuberculosis. Because most physicians believed that these diseases were brought into the United States by immigrants, government health-screening facilities were located at major ports of entry, and the facility on Staten Island was one of them. Dr. Joseph J. Kinyoun was appointed its first director.

NIH historians have documented the Congressional hearing which started the Laboratory of Hygiene on the path to becoming the NIH. In 1888, a Congressional committee hoped to establish a competing bureau, a turf battle which is not uncommon between Congressional committees. Also not uncommonly for administrations, the administration of President Grover Cleveland wished to head off the establishment of additional bureaucratic structure, and sent the Surgeon General to testify against the proposal on February 24, 1988. The Surgeon General testified that the Marine Hospital Service already was performing every function that the proposed new bureau might be asked to carry out. In response to a question about research from one of the committee members, the Surgeon

General replied that the Laboratory of Hygiene already was carrying out research, and, only weeks earlier, had published research reports on cholera (using equipment costing $300, which had been set up in the attic of the lab). This satisfied the committee, thus heading off a new bureau, and making the work of the laboratory public for the first time.

In 1891, the laboratory was moved to Washington, DC, to one floor of a Capitol Hill building, and was renamed the Hygienic Laboratory. The little laboratory soon began to make an impact in research. After 12 years, it was moved to its own building, which it occupied for 35 years. Two million dollars a year was provided in the mid-30's for "diseases and sanitation investigations", and the support of President Roosevelt was obtained. In 1938 the present facility in Bethesda was begun, on 45 acres of land donated by Mr. and Mrs. Luke Wilson of Bethesda, providing the nucleus of what eventually would become the 306.4 acre NIH "reservation". President Roosevelt dedicated the original NIH buildings at the new location on October 31, 1940.

At the time of the dedication of the new facility, the NIH was a free-standing Government laboratory of limited size. With the needs for additional knowledge on dealing with the health problems of the armed forces in World War II, the Government turned to every available resource. Partnerships were formed with academic institutions and independent laboratories for the conduct of biomedical research. Grants and contracts were awarded to these institutions, for the support of projects carried out by scientists employed there. This effort was highly successful, and it resulted in the emergence of a new Federal science policy that emphasized the social and economic value of health research as a Government activity. The partnership with non-Federal scientists was recognized as one of the most effective means for conducting the search for new knowledge.

At the end of World War II, 250 health-related projects in progress at University medical schools and independent laboratories, administered by the Office of Scientific Research and Development, were transferred to the NIH for continuation. This transfer amounted only to $1 million, but it marked the beginning of today's $5 billion NIH extramural research and training program. In house, or intramural, research, continued at NIH, and through it, many of the leaders of the US research effort have been trained. It represents the largest laboratory of its kind in the world. However, the extramural program has grown even more, so that today, nearly 90% of the NIH budget is awarded to investigators through grants, cooperative agreements, and contracts.

MODERN STRUCTURE OF THE NIH

At about the time of the move to Bethesda, the process of establishing categorical Institutes for special purposes began. In August, 1937, the National Cancer Institute was established, the National Heart Institute in June, 1948, and National Institute of Arthritis and Metabolic Diseases in August, 1950. The process continues into the present, with the 1986 reestablishment of the programs of the National Institute of Arthritis, Diabetes, and Digestive and Kidney Diseases (NIADDK) as two Institutes, the National Institute of Arthritis and Musculoskeletal and Skin Diseases (NIAMS) and the National Institute of Diabetes and Digestive and Kidney Diseases (NIDDK). Today, the NIH is organized into an Office of the Director, 12 research Institutes, one research Division, the Library of Medicine, the Clinical Center, and several support Divisions, including the Division of Research Grants.

The Institutes, in general, are organized into intramural and extramural programs, with the extramural programs responsible for awarding and administering support mechanisms of a variety of kinds. Most have review staff, to supplement the review support of the Division of Research Grants, program staff, who participate in the funding process, and are responsible for the scientific administration of the programs, and grants management/contracts management staff, for the budget and business aspects of the awards.

RECENT FUNDING TRENDS

You all are aware of the modern role of the NIH in providing funds for the research projects that you wish to carry out. That funding generally has followed an upward course, by all measures applied. The Director of NIH, Dr. James Wyngaarden, reported in a 1987 article in Science that, in 1970, NIH funded 10,000 research grants, 15,970 in 1982, and 19,300 awards in 1986. The number of new and competing renewal awards also has risen from 5,027 in fiscal year 1982 to more than 6,000 in fiscal year 1987. This has been achieved under a policy of stabilization which has emphasized investigator-initiated research projects, and a reduced emphasis on contracts and centers. Thus, the NIH budget fraction devoted to basic research has risen from 45% in 1970 to 52% in 1980 and 63% in 1986.

The average cost of research grants has risen with inflation, to $151,847 in 1986, but in constant dollars, it is equivalent to that in 1972. Of course, we all recognize that there is much concern about the success rate. Dr. Wyngaarden has pointed out that, until the mid-1970s, 70-75% of submitted applications were recommended for approval by the study sections and councils. Typically, NIH funded 1/2 of these, for a 50% award rate, and a payline of 225-250. In recent years, there has been a steady shift in the behavior of the study sections, with 90% of submitted applications now recommended for approval, an award rate of 35-40%, and the payline 170-180. The number of applications has increased from 7,570 in 1970 to 15,858 in 1986. Many more applications receive high enthusiasm ratings than previously, for whatever reason, making it difficult to compare paylines. The success rate, however, has not changed significantly, since a 38% award rate when 90% of applications are approved (34% success rate) is comparable to the old 50% award rate when 70% of the applications are approved (35% success rate).

MECHANISMS FOR RESEARCH FUNDING

The traditional regular research project grant (RO1) remains the mainstay of the NIH funding mechanisms. The clear intent of the Congress in providing NIH appropriations is to fund as many of these as is possible, consistent with available resources. Additional types of mechanisms have been developed to satisfy special needs. Individual and Institutional National Research Service Awards (F32 and T32, respectively) are offered by NIH training programs. In order to assure that new investigators have appropriate opportunities for funding, the FIRST award was developed, which limits the use of track record in evaluation of the application. The Physician Scientist award and the Clinical Investigator award are designed to give intensive research experience to individuals with clinical training. Research Career Development Awards (RCDA), while somewhat more restricted than previously, still are important mechanisms for the development of both Ph.D. and medical degree holders in the early phases of their research careers. MERIT awards were initiated this past year, in order to provide a way to ease the application-writing burden of selected productive senior investigators in mid-career. MERIT awards allow an additional five years of support, based on satisfactory progress, rather than a complete competing renewal application. An additional, and under-utilized, mechanism is the Senior Fellow award (F33), which is designed to provide experienced scientists an opportunity to change research career direction, to broaden their scientific background, or to acquire new research capabilities.

The research funding for different areas of science, and the availability of funds under the variety of mechanism used, does vary from year to year. Even within each fiscal year, funding varies considerably, particularly in recent years. Therefore, it is essential that you seek out NIH representatives at various stages of your application process, beginning as early as when you are making initial plans to formulate an application. They can advise you of the programs most likely to have an interest in your application, the type of application best suited for the goal you wish to accomplish, and the format which would be most advantageous for presenting your plans. Also, since the policies of different Institutes may differ for certain kinds of applications, such as program projects, or training or career awards, you need to obtain advice from a program official from the Institute most likely to receive the assignment. NIH staff can steer you in the right direction. Also, you need to be in close contact with the appropriate NIH staff all through the application and review process, because current information can be critical to making good decisions on your strategy.

IMPORTANCE OF ACTIVE PARTICIPATION BY THE SCIENTIFIC COMMUNITY

The process of developing and administering support programs, making decisions on funding of applications, and ensuring that programs are maintaining the proper emphasis in changing times, demands much more from the scientific community than simply sending in applications and waiting to hear about funding. This system has functioned as well as it has only through the enthusiastic and dedicated involvement of people like yourselves in the whole process, which includes providing local quality control before an application is submitted, sitting on Institutional Review Boards, monitoring the use of human subjects or animals, and membership on peer review committees of NIH, whether standing study sections or ad hoc groups set up by the Institutes. Some of you may be members of National Advisory Councils of the Institutes, or may assist with other advisory committees, which are essential to the process of development of initiatives to meet new, urgent needs. In these days of rapid scientific advances and enormous opportunities, accompanied by rising costs of doing research and increased competition for the Federal dollar, the active participation of active scientists in helping to plan Program directions is essential. This involvement is vital to NIH's function, and is a form of citizen participation which is a model for the whole government.

Individuals always have had a considerable impact on the way that NIH functions, and on its programs and processes. The Institutes are receptive to the views of individuals, whether solicited or unsolicited, about areas needing emphasis or de-emphasis, new mechanisms which might be needed, such as for training new investigators, or for dissemination of information, through workshops or conferences. It is not uncommon for an individual investigator or group of investigators to work through Institute staff to visit and make a presentation about an area which needs more emphasis. Institute staff are very receptive to this kind of contact, and I would encourage you to make use of opportunities to interact with them.

Finally, meetings such as this one, which highlight areas of rapid advance, present a chance to influence the directions of research funding, and programmatic balance. The work that is presented here, as well as associations and collaborations which will grow out of it, certainly will have an impact later on that will be expressed in directions of NIH programs.

In conclusion, the NIH and the scientific and medical community, through all of their interactions at all levels, have produced improvements in health and medical care over the past century, in a manner which is difficult to match in any other areas of progress. This has been due in many ways to the close cooperation over these years between NIH and the members of the scientific community. Continuation of this cooperation surely will lead to even more spectacular progress in the next century.

REFERENCES

Wyngaarden, J.B. 1987. The National Institutes of Health in its Centennial Year. Science 237: 869-874.

A TRIBUTE TO THE NATIONAL INSTITUTE OF HEALTH

William Curry Moloney

Dr. Abraham, participants in this symposium and guests, I am delighted to be present on this occasion to honor the 100th anniversary of the NIH. This morning Dr. Badman described the history and the great achievements of the Institute and I am not going to repeat or amplify his remarks. When Dr. Abraham invited me to speak at this dinner, I was loathe to accept on the grounds that I am not a bench researcher nor have I been an in house member of the NIH staff. However, I represent a vast group of clinical investigators, teachers of medicine and practicing physicians and all of us owe a tremendous debt of gratitude to the NIH.

In a way I am in a unique position since I have been working in the field of hematology for the past 52 years, during more than half the life span of the NIH, and certainly these past 5 decades have been the most exciting and productive in medical history. Therefore, I would like tonight to speak in a personal way about the tremenous changes that have taken place over the past 50 years and specifically in my own field of hematology. I began to specialize in hematology in 1936 at a time when there were no boards of internal medicine and hematology was not a recognized specialty. Fortunately early in my career I joined the staff of a teaching hospital associated with Tufts Medical School and was able as an outpatient physician to establish a small hematology laboratory. Since I carried out all my own blood studies, including the newly discovered prothrombin test, I was welcomed especially by surgeons and obstetricians. Moreover, I owned and could use the only available bone marrow aspiration needle and this became an important addition to my armamentarium. In the decade prior to WWII progress in hematological research was relatively slow except in the field of the anemias, especially pernicious anemia, and until 1948 there was little opportunity or support for research or research fellowships. It is difficult to realize today that in this period sulfur drugs were just appearing, there were no antibiotics and no blood banks. Under the pressure of the war, penicillin was manufactured and made available in quantity, anticoagulants and developments in blood typing made possible blood banks. Production of plasma for the armed forces led to methods for separation of plasma proteins and paved the way for fundamental research on plasma fractions.

Until after WWII few important contributions had been made toward solving the problems of the pathogenesis and management of hematologic malignancies, an area in which I have had a special interest for the past 40 years. Accidental discovery during the war of the effectiveness of nitrogen mustard was followed by the synthesis and use of antifolic acid compounds by Farber and Subarow. Subsequently 6 MP and other antimetabolites were synthesized to be followed by ARA c and the anthracyclines. Of direct interest to my own relationship to the NIH-NCI were developments in the field of leukemogenesis in the 1960 decade and the establishment of the viral oncology program. Following the invention of the electron microscope, viral particles which were well recognized in poultry and mice as leukemogenic agents, began to be discovered in cows milk and in domestic animals including cats. This led to tremendous activity at NIH and a project (SALES) Special Animal Leukemia Ecology Studies, was established. The steering committee was made up of chiefly virologists, including several distinguished veterinarians. John B. Moloney was chief of the viral oncology section and as you may recall, Dick Rauscher succeeded John Moloney and later became Director of the NCI. I became a member of the SALES committee in 1964 and served on it until shortly before the program was disbanded. Much was expected of viral research and considerable political pressure was exerted to produce quick results in the identification and isolation of human leukemia viruses. The hope was of course to produce a vaccine or other methods of treatment for leukemia in man. Unfortunately these expectations did not materialize, the program came under severe criticism and was disbanded. Failure, however was not due to any lack of effort on either the part of the NCI or the researchers and programs it supported. A great body of excellent research resulted from the SALES program, e.g. identification, classification and characterization of RNA viruses, work on feline leukemia viruses, development of elegant methods such as DNA probes, were among many of the invaluable contributions; not to mention the training of cadres of young scientists who today help populate the fields of viral oncology. Certainly much of the research on T cell lymphoma viruses and AIDS virus sprang from studies initiated by the NCI programs.

At the outset of this talk I said I would not repeat the well earned praises or list the many outstanding contributions the NIH has made not only to the US but also for the benefit of the whole world of medicine. But to me as a teacher of medical students, interns, residents and fellows for over 50 years, the most inspiring influence of the NIH has been in providing a generation of medical scientists as exemplified by many of the participants of this outstanding symposium.

In closing I would like to say that I am very optimistic about the future. The developments in molecular biology, cytogenetics and other basic fields illustrated by the calibre of the presentations at this symposium, make me believe that the problems of cell proliferation and differentiation will be solved in the not too distant future and will lead to our understanding of the pathogenesis of leukemia and control of the disease in man.

RETROVIRAL TRANSFER OF GENES INTO CANINE HEMATOPOIETIC PROGENITOR CELLS

Friedrich Schuening, Rainer Storb, Richard Nash,
Richard B. Stead, William W. Kwok, and A. Dusty Miller

Division of Clinical Research and
the Department of Molecular Medicine
Fred Hutchinson Cancer Research Center, and
the University of Washington School of Medicine
Seattle, Washington

ABSTRACT

Amphotropic retroviral vectors containing either the bacterial neomycin phosphotransferase gene or a mutant dihydrofolate reductase gene (DHFR*) were used to infect canine hematopoietic progenitor cells. Successful transfer and expression of both genes in canine hematopoietic progenitor cells has been achieved as measured by the ability of the viruses to confer resistance to either methotrexate (MTX) or the aminoglycoside G418, respectively. Gene transfer was achieved using helper-free retroviral vectors. The rate of gene expression in canine granulocyte/macrophage colony-forming units (CFU-GM) after cocultivation for 24 hours with virus-producing packaging cells ranged from 6-25%. Autologous marrow cocultivated for 24 hours with virus-producing packaging cells was transplanted into six dogs after lethal total body irradiation. All dogs showed engraftment within two weeks and four dogs survived for 5-7 months without adverse effects. One dog that had been given marrow infected with a DHFR* virus and that received MTX as in vivo selection after marrow transplantation and survived, showed 0.1 and 0.03% MTX-resistant CFU-GM at weeks 3 and 5. The efficiency of gene transfer into canine CFU-GM has been increased threefold by culturing marrow cells for six days in long-term marrow culture after 24 hour cocultivation with virus producing packaging cells.

INTRODUCTION

There are many potential benefits of developing a strategy to transfer genetic material to multipotent hematopoietic stem cells in the canine model. This model is well established in the field of bone marrow transplantation, helping to establish the importance of histocompatibility for the success of engraftment[1] as well as the use of methotrexate (MTX) for prevention of graft versus host disease[2]. It is a logical intermediate step in current investigations of gene transfer techniques, prior to application of these techniques to human subjects.

Retroviral vectors are effective in mediating the transfer of genetic material into mammalian cells. The use of these vectors for gene transfer has several advantages. Proviral DNA integrates in a linear fashion into host genomic DNA and subsequently appears to behave like normal host genes. Retroviral vectors can be designed such that viral replication occurs only in packaging cells[3,4] or in the presence of helper virus but otherwise is prevented in target cells, not allowing spread of virus through host tissues. Vectors can be amphotropic and therefore are able to infect cells of different types and species. Retroviral vectors then would appear to be suitable vehicles for gene transfer into mammalian cells, but more experience is needed with infection of stem-cell derived progeny.

Interest has focused on bone marrow as a target for gene transfer studies because of ease of access and the ability to transplant marrow after manipulation of cells in the laboratory. There are also a number of immunological and hematological hereditary diseases that potentially could be treated with somatic gene transfer techniques. Infection of hematopoietic progenitor cells with retroviral vectors has been demonstrated in different animals and humans[5-8]. Infection rates as determined by growth of resistant granulocyte/macrophage colony-forming units (CFU-GM) were from 3-25%. This granulocyte/macrophage progenitor assay gives no indication of successful infection of pluripotent hematopoietic stem cells, but showed that infection was possible in committed hematopoietic precursor cells.

Infection of pluripotent hematopoietic stem cells was first demonstrated in gene transfer studies in murine marrow transplants[5,9]. Marrow reconstitution occurred in these animals with a high percentage of progeny derived from infected stem cells. Expression of these genes is generally poor and the frequency of cells expressing the gene of interest can decrease over time. In other animal models, transplantation studies to determine the efficiency of infection of stem cells are few[7,10]. Very low infection rates have been demonstrated. The higher rates of infection of reconstituted marrow in mice with retroviral vectors may result from the relatively larger amount of marrow from congenic mice that can be preselected prior to transplant, as well as vectors which derive from murine retroviruses that may infect murine stem cells more readily.

The purpose of the current studies is to ascertain the feasibility of somatic gene transfer into canine hematopoietic precursor cells using retroviral vectors. Different strategies may be needed to increase in vitro infection rates of nonmurine animals to efficiently reconstitute marrow with infected stem cells. These studies have implications for the therapy of human genetic diseases and for investigation of the behavior of pluripotent hematopoietic stem cells and their multilineage progeny after marrow transplantation using the unique proviral integration sites as clonal markers.

METHODS

Vectors

Vector constructs pN2[5] and pSDHT[11] have been described previously. The pN2 vector contains the bacterial neomycin phosphotransferase gene which confers resistance to G418 in mammalian cells. The pSDHT vector contains a mutant dihydrofolate reductase (DHFR*) gene which confers resistance to MTX. Both vectors are replication defective.

Cell Lines

NIH 3T3 (TK$^-$)[12], PA317[4] and PA12[13] packaging cells have been described. PA317 packaging cells have less potential to produce helper virus as compared to PA12 packaging cells[4,11].

Selection

Virus assays for pSDHT were done in 0.1 μM MTX with dialyzed fetal bovine serum. CFU-GM assays used 0.25 μM MTX. For pN2, the concentration of G418 used in virus and CFU-GM assays was 1 mg/ml active compound.

Virus Assay

The technique for assaying virus has been described[14]. Virus titers for these studies ranged from 10^6 - 10^7 cfu/ml for both viruses. Helper virus was assayed using the S^+L^- assay[4]. PA12-N2 contains helper virus, whereas PA12-SDHT and PA317-N2 are helper-free.

CFU-GM Assay

For CFU-GM colony assay[15], 7.5×10^4 mononuclear marrow cells separated over Ficoll-Hypaque gradient were seeded in 2 ml of semisolid medium consisting of DME medium with 20 μg of L-asparagine per ml, 75 μg of DEAE-dextran per ml, 20% heat-inactivated human AB plasma, 5% phytohemagglutininstimulated lymphocyte-conditioned medium and 0.3% agar with or without selective agents. Colonies were counted after 14 days. Selective agents were G418 and MTX as described earlier. Infection rates were determined as a percentage by dividing the number of resistant CFU-GM colonies grown in the presence of selective agents by the number of colonies grown in the same conditions without selective agents.

Long-Term Marrow Culture Assay

The conditions for long-term cultures of canine marrow cells have been described[16]. 6×10^7 mononuclear marrow cells were cultured in 75 cm^2 canted-neck flasks (Corning, Corning, NY) at 2×10^6 cells/ml in RPMI-1640 medium (M.A. Bioproducts, Walkersville, MD) supplemented with 20% prescreened heat-inactivated horse serum (Flow Laboratories, Rockville, MD), 10^{-7} M hydrocortisone, 1% nonessential amino acids, 1% pyruvate, 2% L-glutamine, and 1% penicillin-streptomycin. Cultures were maintained at 37°C in a humidified atmosphere of 5% CO_2 in air. After 1 week of culture, nonadherent cells and medium were removed from the flasks and centrifuged at 400g for 10 minutes. Half of the used medium (15 ml) together with 15 ml of fresh medium and 6×10^7 freshly aspirated mononuclear marrow buffy coat cells (obtained by centrifugation at 800g for 10 minutes) were returned into each flask (marrow boost). At weekly intervals, half the supernatant medium was removed and replaced by the same amount of fresh medium.

Infection of Canine Hematopoietic Progenitor Cells

Canine bone marrow was aspirated from the humeri of anesthetized dogs, diluted 1:3 with Waymouth's medium and separated using Ficoll-Hypaque density-gradient centrifugation (density 1.077). Nucleated marrow cells were cocultivated for 24 hours at a ratio of 2:1 with the vector-producing fibroblasts that had been irradiated with 1500 cGy. Cells were kept in 60mm dishes at 37°C and DME medium supplemented with 20% fetal bovine serum and 2 μg/ml Polybrene was used.

Because of the large amount of canine marrow necessary for autologous marrow transplants, cocultivations with virus-producing packaging cells were done in roller bottles. Conditions for infection of bone marrow and procedures for transplantation have been described[10]. Briefly, bone marrow was aspirated from the humeri and femora of anesthetized dogs, diluted 1:3 with Waymouth's medium and separated using Ficoll-Hypaque density-gradient centrifugation. Nucleated marrow cells were cocultivated with virus-producing packaging cells (PA12-SDHT or PA12-N2) at a ratio of 1:2 for 12-24 hours in 850 cm^2 roller bottles (Corning). Packaging cells were seeded in the roller bottles at $3\text{-}5 \times 10^7$ cells 48 hours prior to the addition of marrow. The roller bottles received 1500 cGy irradiation just prior to the addition of marrow cells, were supplemented with CO_2 and maintained at 37°C, rotating at 1-2 rpm on a Belco roller apparatus. Following cocultivation marrow cells were carefully removed without dislodging the packaging cells, washed, resuspended in serum free medium and reinfused into the marrow donor which had received 500 cGy midline tissue dose of total body irradiation from two opposing ^{60}Co sources at a rate of 10 cGy/minute approximately 2 hours prior to marrow infusion.

To infect canine marrow utilizing long-term marrow cultures, Ficoll separated marrow cells were first cocultivated for 24 hours in roller bottles with virus-producing packaging cells as above. Following cocultivation, marrow cells were used to boost long-term cultures, established one week before. Cultures were fed twice with 15 ml virus-containing fresh medium (2-5 x 10^6 cfu/ml) and 15 ml used medium. Controls were placed directly after Ficoll separation into long-term cultures, which were fed twice with 15 ml virus-free fresh medium and 15 ml used medium. Control and virus flasks were harvested after 6 days of culture and cells were cultured in CFU-GM assay with and without selective agent.

RESULTS

Retroviral Transfer of Genes Into Canine CFU-GM After Short-term Virus Incubation

To establish the CFU-GM assay as a tool for determining the percent of progenitor cells infected with the retroviral vector, the sensitivity of canine CFU-GM had to be determined to the selective agents. G418 completely inhibited CFU-GM colony growth at 1 mg/ml and MTX suppressed all growth at .25 µM. The infection rate obtained by cocultivation of marrow cells with virus-producing packaging cells was compared with the infection rate obtained following exposure of hematopoietic cells to virus-containing medium. Cocultivation resulted in three- to tenfold more efficient gene transfer and was therefore used in the following experiments. The efficiency of infection as determined by the rate of resistance, ranged from 7-16% with PA12-SDHT, from 16-25% with PA12-N2 and was 6% with PA317-N2 (Tables 1 and 2). Phenotypic expression of MTX resistance by CFU-GM containing the DHFR* gene was concentration-dependent but resistant colonies were seen at high concentrations suggesting differences in expression among infected CFU-GM. Of interest was the survival of CFU-GMs in a clinically therapeutic range of MTX (Table 3) suggesting the possibility of in vivo selection for infected cells.

Autologous Transplantation of Marrow Infected With Retroviral Vectors

To assess the ability of the retroviral vectors to infect canine pluripotent hematopoietic stem cells, autologous transplants of infected marrow were done in six dogs. The vector-producing packaging cells used to infect marrow in five of six transplants were PA12-SDHT and in the other PA12-N2. Infection rates of CFU-GM assayed immediately after cocultivation were 2.8-12.9% (Table 4). Since PA12-SDHT confers

MTX resistance, 0.5 mg/kg MTX was administered intravenously on days 1,3,5 posttransplant and twice weekly thereafter to three of the dogs as an in vivo selection step. All dogs engrafted without delay with peripheral polymorphonuclear blood counts >500/mm^3 occurring at an average of 14 days posttransplant. Two dogs died at three weeks after marrow transplantation from severe gastrointestinal toxicity due to MTX, and the remaining four survived >5-7 months with normal grafts. The one dog that survived the in vivo selection with MTX had received the autologous marrow with the higher rate of infection (12.9%).

Posttransplant marrow from the four dogs were studied for evidence of infection of stem cells with retroviral vectors. Resistant CFU-GM

Table 1. MTX-resistance of canine CFU-GM following DHFR* virus infection. CFU-GM Colonies (# MTX resistant/total # colonies).

PA12	PA12-SDHT
0/682	86/551 (16%)
0/853	59/670 (9%)
0/474	62/480 (13%)
0/958	55/486 (11%)
0/439	17/238 (7%)
0/3406	279/2425 (12%)

Ficoll separated marrow cells were cocultivated with packaging cells not producing retroviral vectors (PA12) or those producing DHFR* virus (PA12-SDHT) at a ratio of 1:2 for 24 hours and then assayed for CFU-GM formation in the presence or absence of MTX (500 nM). Results are expressed as the percentage of drug-resistant CFU-GM compared with nonselected CFU-GM. Each row represents data from a separate experiment. Totals are given at the bottom of each column.

Table 2. G-418 resistance of canine CFU-GM following neo-virus infection. CFU-GM Colonies (# G-418 resistant/Total # colonies)

PA12	PA12-N2	PA317	PA317-N2
0/ 642	211/1052 (20%)	0/1560	119/1960 (6%)
0/ 620	217/ 852 (25%)		
0/ 471	257/1325 (19%)		
0/1374	147/ 933 (16%)		
0/3107	832/4162 (20%)	0/1560	119/1960 (6%)

Ficol separated marrow cells were cocultivated with packaging cells not producing retroviral vectors (PA12, PA317) or those producing neomycin phosphotransferase virus (PA12-N2, PA317-N2) at a ratio of 1:2 for 24 hours and then assayed for CFU-GM formation in the presence or absence of G418 (1 mg/ml). Results are expressed as the percentage of drug-resistant CFU-GM compared with nonselected CFU-GM. Each row represents data from a separate experiment. Totals are given at the bottom of each column.

Table 3. Percent MTX-resistant CFU-GM as a function of MTX concentration. % drug resistant colonies.

[MTX]	PA12	PA12-SDHT
250 nM	0%	7.8%
500 nM	0%	7.6%
1 uM	0%	6.4%
5 uM	0%	4.6%
10 uM	0%	3.3%
25 uM	0%	2.4%
50 uM	0%	2.0%

Ficol separated marrow cells were cocultivated with packaging cells not producing retroviral vectors (PA12) or those producing DHFR* virus (PA12-SDHT) at a ratio of 1:2 for 24 hours and then assayed for CFU-GM formation in the absence or presence of increasing concentrations of MTX. Results are expressed as the percentage of drug-resistant CFU-GM compared with nonselected CFU-GM. Data are from a representative experiment with duplicate dishes for each condition.

Table 4. Retroviral infection of autologous canine marrow transplants given after lethal total body irradiation.

Dog #	1	2	3	4	5	6	Avg.
Virus producing packaging cells	PA12-SDHT	PA12-N2	PA12-SDHT	PA12-SDHT	PA12-SDHT	PA12-SDHT	
Marrow cells transplanted ($x10^7$/kg)	4.9	12.0	2.9	8.4	3.4	4.8	6.1
Infection rate (%) after cocultivation	4.5	2.8	5.2	3.4	12.9	3.8	5.4
In vivo selection after bone marrow transplant	none	none	none	MTX	MTX	MTX	
Neutrophils >500/mm^3 (day)	16	12	16	13	14	12	14
Platelets >20K/mm^3 (day)	28	20	40	--	19	--	27
Engraftment(histologic)	No	Yes	Yes	Yes	Yes	Yes	
Survival (weeks)	>28	>28	>19	3	>20	3	

Ficoll separated marrow cells were cocultivated with virus-producing packaging cells (PA12-SDHT or PA12-N2) at a ratio of 1:2 for 12-24 hours. Marrow cells were then reinfused into the marrow donor which had received 500 cGy midline tissue dose of total body irradiation from two opposing ^{60}Co sources at a rate of 10 cGy/minue approximately 2 hours prior to marrow infusion. The percentage of drug-resistant CFU-GM after cocultivation was determined. Dogs 4-6 received MTX 0.5 mg/kg intravenously on days 1, 3, 5 after marrow transplantation and twice weekly thereafter.

were present in only one dog; this dog had a high rate of infection immediately after cocultivation and received MTX postgrafting for in vivo selection (Table 5). The percentage of MTX-resistant CFU-GM was low and transient: 0.1% at week 3 and 0.03% at week 5 after marrow transplantation. At 7 and 10 weeks, there was no growth of resistant CFU-GM. To detect proviral DNA which may not express the selectable marker to allow for the selection of resistant CFU-GM, Southern analysis was performed. No proviral DNA was detected at a level of sensitivity of about 1 copy per 50 cells. No adverse effects on engraftment or on survival were seen as a result of cocultivation of marrow with retroviral vectors.

Retroviral Gene Transfer Into Canine CFU-GM After Long-term Marrow Culture

In an attempt to increase the infection rate of canine hematopoietic progenitor cells, marrow was cocultured overnight with virus-producing packing cells and then kept in long-term marrow culture which was fed with virus supernatant. Infection rates of CFU-GM, using the vector--producing packaging cell PA12-SDHT, were variable ranging from 0 to 82% over seven experiments. The average infection rate was 38% averaged over all the experiments (Table 6). Compared to results obtained with cocultivation only, efficiency of gene transfer has been generally increased threefold by combining cocultivation with long-term marrow culture.

DISCUSSION

Studies of retrovirus-mediated gene transfer with canine marrow have demonstrated successful transfer and expression of two genes, a mutant DHFR gene and the neomycin phosphotransferase gene into canine CFU-GM. This was accomplished with both helper containing and helper-free vectors. The rates of infection in vitro in canine CFU-GM were comparable with that of mouse (10-20%)[5], monkey (7-28%)[7], and human (4-10%)[8].

Table 5. MTX-resistant CFU-GM in dog #5 following autologous transplantation of marrow infected with DHFR*-virus.

week after marrow transplantation	MTX (uM) (per culture dish)	CFU-GM colonies (# MTX resistant / Total # colonies: Dog #5		Control
3	0.25	5/ 3480	(0.14%)	0/ 744
	0.50	2/ 3480	(0.06%)	0/ 744
5	0.25	2/ 6900	(0.03%)	0/3860
	0.50	0/ 6900	(<0.014%)	0/3860
7	0.25	0/11960	(<0.008%)	ND
10	0.25	0/ 276	(<0.36%)	0/2100

Bone marrow cells were obtained 3, 5, 7 and 10 weeks after autologous transplantation of marrow infected with DHFR*-virus. Marrow cells were assayed for CFU-GM formation in the presence or absence of MTX (.25 or .50 μM). Control marrow cells were obtained from dog #1 and #2, which received MTX at the time. Results are expressed as the percentage of MTX-resistant CFU-GM colonies compared with nonselected colonies.

Table 6. MTX resistance of canine CFU-GM following DHFR* virus infection in long-term marrow culture. CFU-GM colonies (#MTX resistant/ total colonies).

Control	PA12-SDHT
0/ 1153	44/101 (44%)
0/ 3385	619/1972 (31%)
0/ 1153	114/228 (50%)
0/ 338	301/365 (82%)
0/ 3103	0/99 (0%)
0/ 649	18/96 (19%)
0/ 338	21/73 (29%)
0/10119	1117/2934 (38%

Ficoll separated marrow cells were cocultivated with DHF* virus producing packaging cells (PA12-SDHT) at a ratio of 1:2 for 24 hours. Following cocultivation, marrow cells were used to boost long-term cultures established one week before. Cultures were fed twice with virus containing supernatant (2 - $5x10^6$ cfu-ml). Controls were placed directly after Ficoll separation into long-term cultures which were fed twice with virus-free medium. Control and virus flasks were harvested after six days of culture and cells were cultured in CFU-GM assay with and without selective agent. Results are expressed as the percentage of MTX-resistant CFU-GM colonies compared with nonselected colonies. Each row represents data from a separate experiment. Totals are given at the bottom of each column.

Successful infection of canine CFU-GM is only of limited interest because for successful gene transfer in vivo, pluripotent hematopoietic stem cells must be infected. This was determined by studying reconstituted marrow grafts after lethal total body irradiation. The 'in vivo' study showed infection with retroviral vectors only in one dog and at a low rate. Within several weeks even this very low level of expression was lost. In a recently reported study with nonhuman primates, low levels of infection by the retroviral vector (SAX) were noted also[7]. Southern blot analysis for SAX vector sequences were negative and in situ hybridization for vector-encoded mRNA in peripheral blood mononuclear cells was positive in 0.8% of cells counted. A higher initial infection rate or a preselection step may therefore be necessary to achieve a large number of marrow cells as at current infection rates only 10-20% of cells would survive. Since the amount of marrow aspirated for autologous transplants in dogs is only slightly in excess of that needed for a successful graft, preselection does not appear to be a feasible method. Therefore, it seems to be necessary to increase the initial infection rate of the marrow transplant in order to achieve a higher expression of retroviral vectors in the reconstituted marrow after marrow transplantation.

Long-term marrow cultures allow a prolonged exposure of marrow cells to retroviral vectors after 24-hour cocultivation by feeding the cultures with virus supernatant. Infection rates as measured by assays of CFU-GM are improved with this method. The ability to infect pluripotent hematopoietic stem cells and their survival with this method of infection is unknown. Stem cell survival in standard canine long-term marrow culture is six days as determined by successful autologous transplants of marrow cells harvested from long-term cultures[16] but that may be shortened by overnight cocultivation and exposure to the retroviral vectors. Further investigations will explore autologous transplantation of marrow cocultivated and then exposed to virus supernatant in long-term cultures.

ACKNOWLEDGEMENTS

The authors wish to thank Drs. Eli Gilboa and David Trauber for supplying the plasmids pN2 and pSDHT, respectively, and Sondra Goehle, Joey Meyer, Theodore Graham, Ray Colby, Greg Davis, Lori Ausburn, Robert Raff, Robin Rowedder, and the Shared Word Processing Department for expert technical assistance. This work was supported by grants number CA 15704, CA 18105, CA 18221, CA 41455 and CA 31787, awarded by the National Cancer Institute, and HL 36444 awarded by the National Heart, Lung and Blood Institute, DHHS.

REFERENCES

1. Storb, R., R.B. Epstein, J. Bryant, H. Ragde, and E.D. Thomas. 1968. Marrow grafts by combined marrow and leukocyte infusions in unrelated dogs selected by histocompatibility typing. Transplantation 6: 587-593.
2. Storb, R., R.B. Epstein, T.C. Graham, and E.D. Thomas. 1970. Methotrexate regimens for control of graft-versus-host disease in dogs with allogeneic marrow grafts. Transplantation 9: 240-246.
3. Mann, R., R.C. Mulligan, and D. Baltimore. 1983. Construction of a retrovirus packaging mutant and its use to produce helper-free defective retrovirus. Cell 33: 153-159.
4. Miller, A.D., and C. Buttimore. 1986. Redesign of retrovirus packaging cell lines to avoid recombination leading to helper virus production. Mol. Cell Biol. 6: 2895-2902.
5. Keller, G., C. Paige, E. Gilboa, and E.F. Wagner. 1985. Expression of a foreign gene in myeloid and lymphoid cells derived from multipotent hematopoietic precursors. Nature (London) 318: 149-154.
6. Kwok, W.W., F. Schuening, R.B. Stead, and A.D. Miller. 1986. Retroviral transfer of genes into canine hematopoietic progenitor cells in culture: A model for human gene therapy. Proc. Natl. Acad. Sci. USA 83: 1-4.
7. Kantoff, P.W., A.P. Gillio, J.R. McLachlin, C. Bardignon, M.A. Eglittis, N.A. Kernan, R.C. Moen, D.B. Kohn, S. Yu, E. Karson, S. Karlsson, J.A. Zwiebel, E. Gilboa, R.M. Blaese, A. Nienhuis, R.J. O'Reilly, and W. French Anderson. 1987. Expression of human adenosine deaminase in nonhuman primate after retrovirus-mediated gene transfer. J. Exp. Med. 166: 219-234.
8. Hock, R.A., and A.D. Miller. 1986. Retrovirus-mediated transfer and expression of drug resistance genes in human hematopoietic progenitor cells. Nature (London) 320: 275-277.
9. Dick, J.E., M.C. Magli, D. Huszar, R.A. Phillips, and A. Bernstein. 1985. Introduction of a selectable gene into primitive stem cells capable of long term reconstitution of the hematopoietic system of W/W^v mice. Cell 42: 71-79.
10. Stead, R.B., W.W. Kwok, R. Storb, and A.D. Miller. 1988. A canine model for gene therapy: reconstitution of dogs with autologous marrow infected with retroviral vectors. Blood 71: 742-747.
11. Miller, A.D., D.R. Trauber, and C. Buttimore. 1986. Factors involved

in production of helper virus-free retrovirus vectors. Somatic Cell Mol. Genet. 12: 175-183.
12. Wei, C., M. Gibson, P.G. Spear, and E.M. Seolmick. 1981. Construction and isolation of a transmissible retrovirus containing the src gene of Harvey murine sarcoma virus and the thymidine kinase gene of herpes simplex virus type I. J. Virol. 39: 935-944.
13. Miller, A.D., M.-F. Law, and I.M. Verma. 1985. Generation of helper-free amphotropic retroviruses that transduce a dominant-acting methotrexate-resistant DHFR gene. Mol. Cell. Biol. 5: 431-437.
14. Bender, M.A., T.D. Palmer, R.E. Gelinas, and A.D. Miller. 1987. Evidence that the packaging signal of Moloney murine leukemia virus extends into the "gag" region. J. Virology 61: 1639-1646.
15. Schuening, F., C. Emde, and U.W. Schaeffer. 1983. Improved culture conditions for granulocyte-macrophage progenitor cells. Exp. Hematol. 11: 205: (suppl 14) (Abstr.)
16. Schuening, F., R. Storb, J. Meyer, and S. Goehle. Long-term culture of canine bone marrow cells. Submitted for publication.

RETROVIRAL-MEDIATED GENE TRANSFER INTO HEMOPOIETIC CELLS

Martin A. Eglitis[1], Philip W. Kantoff[1], Donald B. Kohn[2*], Evelyn Karson[1], Robert C. Moen[1], Clinton D. Lothrop, Jr.[3], R. Michael Blaese[2] and W. French Anderson[1]

[1]Laboratory of Molecular Hematology, NHLBI, NIH, Bethesda, MD,
[2]Metabolism Branch, NCI, NIH, Bethesda, MD,
[3]Dept. of Environmental Practice, College of Veterinary Medicine, University of Tennessee, Knoxville, TN

ABSTRACT

Retroviral vectors have provided a means for the introduction of functioning exogenous genes into the hematopoietic system of whole animals. Although these vectors are quite efficient in the mouse model, when applied to non-murine _in vivo_ systems, the efficiency of gene transfer has diminished to impractical levels. Since _in vivo_ analyses are expensive and time consuming, _in vitro_ models have been developed to speed the evaluation of alternative protocols. Using _in vitro_ colony assays, three approaches were evaluated for their ability to improve the infectivity of hematopoietic progenitor cells with retroviral vectors. Exogenously applied hematopoietic growth factors increased the proportion of hematopoietic colonies _in vitro_ up to an average of 5 fold. When alternative sources of progenitors, such as fetal cord blood, were used, improvements in infection efficiency were also obtained. Finally, evidence was acquired suggesting that xenotropic packaging of vectors also improved infection efficiency.

INTRODUCTION

Recently, retroviruses have been altered to permit their use as vehicles for the transfer of exogenous genes into a variety of mammalian target cells[1-3]. They have been particulary useful in transferring genes into cells which are refractory to the uptake of exogenous genes by alternative techniques. Such retroviral vectors have been used to successfully transfer functioning genes into the bone marrow of mice and monkeys _in vivo_[4-7].

The retroviral vector N2 has been used as the backbone for the transfer of a human adenosine deaminase (ADA) gene into both mice and monkeys (Figure 1). There are, however, a number of differences between species in the behavior of these vectors (Table I). In mice, N2 and its derivatives are very efficiently integrated into the genome of hematopoietic progenitors. However, only the parental N2 vector has been shown to function in murine hematopoietic cells. For example, the N2

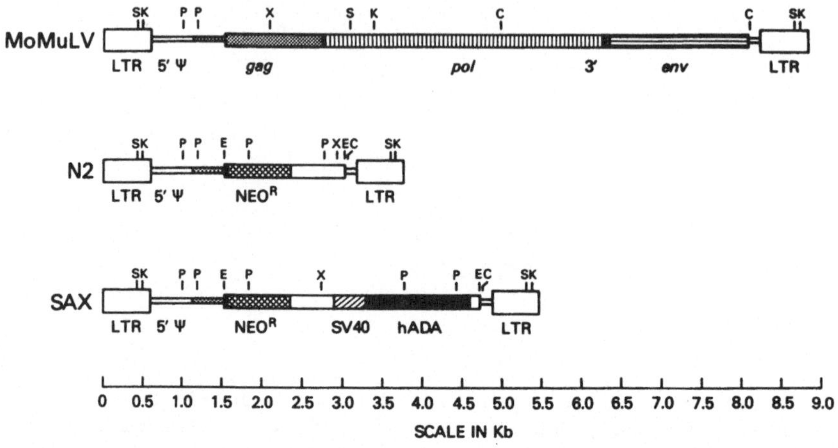

Fig.1. Structure of the proviral forms of Moloney Murine Leukemia Virus (MoMuLV), and the derivative vectors N2 and SAX. LTR: long terminal repeates, containing retroviral enhancer and promoter sequences. 5' : 5' splice donor site. U: Packaging signal for genomic RNA. gag: core antigen gene, expressed on a single unspliced message together with pol (reverse transcriptase gene). 3' : 3' splice acceptor site. env: envelope gene, expressed from the spliced transcript. NEO^R: the bacterial gene for neomycin phosphotransferase, a dominant drug resistance selectable marker. SV40: the early SV40 promoter. hADA: the cDNA for human adenosine deaminase. Restriction enzyme sites: S = SacI, K = KpnI, P = PstI, C = ClaI, E = EcoRI, X = XhoI.

derivative SAX expresses neither the bacterial neo^R nor the SV40-promoted human ADA genes in murine hematopoietic cells _in vivo_. In contrast, SAX infects primate hematopoietic progenitor cells with substantially reduced efficiency. Nevertheless, both the neo^R and human ADA genes within this vector function efficiently in primate cells _in vivo_.

In summary, extrapolations in the behavior of retroviral vectors from one species to a distantly related species must be done with caution. Vectors which transfer genes efficiently in one species may perform much less efficiently in another. Likewise, genes which may function in one species _in vivo_ may function very poorly or not at all in another. Since a goal of our studies is to establish a clinically applicable protocol by which to use retroviral vectors to transfer genes into individuals suffering from inborn errors of metabolism, we are interested in improving the efficiency of gene transfer in primate hematopoietic progenitors. However, performing such studies on whole animals would be both time consuming and inefficient, since each animal could be used to evaluate

Table I. SUMMARY OF _IN VIVO_ EXPRESSION

Species:	Vector:	Infection Efficiency:	Expression? (gene)
Mus	N2	86%	Yes (Neo^R)
	SAX	75%	No (Neo^R, ADA)
Cynomolgus	SAX	< 1%	Yes(Neo^R, ADA)

only a single methodology at a time. Therefore, we have turned to _in vitro_ model systems to evaluate the ability of retroviral vectors to transfer functioning genes into hematopoietic cells. Earlier studies of expression of a transferred ADA gene have used ADA-deficient T and B cell lines derived from ADA-deficient patients[8]. Here we will also describe our use of primary hematopoietic colony culture systems to evaluate the impact of a variety of approaches on the efficiency with which retroviral vectors transfer a functioning selectable marker into human bone marrow cells.

Materials and Methods

The retroviral vectors used in this study are all derived from the Moloney murine leukemia virus (Figure 1). The construction of both N2 and its derivative SAX has been described previously[8,9].

The isolation and culture of the ADA-deficient T cell line TJF-2 has been described previously[8]. After incubation for 2 hr in vector-containing medium, TJF-2 cells were cultured in 400 µg G418 (BRL/Gibco, active weight)/ml to select for a population of cells containing integrated vector. Sensitivity to increasing amounts of deoxyadenosine in the culture medium was evaluated by measuring the amount of ^{3}H-thymidine incorporated into DNA in a 4 hr period.

Human bone marrow was obtained from normal volunteers after informed consent. Fetal hematopoietic progenitors were grown from fetal cord blood obtained, with informed consent, at the time of premature delivery at the times indicated. Culture of erythroid and granulocyte/macrophage colonies was as previously described[10]. Efficiency of gene transfer was determined by measuring the proportion of colonies which grew in the presence of 1200 µg G418 (active weight)/ml, a concentration which completely suppressed the growth of such colonies in the absence of gene transfer.

Dog bone marrow was obtained either from normal, healthy dogs or from cyclic hematopoiesis dogs[11]. Granulocyte/macrophage colonies derived from such marrow were cultured as described previously[12].

For studies on the effects of growth factors on infection efficiency of human marrow, granulocyte/macrophage-colony stimulating factor (GM-CSF) was provided by Genetics Institute, Cambridge, MA and interleukin-1a (IL-1a) was provided by Immunex, Seattle, WA. For a general source of growth factors, medium was conditioned by culture of phytohemagglutinin-stimulated peripheral lymphocytes (PHA-LCM). Nucleated cells from peripheral blood were isolated by centrifugation through LSM (Bionetics), then grown for 7 days in Iscove's Modified Dulbecco's Minimal Essential Medium with added PHA (1 µg/ml final concentration).

For studies on the effect of retroviral envelope on the infectivity of human bone marrow, 2776X xenotropic retrovirus and 4070A amphotropic retrovirus were kindly provided by Dr. Janet Hartley. Pseudotyped N2 vector was generated by infecting mink cells (Mv 1 Lu) which had been previously transfected with the pN2 plasmid containing the cloned N2 provirus.

RESULTS AND DISCUSSION

T Cell Studies

Earlier studies demonstrated the utility of T cell lines derived from ADA-deficient patients for the evaluation of ADA vectors[8]. Normal T cells exhibit a sensitivity to deoxyadenosine in the culture medium at concentrations above 2000 μg/ml (Figure 2). T cells derived from ADA-deficient patients are more sensitive to deoxyadenosine by nearly two orders of magnitude, with an LD_{50} of around 100 μg/ml. After introduction of normal ADA gene by using the vector SAX, the sensitivity of the TJF-2 cell line is restored to the concentration range exhibited by normal T cell lines. Thus, at least in this model system, retroviral vectors can transfer a normal ADA gene to ADA deficient T cells and allow sufficient expression that by one assay, sensitivity to deoxyadenosine, normal physiological function is restored.

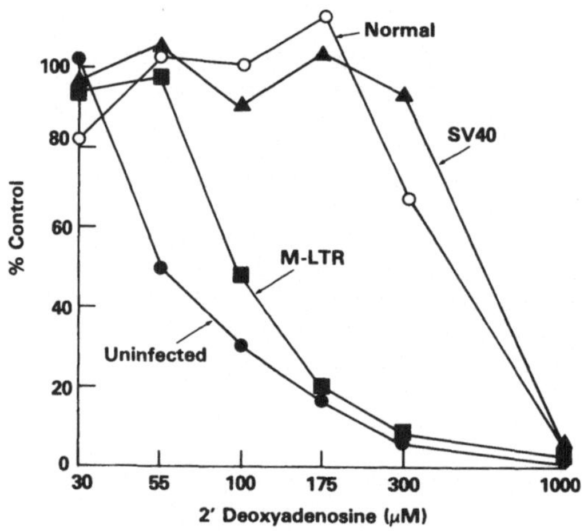

Fig.2. Correction of Deoxyadenosine Induced Inhibition of Human ADA T cell Growth by ADA Containing Retroviral Vectors. Cells were grown at the indicated concentrations of deoxyadenosine, then viability determined by uptake of ^{3}H-thymidine into DNA. 100% control represents the level of uptake in T cells derived from idividuals with normal ADA activity. Normal = sensitivity curve or normal T cells. Uninfected = sensitivity curve of ADA-deficient T cells derived from an ADA-deficient T patient. M-Ltr=sensitivity curve of ADA-deficient T cells after infection with a retroviral vector in which the hADA is expressed from the spliced message generated off the MoMuLV promoter. SV40 = sensitivity curve of ADA-deficient T cells after infection with the retroviral vector SAX, in which the hADA is expressed off an internal SV40 early promoter.

In Vitro Colony Assays

These T cell studies provided encouraging evidence that retroviral vectors could function in appropriate target cells. To evaluate procedures that could improve infection efficiency, an in vitro model system more closely analogous to the actual hematopoietic progenitor target cell would be necessary. Although in vitro colony assays do not characterize the stem cell compartment which will actually reconstitutes an animal, they do permit analysis of relatively immature derivatives of those cells, namely the erythroid BFU-E and myeloid CFU-GM. When nucleated bone marrow cells are infected with the N2 vector by culturing the marrow cells overnight together with irradiated packaging cells producing the vector (PA317-N2), infection efficiencies of between 1 and 2% are obtained (Table II). Although these low efficiencies are discouraging, they at least suggest that infection of such progenitors in primary culture could provide a useful model for improving infection efficiency.

Hematopoietic Growth Factors

The integration of a retrovirus into the host genome requires that the target genomic DNA undergoes at least one round of replication. By their nature, progenitor cells are relatively quiescent in terms of their cell cycle. However, a variety of colony stimulating factors have been recently described which can induce cell division in progenitor cells. Since some of these have now been cloned and are available in pure form, we undertook to evaluate their effect on the efficiency of infection of hematopoietic progenitors with retroviral vectors. As part of an initial screen, we studied IL-1α and GM-CSF. A third source of growth factors was the mix of factors present in PHA-LCM.

In these experiments, bone marrow cells were either infected directly after collection (as before) or after a 40-48 hr period of cultivation in culture medium enriched with either PHA-LCM, IL-1α, or GM-CSF. As shown in Table III, all of these sources of growth factor had some effect in improving infection of progenitors with N2. The greatest effect was observed with the mix of growth factors present in PHA-LCM. Although the average enhancement in infection efficiency was around 5 fold, this represents infections where the improvement ranged as high as 10.2 fold. Likewise, with GM-CSF, the maximal improvement in infection efficiency could be as much as 18 fold. In general, IL-1α did not result in as great a degree of stimulation, which may reflect its role more as an inducer of differentiation than as an overall stimulator of bone marrow cell growth, at least in vitro[13].

Other Sources of Progenitor Cells

In addition to the direct exposure of adult bone marrow cells to growth factors, it was reasoned that progenitor cells derived from earlier stages of development may be stimulated to a greater degree as part

Table II. SUMMARY OF COLONY ASSAYS OF N2 INFECTED HUMAN BONE MARROW

Colony Type	Number Resistant/ Number Assayed
BFU-E	296/22449 (1.3%)
CFU-GM	313/18181 (1.7%)

Table III. EFFECT OF PRESTIMULATION IN VITRO WITH GROWTH FACTORS ON INFECTIVITY OF HUMAN HEMATOPOIETIC PROGENITORS

Colony Type	Without Prestimulation	Prestimulation with LCM	GM-CSF	IL-1a
CFU-GM	313/18181 (1.7%)	395/4650 (8.5%)	281/5654 (5.0%)	217/7245 (3.0%)
BFU-E	296/22449 (1.3%)	252/5141 (5.0%)	263/6870 (3.8%)	210/9919 (2.1%)

of the process of seeding newly developing marrow space in the fetus. Therefore, the infection efficiency of fetal progenitors obtained from cord blood was also determined. Progenitors derived from a 29 wk fetus did infect at roughly 3 times greater efficiency than adult cells infected simultaneously (Table IV). Still greater improvements were obtained with progenitors obtained from a 24 wk fetus. Because of the scarcity of material, these results must be looked upon as very preliminary. Nonetheless, they do suggest that the physiological milieu from whence progenitors are derived may affect their infectivity with retroviral vectors.

Rate of proliferation may not, however, be the simple explanation for improvements in infection efficiency. A dog model exists for cyclic hematopoiesis, a blood disorder where, rather than being generated by a steady proliferation of progenitors, the hematopoietic system is instead generated cyclically[11]. At particular times in this cycle, the marrow of afflicted dogs is greatly enriched for hematopoietic progenitors. It was reasoned that, by infecting marrow aspirated at such times, improvements in the infection efficiency of progenitors could be detected. The results of such infections (Table V), however, show no such improvement. This may be the result of sampling at an inappropriate time; perhaps the actual time in the proliferative cycle when progenitor cells are the most receptive to vector infection is distinct from the time at which they are present in the greatest abundance. Alternatively, the block to increase in infection efficiency may be an inherent characteristic of dog marrow cells, completely unrelated to the rate of cell cycling.

This result highlights the fact that the underlying feature of any of the stimulated progenitor cells in which improvements in infection efficiency were detected is not understood. Rather than being the product of DNA replication, perhaps infection efficiency is increased by changes in the level and type of receptor expressed by the potential target cells. The mechanism of improvement in infection efficiency will need further analysis.

Alternative Envelopes

Since infection with a retroviral vector is mediated through a

Table IV. INFECTIVITY OF CFU-GM FROM ADULT MARROW AND FETAL CORD BLOOD

Cell Type	$G418^R$ Colonies/Total Plated Uninfected	Infected	% $G418^R$
Adult	0/405	2/184	1.1
29 wk Fetal	2/251	10/251	3.2
24 wk Fetal	2/176	20/175	10.3

Table V. RECOVERY OF CFU-GM FROM NORMAL AND CYCLIC HEMATOPOIETIC DOGS AFTER INFECTION N2 VECTOR

SOURCE OF MARROW	G418 Concentration (ug/ml) 0	1500	1750	2000
Normal Dogs	109	9 (8.3%)	5 (4.6%)	3 (2.8%)
CH Dogs	172	3 (1.7%)	8 (4.7%)	6 (3.5%)

cell surface receptor, perhaps some of the problem in the reduced efficiencies of infection observed with non-murine species is a result of rarity or absence of the receptor of the particular envelope used to package the vector. Murine retroviruses may be divided into three classes according to their infection tropisms: ecotropic, amphotropic and xenotropic[14]. Ecotropic envelopes restrict infection to only mouse or rat cells, whereas amphotropic envelopes permit infection of a wide range of species, including primates. Xenotropic envelopes permit infection only of non-murine cells. Each of these envelopes acts through its own distinctive receptor. It is possible that, by changing to a xenotropic envelope, the infection efficiency of retroviral vectors could be improved.

To analyze this possibility, we utilized pseudotyping to alter the envelope of the N2 vector. Although this procedure requires that helper virus be present, earlier studies have suggested that the presence of such helper virus need not interfere with the infectivity of vectors themselves[15]. For these studies, we compared the 2776X xenotropic envelope to the 4070A amphotropic envelope, which had been used in the PA317 packaging system[16] that normally generates the N2 vector. As shown in Table VI, such xenotropically pseudotyped vector exhibited the appropriate species tropisms. In addition, expression of an amphotropic envelope did not interfere with its infection of mink cells[17].

When amphotropically pseudotyped N2 (4070A-N2) was used to infect human hematopoietic progenitors, no G418 resistant colonies were detected (Table VII). Amphotropically packaged N2 (PA317-N2) did generate somewhat less than 1% G418 resistant colonies. This discrepancy could be a consequence of the titre of 4070A-N2 being reduced by nearly two orders of magnitude relative to PA317-N2. However, xenotropically packaged N2 (2776X-N2), despite having a titer on mink cells identical to 4070A-N2, did generate G418 resistant colonies. Although a thorough evaluation of the role of the envelope in vector infectivity requires the development of a xenotropic packaging cell line analogous to PA317, these results do suggest that a xenotropic envelope, perhaps by having greater affinity to receptors present on human hematopoietic progenitors, is capable of overcoming a low titer and demonstrating infection of human progenitors.

CONCLUSIONS

These studies show that _in vitro_ assays provide a useful means by which to screen vectors and infection protocols. However, they suffer from the failing that in no case is the true engrafting stem cell characterized for its infectivity with retroviral vectors. Therefore, after such initial _in vitro_ screens, vectors and infection protocols will still need evaluation _in vivo_. In addition to the mouse and monkey studies mentioned earlier, we are using other larger animals such as dogs, or, for _in utero_ studies as described later in this Symposium by Dr. Zanjani, with sheep. All of this work is directed toward fulfilling the ultimate promise of retroviral-mediated gene transfer, namely the therapy of human genetic disease.

Table VI. HOST RANGE AND INTERFERENCE PATTERN OF PSEUDOTYPED N2 VECTOR AS DETERMINED BY TITERING ON VARIOUS TARGET CELLS

	Target Cell[a]				
	A. Host Range		B. Interference Pattern		
Viral Supernatant[b]	NIH-3T3	Mink	PA317 (NIH-3T3)	4070A (Mink)	2776X (Mink)
N2-Mink	0[c]	0	0	0	0
PA317-N2	6.70	5.70	3.70	2.60	6.28
4070A-N2 Mink	3.28	4.00	2.00	1.18	4.30
2776X-N2 Mink	0	4.00	0	3.60	0.90

[a] Target Cells:

NIH-3T3 = transformed murine fibroblast line
Mink = Mu 1 Lv mink lung cells
PA317 = 4070A amphotropic retroviral packaging line derived from NIH-3T3 cells
4070A Mink = mink cells infected with 4070A virus and passaged for over one month
2776X Mink = mink cells infected with 2776X virus and passaged for over one month

[b] Supernatants:

PA317-N2 = PA317 helper cells transfected with pN2
N2-Mink = Mink cells transfected with pN2, G418-selected population
4070A-N2 Mink = N2-Mink cells infected with 4070A virus and passaged for over one month
2776X-N2 Mink = N2-Mink cells infected with 2776X virus and passaged for over one month

[c] Titers expressed as the logarithm

Table VII. PSEUDOTYPED INFECTION OF HUMAN MARROW PROGENITORS

	N2 Packaged using:		
Colony Type	PA317	2776X	4070A
CFU-E	6/1199 (0.5%) n=1	1/961 (0.1%) n=1	0/2095 (0%) n=2
BFU-E	28/3548 (0.8%) n=2	7/4365 (0.2%) n=2	0/4014 (0%) n=2
CFU-GM	6/4234 (0.2%) n=2	4/3675 (0.1%) n=2	0/3251 (0%) n=2

REFERENCES

1. Doehmer, J., M. Baringa, W. Vale, M.G. Rosenfield, I.M. Verma, and R.M. Evans. 1982. Introduction of rat growth hormone gene into mouse fibroblasts via a retroviral DNA vector: expression and regulation. Proc. Natl. Acad. Sci. USA 79: 2268-2272.
2. Perkins, A.S., P.T. Kirschmeier, S. Gattoni-Celli, and I.B. Weinstein. 1983. Design of a retrovirus-derived vector for expression and transduction of exogenous genes in mammalian cells. Mol. Cell Biol. 3: 1123-1132.
3. Hwang, L.H. and E. Gilboa. 1984. Expression of genes introduced into cells by retriviral infection is more efficient than that of genes introduced into cells by DNA transfection. J. Virol. 50: 417-424.
4. Dick, J.E., M.C. Magli, D. Huszar, R.A. Phillips, and A. Bernstein. 1985. Introduction of a selectable gene into primitive stem cells capable of long-term reconstitution of the hemopoietic system of W/Wv mice. Cell. 42: 71-79.
5. Keller, G., C. Paige, E. Gilboa, and E.F. Wagner. 1985. Expression of a foreign gene in myeloid and lymphoid cells derived from multipotent haematopoietic precursors. Nature 318: 149-154.
6. Eglitis, M.A., P. Kantoff, E. Gilboa, and W.F. Anderson. 1985. Gene expression in mice after high efficiency retroviral-mediated gene transfer. Science 230: 1395-1398.
7. Kantoff, P.W., A.P. Gillio, J.R. McLachlin, C. Bordignon, M.A. Eglitis, N.A. Kernan, R.C. Moen, D.B. Kohn, S.-F. Yu, E. Karson, S. Karlsson, J.A. Zwiebel, E. Gilboa, R.M. Blaese, A. Nienhuis, R.J. O'Reilly, and W.F. Anderson. 1987. Expression of human adenosine deaminase in nonhuman primates after retrovirus-mediated gene transfer. J. Exp. Med. 166: 219-234.
8. Kantoff, P.W., D.B. Kohn, H. Mitsuya, D. Armentano, M. Sieberg, J.A. Zwiebel, M.A. Eglitis, J.R. McLachlin, D.A. Wiginton, J.J. Hutton, S.D. Horowitz, E. Gilboa, R.M. Blaese, and W.F. Anderson. 1986. Correction of adenosine deaminase deficiency in cultured human T and B cells by retrovirus-mediated gene transfer. Proc. Natl. Acad. Sci. USA 83: 6563-6567.
9. Armentano, D., S.-F. Yu, P.W. Kantoff, T. von Ruden, W.F. Anderson, and E. Gilboa. 1986. Effect of internal viral sequences on the utility of retroviral vectors. J. Virol. 61: 1647-1650.
10. Gregory, C.J., and A.C. Eaves. 1977. Human marrow cells capable of erythropoietic differentiation in vitro: Definition of three erythroid colony responses. Blood 49: 855-864.
11. Jones, J.B., R.D. Lange, and E.S. Jones. 1975. Cyclic hematopoiesis in a colony of dogs affected with cyclic neutropenia. J. Am. Vet. Med. Assoc. 166: 365-367.
12. Dunn, C.D.R., J.B. Jones, J.D. Jolly, and R.D. Lange. 1977. Progenitor cells in canine cyclic hematopoiesis. Blood 50: 1111-1120.
13. Neta, R., M.B. Sztein, J.J. Oppenheim, S. Gillis, and S.D. Douches. 1987. The in vivo effects of Interleukin I. I. Bone marrow cells are induced to cycle after administration of interleukin I. J. Immunol. 139: 1861-1866.
14. Teich, N. 1982. Taxonomy of retroviruses. in RNA Tumor Viruses. R. Weiss, N. Teich, H. Varmus, and J. Coffin, editors. Cold Spring Harbor Laboratory. 25-208.
15. Hogge, D.E. and R.K. Humphries. 1987. Gene Transfer to primary normal and malignant human hemopoietic progenitors using recombinant retroviruses. Blood 69: 611-617.
16. Miller, A.D. and C. Buttimore. 1986. Redesign of retrovirus packaging cell lines to avoid recombination leading to helper virus production. Mol. Cell Biol. 6: 2895-2902.

RETROVIRAL GENE TRANSFER:

APPLICATIONS TO HUMAN THERAPY

Eli Gilboa

Memorial Sloan-Kettering Cancer Center
1275 York Avenue
New York, NY 10021, USA

The understanding of gene expression has been greatly enhanced by the ability to transfer cloned genes into cells and to study the mechanism of their regulation. For the past several years it has been recognized that retroviruses are good candidates as vehicles, or vectors, to introduce genes into eucaryotic cells. Retrovirus-derived vectors utilize the biochemical processes unique to this group of viruses, to transfer genes with high efficiency into a wide variety of cell types, in vitro and in vivo. By using retrovirus derived vectors, the effect of newly introduced genes and the mechanism of gene expression can be studied in cell types so far refractory to gene transfer. The special features of this new gene transfer technology have provided for the first time the opportunity of introducing genes into the somatic cells of live animals. Although at present limited to gene transfer into hemopoietic cells, its potential in general studies and applications to human therapy is beginning to be recognized.

Retroviral gene transfer is used for the purpose of introducing functional genes into cells at one copy per cell, without affecting the proliferative capacity of the recipient cell. The suitability of retroviruses for gene transfer stems from their mode of replication. By "simply" replacing the viral genes with the gene of interest and utilizing the efficient viral infection process, the gene is transferred into the target cell as if it were a viral gene. As with many new emerging technologies, the potential of retroviral gene transfer created great expectations. However, it was soon realized that there are certain limitations and difficulties associated with this technology. With numerous refinements it is now quite simple to insert a gene into a retrovirus vector, obtain recombinant virus, infect target cells and express the foreign gene. What is however more difficult and elusive at this moment is to maximize the efficiency of the process. The nature of the retrovirus vector will determine to a large extent those parameters. The reason why the development of "all--purpose", "super-efficient", or "highly transmissible" retrovirus vectors is elusive, stems from the simple fact that we don't understand yet some of the more subtle details of the structure and biology of retroviruses, nor that of mammalian genes.

Fig. 1 (top) shows the structure of a retroviral vector in which the transduced gene is expressed from an internal promoter, hence the name, vectors with internal promoters (VIP). In these vectors, the selectable gene is linked to the left end of the viral DNA and is expressed

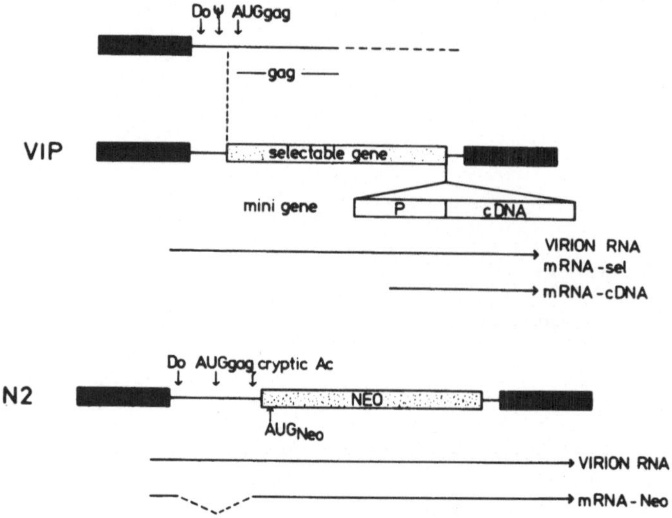

Fig. 1.

from the viral promoter. A minigene consisting of a DNA fragment encoding a promoter linked to a cDNA copy of the gene of interest is inserted downstream of the selectable gene[1]. The promoter-encoding DNA fragment, which is responsible for the expression of the transduced gene can be derived from any gene, and therefore in using this type of vector one has the flexibility of choosing the promoters to express the transduced gene in a manner most appropriate for a particular experimental design. The main drawback of this strategy of vector design is that an internal promoter is placed within the retroviral transcriptional unit which affects the activity of the promoter, and results in variable and often low level of expression of the transduced gene. Fig. 1 (bottom) shows the structure of a VIP vector called N2[2]. The unique feature of this vector is the generation of significantly higher titers of virus, as compared to other retroviral based vectors. As shown in Fig. 1 (bottom), in the N2 vector the region downstream from the 5' LTR extends beyond the gag AUG initiation codon and includes 418 base pairs of the gag coding sequences to which the bacterial Neo gene is fused. It appears that this extra region present in N2 is responsible for the production of 10- to 50- fold higher titers of virus as compared to similar vectors lacking these sequences. In N2 the functional AUG and 418 base pairs of gag coding sequence, are out of frame with the Neo gene coding sequences. How is the neo gene expressed? As illustrated in Fig. 1, it appears that a cryptic 3' splice site is activated in the gag coding sequences upstream of the Neo gene, generating a spliced RNA form which serves as the mRNA for the Neo gene[2]. The usefulness of N2 based vectors was demonstrated in several laboratories for the introduction of genes into cultuted hemopoietic cell lines and fresh bone marrow cells.

In 1978, Zamecnik and his colleagues[3] have shown that it is possible to inhibit the replication of an avian retrovirus by using a short oligonucleotide which is complementary to the viral RNA. This approach was recently revived by the studies of Weintraub, Inoue, Pestka and others who have shown that it is possible to inhibit specific genes in diploid cells, by introducing into the cell an RNA species which is complementary to the RNA transcribed of the gene in question[4-6]. This RNA is called antisense RNA. Although initial observations have created an undue sense of pessimism as to the usefulness of this approach, recent studies are more encouraging. In particular, it was shown that effective inhibition can be achieved by using DNA templates for the synthesis of antisense

RNA, which were introduced into cells in a genetically stable manner at one or a few copies per cell. The usefulness of this approach to eucaryotic genetics cannot be overstated. It is also tempting to speculate that coupling of antisense RNA inhibition to an effective in vivo gene transfer procedure will enable the inhibition of specific genes in the live patient, and the treatment of diseases caused by the activity of deleterious genes.

The objective of our studies was to use this approach to inhibit the replication of a virus, HTLV-I, in human lymphoid cells. Introduction of an effective antisense DNA template into the cell prior to viral infection will confer resistance against viral replication and thus, will protect the cells from viral infection.

Human T Cell leukemia virus, HTLV-I, is a human retrovirus which is the etiological agent of a particularly aggressive form of T cell leukemia in adult patients. HTLV-I is poorly infectious in established human cell lines but it readily transforms primary T-lymphocytes in vitro. The infected cells become "immortalized" and can be propagated indefinitely in culture. We wanted to see whether antisense templates, specific to HTLV-I, will protect the cells from viral mediated transformation. A series of retroviral vectors were constructed, termed anti-sense vectors, carrying different regions of the HTLV-I genome in reversed transcriptional orientation under the control of heterologous internal promoters (Fig. 2). Two segments of the HTLV-I genome were chosen. The sequence spanning the 5' end of the viral mRNA species ("C") harbors cis acting elements, essential for viral gene expression (5' splice site) and virus replication (t-RNA primer binding site) and perhaps, signals for packaging of the genomic virus RNA. The second sequence ("X") corresponds to the first 1kb of the pX gene c-DNA. The expression of this gene is required for virus replication and has been also implicated in host cell transformation. The HTLV-I derived sequences were fused either to the early SV40 promoter or to the immediate early CMV promoter and inserted into the retroviral vector N2 to generate anti-sense vectors, as shown in Fig. 2. PHA-stimulated mononuclear cells (MNC) derived from umbilical cord blood, a source of primary T-lymphocytes, were transduced with the recombinant murine retroviruses carrying the anti-sense sequences and the parent vector. After allowing the vectors to integrate and express anti-sense RNA, the cells were infected with HTLV-I secreted from a productively infected cell line, c10/MJ2[7]. All cultures were grown in the presence of exogenous T-cell growth factor, interleukin-2 (IL-2), at a concentration which supported maximal proliferation of stimulated T-lymphocytes. Cell proliferation was measured directly by counting the cells as shown in Figure 3.

Uninfected cells proliferate transiently in response to PHA-stimulation, and transduction with anti-sense vectors has no effect on cell

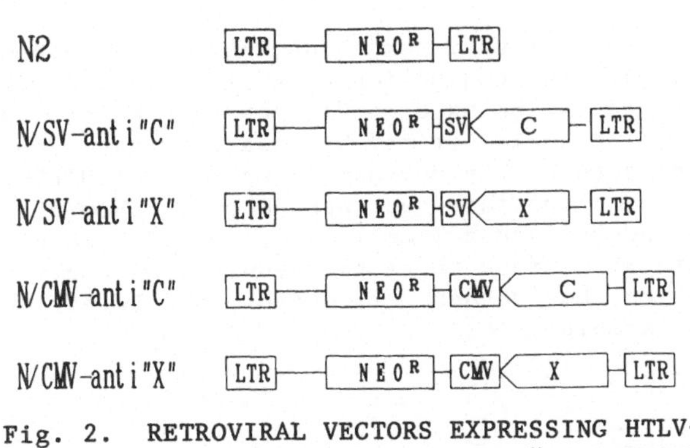

Fig. 2. RETROVIRAL VECTORS EXPRESSING HTLV-I SEQUENCES IN "ANTI-SENSE" ORIENTATION

proliferation. On the other hand, infection of cells with HTLV-I causes the reappearance of blast like cells after approximately two weeks of cocultivation. In the presence of G418 (Figure 3), an initial decline in total cell number was observed in all cultures during the first days in the selective media. Subsequently, cell numbers steadily increased in the cultures carrying the N2 control vector (N2) and the anti-sense vectors, SC and SX. In contrast, the cell count continued to decline in cultures infected with the CC and CX vectors, until only a few living cells could be detected. No proliferation was observed over a period of more than two weeks, when blast like cells could be detected, and proliferation resumed in these cultures as well.

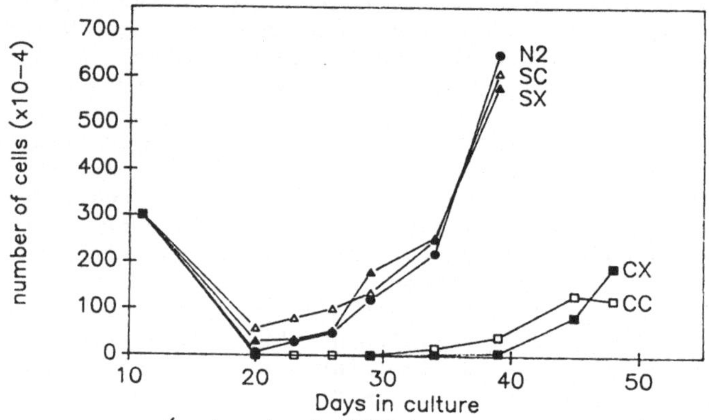

Fig. 3.

In this study we have shown that anti-sense RNA expressed from recombinant murine retroviruses in primary human T-cells leads to significant, though not complete, inhibition of HTLV-I replication and viral mediated cell transformation. One to several copies of vector, stably integrated in the chromosome, are responsible for the synthesis of sufficient amounts of anti-sense RNA in the cell to mediate this effect. Use of a retroviral gene transfer system was responsible for the efficient transfer of anti-sense templates into primary human T-cells

Possible applications of anti-sense RNA inhibition to human medicine have not gone unnoticed. In conjunction with effective in vivo gene transfer techniques this approach may be used to "switch off" harmful genes in the live patient". For example, it may be used to treat or prevent virus induced diseases. Because of the dominant phenotype of HTLV-I induced malignant cell transformation this approach does not show great promise in preventing viral induced acute T-cell leukemia. At present, this approach may be more useful in situations where the result of virus infection is not cell transformation, but rather cell death. HIV infection leading to AIDS is one such example.

REFERENCES

1. Miller, D.A., E.S. Ong, M.G. Rosenfeld, I.M. Verma, and R.M. Evans. 1984. Infectious and Selectable retrovirus containing an inducible rat growth hormone minigene. Science 225: 993-997.
2. Armentano, D., S.F. Yu, P.W. Kantoff, T. von Ruden, W.F. Anderson and E. Gilboa. 1987. Effect of Internal Viral Sequences on the Utility of Retroviral Vectors. J. Virol. 61: 1647-1650.
3. Zamecnik, P.C., and M.L. Stephenson. 1978. Inhibition of Rous sarcoma virus replication and cell transformation by a specific oligodeoxynucleotide. Proc. Natl. Acad. Sci. 75: 280-284.
4. Izant, J.G., and H. Weintraub. 1984. Inhibition of thymidine kinase gene expression by anti-sense RNA: A molecular approach to genetic analysis. Cell 36: 1007-1015.
5. Coleman, J., Green, P.J. and Inouye, M. 1984. The use of RNAs complementary to specific mRNAs to regulate the expression of individual bacterial genes. Cell 37: 429-436.
6. Pestka, S., B.L. Dougherty, V. Jung, K. Hoffa, and R.K. Pestka. 1984. Anti-mRNA: specific inhibition of translation of single mRNA molecules. Proc. Natl. Acad. Sci. 81: 7525-7528.
7. Markham, P.D., S.Z. Salahuddin, V.S. Kalyanaraman, M. Popovic, P. Sarin, and R. Gallo. 1983. Infection and transformation of fresh human umbilical cord blood cells by multiple sources of Human T-Cell Leukemia Virus (HTLV). Int. J. Cancer. 31: 413-420.

CONSTRUCTION OF A SAFE AND EFFICIENT RETROVIRUS PACKAGING CELL LINE

Dina Markowitz, Stephen Goff and Arthur Bank

Columbia University
College of Physicians and Surgeons
Department of Genetics and Development
Department of Biochemistry and Molecular Biophysics, and
Department of Medicine
New York, NY

ABSTRACT

Ecotropic and amphotropic retrovirus packaging cell lines have been constructed in which the helper virus genome have been separated onto two plasmids, and the ψ packaging signal and 3' LTR have been removed. The <u>gag</u> and <u>pol</u> genes on one plasmid and the <u>env</u> gene on another plasmid were transfected into NIH 3T3 cells. Packaging cell lines produced by these transfected genes released titers of replication-defective retroviral vectors which were comparable to titers produced by packaging cell lines containing the helper virus genome on one plasmid. There has been no evidence of recombination events between the ecotropic helper virus plasmids and the vector virus plasmid that would result in the generation of intact replication-competent virus. These results suggest that a packaging cell line containing <u>gag</u>, <u>pol</u> and <u>env</u> on different plasmids is efficient and safe for use in retroviral gene gransfer.

KEY WORDS:

Retroviral gene transfer -- packaging cell line.

INTRODUCTION

Retroviruses appear to be the method of choice as vectors for the transfer of exogenous genes into humans. In particular, the cloning, transfer, and expression of human globin genes into erythroid cells in culture has raised the possibility of autotransplantation of bone marrow cells with normal β-globin genes as an approach to the therapy of β-thalassemia and sickle cell anemia in humans[1].

A major prerequisite for the use of retroviruses is to insure the safety of their use[2]. The major danger of the use of retroviruses for gene therapy is the possibility of the spread of wild-type retrovirus in the cell population. The proliferation of wild-type virus can lead to multiple integrations of the retrovirus into the genome which may result in the activation of potentially harmful genes such as oncogenes[3,4]. The development of packaging cell lines that produce only replication-defective retroviruses has increased the utility of retroviruses for gene therapy[5-9]. In these cell lines, the sequence required for packaging of the viral RNA (ψ sequence) has been deleted, therefore the packaging cell produces viral proteins but is unable to package the viral RNA genome into infectious virions. When these packaging lines are transfected with a replication-defective retroviral vector containing an intact ψ sequence required for its own packaging, wild-type retrovirus has been shown to arise[6,10,11] presumably due to recombination events between the helper virus genome and the vector virus.

We have created novel ecotropic and amphotropic retrovirus packaging cell lines which should virtually eliminate the possibility of recombination between the helper virus and the vector virus leading to wild-type retrovirus. In our cell lines, the helper virus DNA has been separated onto two plasmids; the gag and the pol genes are on one plasmid and the env gene is on another plasmid. In addition, the packaging sequence and the 3' LTR have been deleted in both plasmids. With this type of strategy at least three recombination events between the two helper plasmids and the vector virus are necessary to generate a wild-type virus. We will describe the development of both a stable ecotropic and a stable amphotropic packaging line that are both efficient and safe for use in gene transfer experiments.

Generation of the Ecotropic Packaging Line

To generate an ecotropic packaging cell line two helper virus plasmids, pgag-polgpt and penv, were constructed using Mo-MULV proviral DNA from the plasmid 3PO (Figure 1). 3PO contains a deletion of the ψ packaging signal. pgag-polgpt (Figure 1) was constructed by isolating a fragment containing the 5' LTR and the gag and pol DNA and inserting this fragment into the plasmid pSV2gpt[12]. penv (Figure 1) was constructed by isolating a fragment from 3PO that contains the 3' acceptor splice site and the env gene and ligating it to another fragment from 3PO containing the 5' LTR and 5' donor splice site.

3T3 cells were co-transfected, by electroporation[13], with pgag-polgpt and penv DNAs. Recipient cells were selected for the presence of the gpt gene with media containing mycophenolic acid (MA). MA-resistant (GP+E) clones were then tested for their ability to produce reverse transcriptase (RT), the pol gene protein[14]. One third of the clones

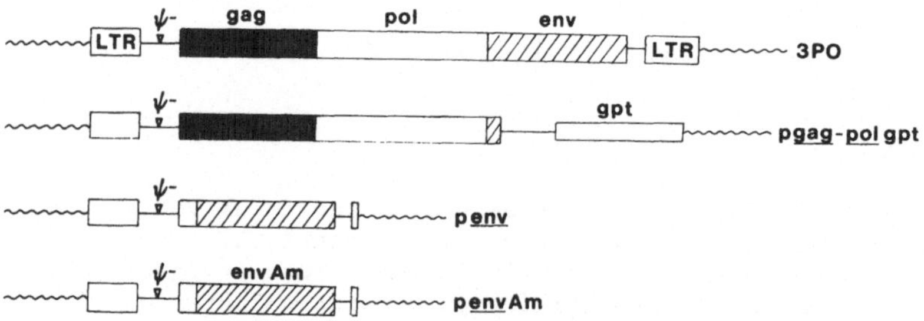

Fig. 1. Comparison of viral sequences contained in parent plasmid 3PO and constructs pgag-polgpt, penv and penvAm. Mo-MULV LTRs and the ψ deletion are indicated. Solid regions represent pol sequences; hatched regions represent env or envAm sequences; wavy lines represent pBR322 sequences.

tested produced levels of reverse transcriptase which were comparable to the levels of RT produced by 3T3 cells containing the parent plasmid 3PO. The high RT-producing clones were then tested for env protein production by immunoprecipitating labelled cellular proteins with env antisera[15].

Five of the GP+E cell lines which expressed high RT and medium-to-high env protein were tested for their ability to package the replication-defective retroviral vector Δneo. Δneo is a 6.6 kb replication-defective retroviral plasmid containing a neomycin resistance gene driven by a SV40 promoter (Figure 2). Cell lines were transfected with Δneo and G418-resistant clones were collected. Supernatants from G418-resistant clones were then tested for Δneo viral particles. The titers of GP+E+Δneo clones ranged from 2×10^2 to 1.7×10^5 CFU/ml (Table 1). These titers were comparable to Δneo titers released from the 3PO-18 packaging line, which was constructed by transfecting 3PO into NIH 3T3 cells (8×10^2 to 6.5×10^4 CFU/ml) as well as from the $\psi 2$ packaging line[6] (4.6×10^4 - 5.4×10^4 CFU/ml). The GP+E-86 packaging line produced titers that were consistently higher than the other four GP+E lines, was therefore used in subsequent experiments.

To test the effect of changing the structure of the retroviral vector containing the exogenous gene, in this case the neomycin-resistance gene, GP+E-86 cells were transfected with the N2 retroviral vector[16] in which neomycin-resistance expression is controlled by the viral LTR (Figure 2). G418-resistant clones were tested for N2 virus titer, and titers of 5.0×10^3 to 4×10^6 CFU/ml were obtained (Table 1). These titers were comparable to N2 titers released from the 3PO-18 cell line (1.85×10^4 to 5.0×10^5 CFU/ml). Thus, N2 produced titers that were 1-2 logs higher than Δneo, and the packaging line we constructed was as efficient as one in which all three retroviral components are on the same plasmid.

Analysis for Recombinant Infection Retrovirus

As a stringent test for infectious retrovirus which may have been generated through recombination events between the two helper plasmids and the Δneo retroviral vector, 3T3 cells were infected with supernatant from pooled clones containing GP+E-86+Δneo. The infected cells were passaged continuously for one month without G418 selection. This treatment would have allowed a rare wild-type virus to spread throughout the population of 3T3 cells (this population of cells should contain cells successfully infected with Δneo virus as well as uninfected 3T3 cells that are not G418-resistant), and therefore led to the spread of infectious Δneo particles. After one month in culture, supernatant was assayed for Δneo virus production by infecting fresh 3T3 cells and testing for G418

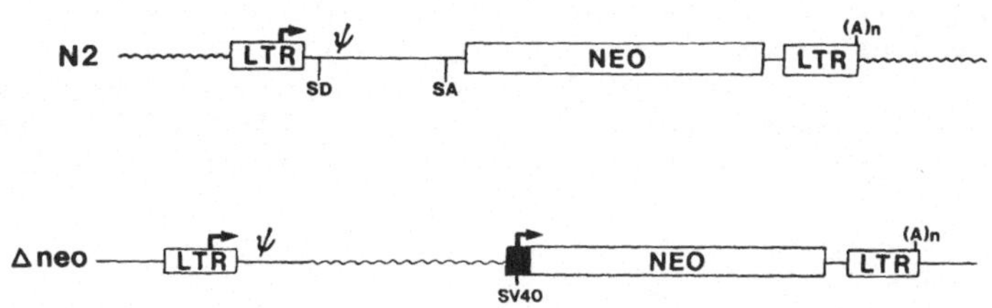

Fig. 2. Replication-defective retroviral vectors N2 and Δneo, ψ, packaging sequence; wavy lines, pBR322 sequences; solid box, SV40 promoter and origin at replication.

Table 1. Virus Production From Packaging Cells Containing Retroviral Vectors

Packaging Cell Line	Vector	# Clones Tested	Titer (CFU/ml) Range	Median	Mean
GP+E-86	Δneo	9	$2x10^2$-$1.7x10^5$	$3.3x10^4$	$5.4x10^4$
ψ^2	Δneo	2	$4.6x10^4$-$5.4x10^4$	NA	NA
3PO-18	Δneo	8	$8x10^2$-$6.5x10^4$	$6.2x10^3$	$1.47x10^4$
GP+E-18	N2	22	$5x10^3$-$4x10^6$	$7.5x10^5$	$1.38x10^6$
3PO-18	N2	9	$1.85x10^4$-$5x10^5$	$1x10^5$	$2.14x10^5$
GP+envAm-12	N2	22	$3x10^3$-$1x10^6$	$6.75x10^4$	$1.45x10^5$
PA317	N2	6	$2x10^3$-$2x10^5$	$1.15x10^5$	$1.01x10^5$

resistance. The infected 3T3 cells yielded no G418-resistant clones, indicating that there was no viral rescue of Δneo from the initial 3T3 cells that were infected with GP+E-86+Δneo supernatant.

In a different test of the safety of the GP+E-86 packaging line, supernatant from cells containing GP+E-86 was used to infect pools of N2-transfected 3T3 cells (3T3-N2 pools). If the 3T3-N2 cells became infected with wild-type virus secreted from GP+E-86 cells, the 3T3-N2 cells would begin to secrete N2 virus. Supernatant from the GP+E-86-infected 3T3-N2 pools was harvested and used to infect fresh 3T3 cells. These 3T3 cells were then tested for the presence of N2 virus by G418 selection. Using this assay we were unable to detect G418-resistant cells, demonstrating that GP+E-86 cells are unable to transfer the packaging function or to rescue N2 virus from 3T3 cells.

Construction of an Amphotropic Packaging Line

To generate a safe amphotropic packaging line we constructed the plasmid penvAm using DNA from pL1[5], a plasmid containing the 4070A amphotropic murine leukemia virus proviral DNA. A fragment containing the env gene and 3' acceptor splice site was isolated and ligated to a fragment from 3PO containing the Mo-MULV 5' LTR and 5' donor splice site (Figure 1). The plasmids penvAm and pRSVhyg[17] were co-transfected into a clone of 3T3 cells that had been transfected with the pgag-polgpt and shown to produce a high level of RT. Hygromycin B-resistant clones were isolated and tested for amphotropic env protein production by metabolic labelling followed by immunoprecipitation with env antiserum. The cell line GP+envAm--12 was selected as a clone producing high levels of both RT and amphotropic env protein. To test packaging ability, GP+envAm-12 cells were transfected with N2. G418-resistant clones were isolated and the titers of released N2 virus were determined by infecting 3T3 cells with harvested supernatants. Titers of GP+envAm-12+N2 clones ranged for 3×10^3 to $>1 \times 10^6$ CFU/ml (Table 1). In a control experiment, N2 was transfected into the amphotropic packaging line PA317[11]. Titers of G418-resistant clones, when used to infect 3T3 cells, ranged from 2×10^3 to 2.0×10^5 CFU/ml. The results indicate that GP+envAm-12 is as efficient in retroviral gene transfer as PA317.

SUMMARY

One of the requirements for the use of retroviral vectors in human gene therapy is the use of a packaging line which is incapable of producing wild-type virus. While recently-designed packaging lines are relatively safe, wild-type virus may be produced through two recombinational events between the helper virus and a replication-defective retroviral vector even with the most frequently used amphotropic line PA317. In order to create a safer packaging line, we have separated the gag and pol genes on one plasmid, and the ecotropic or amphotropic env gene on another plasmid. These plasmids contain deletions of the packaging(ψ) signal and the 3' LTR. We have developed an ecotropic packaging line (GP+E-86) and an amphotropic packaging line (GP+envAm-12) that produce titers of retroviral particles comparable to those of packaging lines containing the helper virus genome on a single plasmid.

We have found no evidence for the generation of wild-type retrovirus using the GP+E-86 packaging line, either alone or in combination with the replication-defective retroviral vectors Δneo and N2. Preliminary experiments also demonstrate that the GP+envAm-12 packaging line appears to be equally safe and therefore appropriate for use in experiments designed for human gene therapy.

ACKNOWLEDGEMENTS

We wish to thank Dr. Norma Lerner for providing the Δneo plasmid, Dr. Leslie Lobel for providing the 3PO plasmid, Dr. Eli Gilboa for providing the N2 plasmid, and Dr. Andrew Murphy for providing the pRSVhyg plasmid.

This work is supported by Public Health Service grants DK-25274, HL-37069, and HL-07230 from the National Institutes of Health; and by the March of Dimes Birth Defects Foundation.

REFERENCES

1. Cone, R.D., A. Webber-Benarous, D. Baorto, and R.C. Mulligan. 1987. Regulated expression of a complete human β-globin gene encoded by a transmissible retrovirus vector. Mol. Cell. Biol. 7: 887-897.
2. Anderson, W.F. 1984. Prospects for human gene therapy. Science 226: 401-409.
3. Neel, B.B., W.S. Hayward, H.L. Robinson, J. Fang, and S. M. Astrin. 1981. Avian leukosis virus-induced tumors have common proviral integration sites and synthesize discrete new RNAs: Oncogenesis by promoter insertion. Cell 23: 323-334.
4. Varmus, H.E., N. Quintrell, and S. Ortiz. 1981. Retroviruses as mutagens: Insertion and exision of a nontransforming provirus alter expression of a resident transforming provirus. Cell 25: 23-36.
5. Cone, R.D., and R.C. Mulligan. 1984. High efficiency gene transfer into mammalian cells: Generation of helper-free retrovirus with broad mammalian host range. Proc. Natl. Acad. Sci. 81: 6349-6353.
6. Mann, R., F.C. Mulligan, and D. Baltimore. 1983. Construction of a retrovirus packaging mutant and its use to produce helper-free defective retrovirus. Cell 33: 153-159.
7. Miller, A.D., M.-F. Law, and I.M. Verma. 1985. Generation of helper-free amphotropic retroviruses that transduce a dominant-acting methotrexate-resistant dihydrofolate reductase gene. Mol. Cell Biol. 5: 431-437.
8. Sorge, J., D. Wright, V.D. Erdman, and A. Cutting. 1984. Amphotropic retrovirus system for human cell gene transfer. Mol. Cell. Biol. 4: 1730-1737.
9. Watanabe, S., and H.M. Temin. 1983. Construction of a helper

cell line for Avian reticuloendotheliosis virus cloning factors. Mol. Cell. Biol. 3: 2241-2249.

10. Hock, R.A., and A.D. Miller. 1986. Retrovirus-mediated transfer and expression of drug-resistant genes in human haematopoietic progenitor cells. Nature 320: 257-277.

11. Miller, A.D., and C. Buttimore. 1986. Redesign of retrovirus packaging cell lines to avoid recombination leading to helper virus production. Mol. Cell. Biol. 6: 2895-2902.

12. Mulligan, R.C., and P. Berg. 1980. Expression of a bacterial gene in mammalian cells. Science 209: 175-183.

13. Potter, H., L.W. Weir, and P. Leder. 1984. Enhancer-dependent expression of human K immunoglobulin genes introduced into mouse pre-B lymphocytes by electroporation. Proc. Natl. Acad. Sci. 81: 7161-7165.

14. Goff, S.P., P. Traktman, and D. Baltimore. 1981. Isolation and properties of Molony murine leukemia virus mutants: Use of a rapid assay for release of virion reverse transcriptase. J. Virol. 38: 239-248.

15. Schwartzberg, P., J. Colicelli, and S. Goff. 1983. Deletion mutants of Molony murine leukemia virus which lack glycosylated gag protein are replication competent. J. Virol. 46: 538-546.

16. Keller, G., C. Paige, E. Gilboa, and E.F. Wagner. 1985. Expression of a foreign gene in myeloid and lymphoid cells derived from multipotent haematopoietic precursors. Nature 318: 149-154.

17. Murphy, A.J. 1987. Molecular techniques for the isolation of transcriptional transacting genes. Doctoral Thesis, Columbia University 133-139.

LINEAGE SPECIFIC EXPRESSION OF A HUMAN β-GLOBIN GENE IN MURINE BONE MARROW TRANSPLANT RECIPIENTS

Elaine A. Dzierzak and Richard C. Mulligan

INTRODUCTION

Introduction of foreign genes into the hematopoietic system of adult animals has been accomplished by recombinant retroviral infection of the pluripotent bone marrow stem cell[1-3]. Retroviral vectors are highly efficient for gene transduction of this rare cell type which is present in bone marrow at a frequency of approximately 1 in 10^5. Engraftment of lethally irradiated recipients with infected bone marrow results in the reconstitution of all mature blood cell lineages from the transduced stem cell(s). The unique integration property of the provirus allows for the lineage mapping of individually transduced stem cell(s) and the assessment of the developmental and dynamic nature of hematopoiesis *in vivo*[3]. While retroviruses have been demonstrated to serve as efficient vectors of the tranduction of exogenous genes into the hematopoietic system, there have been only limited demonstrations of expression of such inserted sequences[1,2,4,5].

Distinct strategies for transcription of gene sequences inserted in retroviruses can be applied for expression in either a lineage specific or constitutive fashion within the developing blood system. Most studies have relied on viral transcription signals or other constitutive promoter sequences to drive transcription of inserted cDNAs. Alternatively, we have constructed retroviruses containing a normal cellular human β-globin gene[6,7] to determine whether genomic sequences can act normally in the context of a retrovirus. We have tested for levels of human β-globin expression as well as lineage specific expression in murine bone marrow transplant recipients reconstituted with retrovirally transduced stem cells[7].

RESULTS

Four to nine months after transplantation of male recipient mice with infected female bone marrow, 18 out of 104 animals were found to harbor the retrovirally transduced gene. Eight mice were chosen for further analysis. Proviral sequences in the spleen and bone marrow DNA of these animals were observed by Southern blot analysis to be of the correct structure and to be present at a frequency of 0.02 to 0.4 copies per genome. When percent engraftment was determined by probing for Y-chromosome specific DNA sequences, the average efficiency of infection was 10% and ranged in individual animals from 2% to 100% of the engrafted bone marrow.

To assess whether the transduced human β-globin gene was expressed in the transplanted animals, SP6 RNA protection analysis was performed on blood RNA. All eight mice were found to express human β-globin but to varying levels. In each animal the level of human β-globin mRNA was compared to the level of endogenous mouse β-globin mRNA. Since only a fraction of stem cells contributing to the mature blood population were infected with the human β-globin recombinant retrovirus, the absolute level of human β-globin expression was determined by correcting for proviral copy number in each transplant recipient. The level of human β-globin expression ranged from 0.4% to 4.0% of the endogenous mouse β-globin level when considered at a single copy per diploid genome. This level of expression is consistent with the levels of expression (at a single copy per genome) observed in germline transgenic mice or in MEL cells when the human β-globin gene is transduced by conventional methods and suggests that retroviral sequences have no effect on the level of expression of this gene. Furthermore, human β-globin protein is produced in the red blood cells of these animals as observed by immunofluorescence with a human β-globin specific monoclonal antibody. The percentage of stained cells in the blood correctly reflects the proviral copy number, suggesting expression in all erythroid cells harboring proviral sequences.

The developmental fate of the pluripotent bone marrow stem cell is to replenish all components of the erythroid, myeloid and lymphoid lineages of the mature blood system. Within each individual animal infected with a human β-globin virus, we observed by Southern blot analysis the same unique proviral integrant in each of these lineages demonstrating reconstitution by an infected stem cell. When RNA from each of these lineages was tested by SP6 RNA protection analysis, high levels of human β-globin mRNA were found only in the erythroid lineage. B and T lymphocytes and macrophages expressed only very low or undetectable levels of human β-globin mRNA. Interestingly, an even more strict lineage specific expression of human β-globin was observed in two mice engrafted with cells infected by a virus in which the enhancer element normally present in the retroviral LTR was deleted. Thus, human β-globin regulatory sequences function appropriately to give only erythroid specific expression in vivo when in context of a retroviral vector.

The results presented here demonstrate that it is possible to obtain the stable, regulated and quantitatively appropriate levels of expression in vivo of a normal cellular gene transduced via retroviral infection of a bone marrow stem cell. These experiments suggest that it may be important to use gene sequences that contain transcriptional signals normally active in hematopoietic cells for efficient and appropriate expression of the retrovirally transduced gene. Although the levels of human β-globin expression have not reached the levels of expression of the endogenous mouse gene, it is most likely due to the segment of the human gene used for these experiments. Additional regulatory sequences have been recently described[8] and will be important for future studies, especially with applications to potential human gene therapy. Finally, with improvements in the efficiency of stem cell infection and higher levels of expression it may be appropriate to reconsider globin disorders as potential candidates for gene therapy.

REFERENCES

1. Dick, J.E., et al. 1985. Cell 42: 71-79.
2. Keller, G., et al. 1985. Nature 318: 149-154.
3. Lemischka, I.R., et al. 1986. Cell 45: 917-927.
4. Eglitis, M.A., et al. 1985. Science 230: 1395-1398.
5. Bowtel, D.D.L., et al. 1987. Mol. Biol. Med. 4: 229-250.
6. Cone, R.D., et al. Mol. Cell. Biol. 7: 887-897.
7. Dzierzak, E.A., et al. 1988. Nature 331: 35-41.
8. Grosveld, F., et al. 1987. Cell 51: 975-985.

A COMPARISON OF METHODS FOR ANALYSIS OF mRNA IN HEMATOPOIETIC CELLS: CONVENTIONAL AND CLONAL NORTHERN ANALYSIS AND IN SITU HYBRIDIZATION

Cathy L. Castiglia

Sloan-Kettering Institute
New York, N.Y.

ABSTRACT

Three methods for the detection of mRNA in hematopoietic cells are addressed, and examples using these techniques to examine human leukemic cells' genetic response to chemical and biological response modifiers shown. The limitations and specific applications of the three techniques are discussed.

INTRODUCTION

A dramatic change in the utilization of a specific gene can serve as a useful marker of cellular activity and response. The study of gene expression will lead to a more detailed understanding of events occurring during development of the hematopoietic system, which contains many cell types at various stages of differentiation. This heterogeneity complicates molecular biological analyses, as signals from members of the population that are few in number are diluted out, making them difficult to identify. Even clonally derived cell lines behave heterogeneously in response to growth and differentiation factors. While the various components of blood and bone marrow may be separated prior to analysis, the diminishing amount of material available for study creates technical problems. Together with the necessity for time consuming and often difficult purification of cellular mRNA, these constitute the main limitations of conventional Northern analysis, using either gel electrophoretic or dot blot techniques.

One resolution has been to probe the mRNA content of hematopoietic colonies grown in semi-solid medium[1]. A population enriched for progenitors is suspended in soft agar and grown under defined conditions after which the culture is harvested, the cells lysed _in situ_ and their contents transferred to a nylon membrane. The blot is then probed using standard hybridization protocols. Our understanding of leukemia and its therapy will be enhanced by the ability to analyze the clonal derivatives of clinical samples, subject them to environmental manipulation, and examine their genetic response. However, relatively few primary AML samples form discrete colonies in agar; a compact morphology is important for optimal sensitivity. Moreover, a high proportion of AML fail to form colonies at all. For these reasons, I have adapted _in situ_ hybridization for examination of hematopoietic samples.

In this method, individual cells are fixed by acidic dehydration and cellular transcripts hybridized to labelled nucleic acid probes. No post-fixation digestion by protease, usually required to increase accessibility of the mRNA bound to the cellular substrata, is necessary, making co-detection of cellular antigens by monoclonal antibodies possible. The resulting specimen can be analyzed by autoradiography or histochemical methods. Clearly, due to its convenience and its universal applicability, in situ hybridization is the method of choice, and should contribute greatly to the diagnosis, treatment and basic understanding of disease.

MATERIALS AND METHODS

A. Conventional Northern Analysis

Cellular mRNA was prepared from no less than 1 x 10^7 cells by cell lysis in guanidinium isothiocyanate buffer followed by centrifugation through CsCl to separate the RNA from DNA and protein[2]. Poly(A)-containing RNA was isolated by chromatography using oligo (dT) cellulose[3] and glyoxalated to generate fully denatured molecules which could be resolved according to molecular weight by agarose gel electrophoresis[4]. Nonpolyadenylated RNA was glyoxalated and run in the last lane of each gel; ribosomal RNA serve as 4.9 kb and 1.8 kb molecular weight markers. The RNA was then transferred to Nytran hybridization paper (Schleicher and Schuell; Nashua, N.H.) by capillary action using 10x SSC, and the resulting blot baked for 2 hours at 80°C in vacuo to remove the glyoxal adduct.

B. Clonal Northern Analysis

The agar disk containing hematopoietic colonies was generated by culturing leukemic cells or normal progenitors under conditions which promote CFU-GM growth. An underlayer of agar containing all the components of the culture except for the cells is used in order to prevent the disk from being too thin in the center for easy manipulation. The sample is harvested by floating the two layers into a dish of PBS, maintaining orientation with the colony-containing disk on top. After picking up the agar with a 75mm x 50mm glass slide, the upper disk is made to adhere to a 50 mm^2 square of 3MM filter paper (Whatman; Maidstone, England) and the lower layer allowed to slough off. The paper is trimmed to slightly larger than the size of the disk and the disk inverted onto Gene-Screen Plus hybridization membrane (New England Nuclear; Boston, MA), which has been prewetted according to the manufacturer's directions and placed atop a porous polypropylene filter fitted into a vacuum chamber (Figure 1), (Kwik--Ezee Inc.; Westbury, N.Y.). Vacuum is applied, and a solution of filtered 10x SSC wicked through the agar, using the 3MM paper to direct the flow. If a replicate filter is desired, a second sheet of Gene-Screen Plus can be placed beneath the first, and 2x SSC used to reduce binding of RNA to the top filter. Cellular RNA, DNA and protein are then transferred to the membrane by removing the 3MM paper and dissolving the agar with formamide heated in a boiling water bath. This destroys the hydrogen bonding which holds agar together, and denatures the RNA such that it binds more efficiently to the nylon membrane. It is imperative that nitrocellulose not be used with this method; it will dissolve. Additionally, Nytran is not as efficient as Gene-Screen Plus at retaining RNA under these transfer conditions.

After hot formamide treatment, which is monitored visually, the blot is thoroughly rinsed with 10x SSC, the vacuum broken, and the blot air dried. No baking is necessary, and storage between sheets of 3MM paper at room termperature under dessication renders the blot stable for long periods of time.

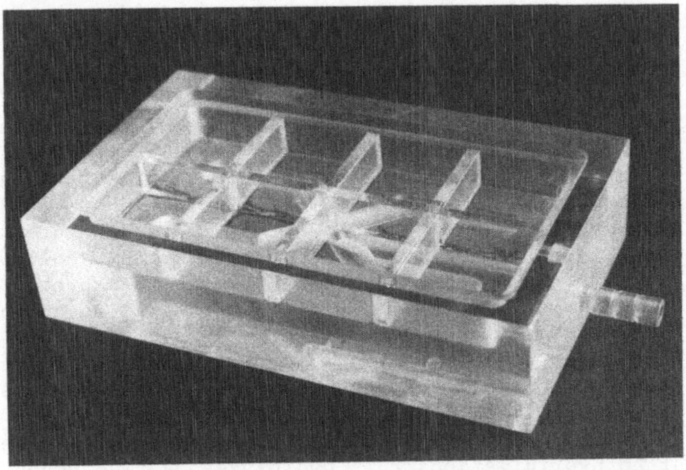

Fig. 1. The vacuum chamber used for transfer of the mRNA contained in hematopoietic colonies to hybridization membrane. Eight cultures may be harvested simultaneously.

C. In Situ Hybridization

Cells to be analyzed are suspended in 10% FCS and either cytocentrifuged onto clean glass slides or allowed to adhere to Lab Tek chambers (Miles; Naperville, IL) for 2 hours at 37°C, after which they are briefly air dried. The fixation procedure takes place at room temperature, beginning with 2-3 minutes in 70% ethanol/10% acetic acid, and then proceeding for four or more minutes each through A) 50% ethanol/50% PBS, B) 70% ethanol/ 30% PBS, C) 85% ethanol, D) 95% ethanol, after which the slides are air dried and stored in a dust-free environment awaiting analysis. I have found the PBS is essential for neutralization of the acetic acid, which over time would otherwise degrade the cellular mRNA.

The hybridization mix consists of 10mM Tris-HCl, pH 7.4, 0.3M NaCl, 1mM EDTA, 1X Denhardt's solution, 50mM DTT, 50% formamide, 0.5 mg/ml poly(A), 0.5 mg/ml tRNA, and 1-2 x 10^6 cpm/ml heat-denatured ^{35}S-labelled DNA. Enough mix is applied to cover the sample, on top of which a siliconized cover slip is placed. Slides are incubated at 37°C in chambers humidified with 50% formamide, such that neither evaporation nor dilution of the hybridization mix occurs.

After an overnight incubation, the reaction is quenched by dunking the slide into an excess of 2x SCC at room temperature, at which time the cover slip floats off. Slides are transferred to fresh 2x SSC and placed at 37°C for 24 hours without agitation. The solution is changed to 1x SSC and the wash continued at 37°C for an additional 4-8 hours. Slides are then dried by passing through an ethanol series, and dipped in Kodak NTB-2 nuclear track emulsion. After exposure at 4°C in the presence of a dessicating agent for the appropriate period of time, which is determined by monitoring "test" slides that are developed every few days until the desired grain density is achieved, the samples are developed using Kodak D-19 developer for 5 minutes, H_2O as a stop bath for 30 seconds, Kodak fixer for 5 minutes, and H_2O as a rinse for 15 minutes. Morphology can be determined by Giemsa stain at pH 7.0 for 1-2 hours (M. Griffin, personal communication).

D. Probe Preparation

DNA probes were generated by gel purification of restriction fragments containing the sequences of interest and extension of hexanucleotide primers randomly hybridized to the heat-denatured DNA by the Klenow fragment of DNA polymerase in a reaction containing ^{32}P- or ^{35}S-dCTP[5]. The reaction was allowed to proceed for 5 hours at room temperature and terminated by the addition of an equal volume of 2X urea buffer (7.5M urea, 2% SDS, 5mM EDTA, 0.4M NaAc, pH 7.0) (Modification of 6). This denatures the DNA-binding proteins in the reaction mix and prevents their collapse onto the nucleic acid during the subsequent phenol/chloroform extraction. In the absence of such preventative measures, the residual peptide bound to the probe may cause a bridge to form between the probe and the hybridization paper or cellular substrata, leading to non-specific signals that are difficult, if not impossible, to prevent by the use of stringent hybridization conditions or eliminate by harsh washing protocols. Many researchers have circumvented this problem inadvertantly by using alkali to denature their probe prior to hybridization; this hydrolyzes peptide bonds and thereby removes residual proteins. This method, however, is not compatible with the use of riboprobes, which have proven to be more sensitive due to the inherently higher stability of an RNA-RNA hybrid.

Following treatment with urea buffer, the probe is separated from unincorporated nucleotides by Sephadex G-50 chromatography, extracted with phenol/chloroform and precipitated by the addition of 2.5 volumes of ice cold 95% ethanol. It should be noted that tubes containing radioactive materials in volatile solvents should be sealed with Parafilm to prevent the aerosol created during centrifugation from contaminating equipment.

RESULTS

Clonal Northern Analysis

Figure 2 shows the signal obtained from colonies of HL-60 cells grown in 0.3% agar and exposed to either chemical or biological agents. The HL-60 cell line was derived from the peripheral blood leukocytes of a patient with acute promyelocytic leukemia and differentiates along the granulocytic pathway, albeit defectively, after treatment with DMSO[7]. Using a tubulin probe as a measure of baseline mRNA content in these colonies, it is clear that treatment with 1.2% DMSO leads to less signal/colony than untreated controls, as would be expected since the cells cease division as they terminally differentiate, and fewer cells/colony means less mRNA as well.

Human granulocyte colony stimulating factor (hG-CSF) has also been shown to cause granulocytic differentiation of HL-60 cells[8,9] although the process is slower than when DMSO is used (data not shown). Since proliferation was inhibited at a later time, the tubulin mRNA content of HL-60 colonies grown in saturating concentrations of recombinant hG-CSF (10,000 U/ml, product of Amgen Biologicals) was less than that of control, but greater than that of DMSO-treated samples. The variable sensitivity of cells within a cloned line to biological response modifiers is also evident. Their heterogeneous response leads to different degrees of proliferation in each colony, and as a result the signals are of varying intensity.

Having a reasonable measure of mRNA/colony under these conditions, it was possible to examine the change in genetic expression of a cell surface antigen which had been studied at the protein level. As there were indications that rhG-CSF up-regulated the IL-2 receptor in HL-60 cells (J. Gabrilove, M. Firpo, personal communication) I undertook a study

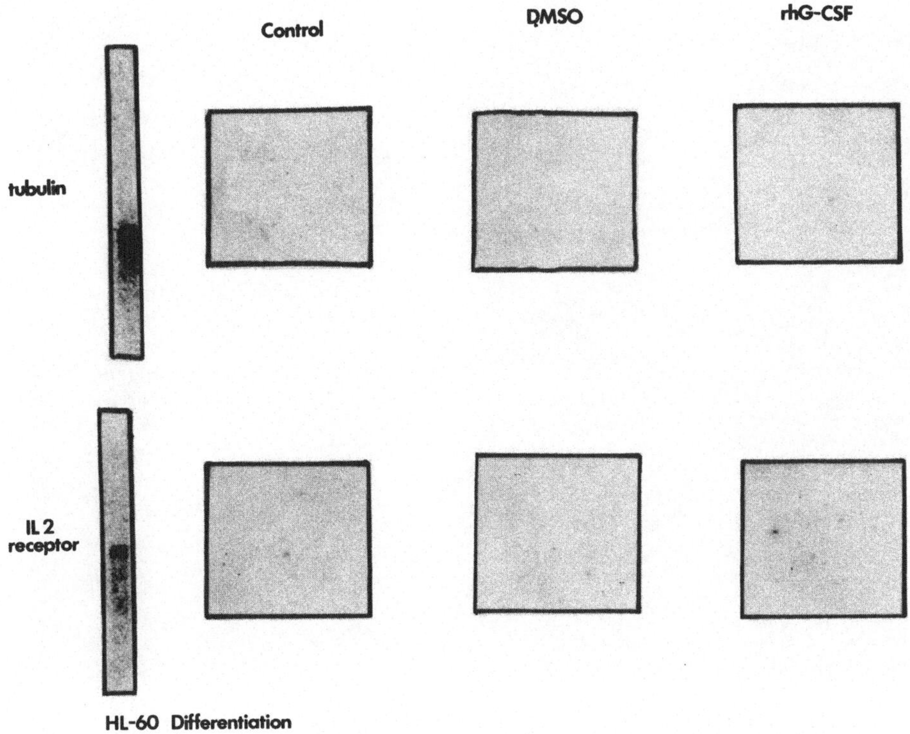

Fig. 2. Clonal northern analysis of HL-60 human promyelocytic leukemic cell grown in semi-solid medium for 5 days and then treated with 10,000 U/ml rhG-CSF (amgen, Thousand Oaks, CA) or 1.2% DMSO for 5 days. The dots on the autoradiogram represent signals from colonies expressing tubulin and IL 2 receptor mRNA. An autoradiogram from convential northern analysis of mRNA isolated from HL-60 cells and hybridized in parallel with the RNA colony blot to ensure the appropriate stringency has been used is shown to the left.

at the molecular level using a probe complementary to c-Tac. Compared to control colonies, rhG-CSF-treated samples do indeed contain more IL-2 receptor mRNA. Unfortunately, signal from the DMSO-treated sample was too low to be conclusive.

In Situ Hybridization

It is a of great scientific and medical importance to understand the genetic states that are associated with leukemia. Although this can be accomplished in some cases by Northern analysis, the ability to simply and rapidly analyze individual cells will facilitate research of this kind. The results of one such study are shown in Figure 3. Bone marrow from a patient presenting with high blast count M-5 leukemia was depleted of accessory cells and incubated in IMDM/10% FCS; 5,000 U/ml rhG-CSF was added to half the culture. After 36 hours at 37°C, the cells were cytocentrifuged and fixed as described in materials and methods. An ^{35}S-labelled DNA probe was used to detect c-myc transcripts within the blasts; very few were positive, and this pattern did not change with G-CSF treatment. It is likely that the few cells expressing high levels of myc mRNA are not part of the leukemic clone.

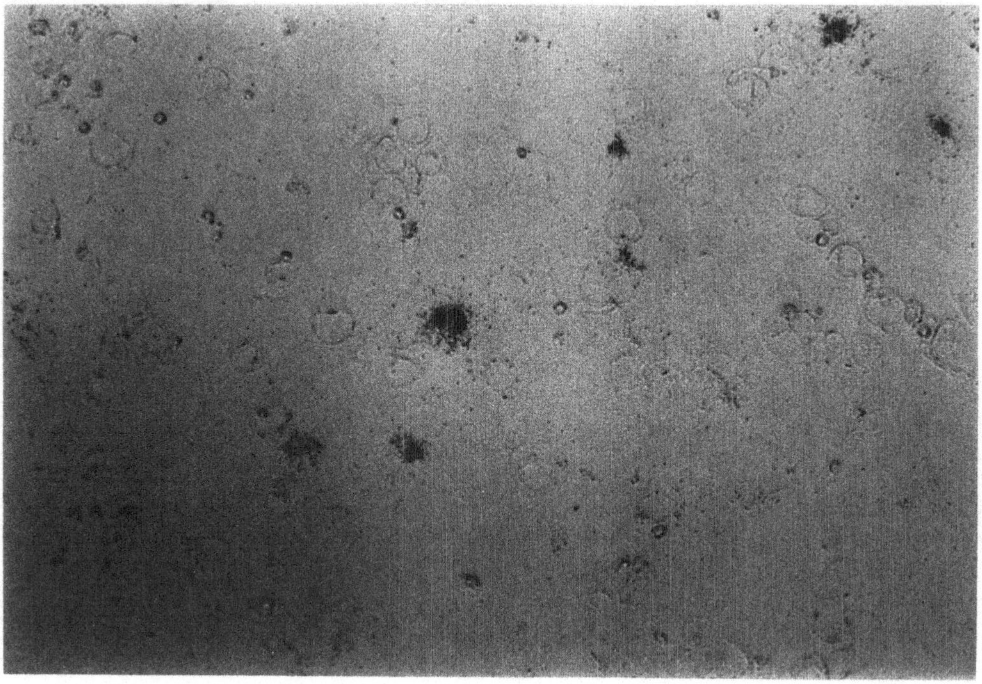

Fig. 3. In situ hybridization of acessory cell-depleted bone marrow from a patient with advanced monymyelocytic leukemia. This photomicrograph (400X magnification) shows the autoradiographic signal produced by the decay of ^{35}S-labelled DNA probe complementary to c-myc hybridized to cellular mRNA. Exposure: 9 days.

Figure 4 depicts the change in expression of c-myc and the interferon-inducible 2'5'-oligo(A) synthetase gene in HL-60 cells exposed to high doses of rhG-CSF (10,000 U/ml) or recombinant human beta-1 interferon (rhB-1 IFN, 500 U/ml; kind gift of M. Revel). As expected from an agent that induces terminal differentiation, c-myc levels dropped significantly after incubation of HL-60 in rhG-CSF for 7 days; rhB-1 IFN, however, has no effect. This agrees with growth curve data (not shown); although the other human leukemic cell lines tested (HEL, U937) are exquisitely sensitive to growth inhibition by rhB-1 IFN (data not shown), HL-60 cells are resistant. The reason why becomes clear upon inspection of synthetase mRNA in HL-60 control cells; this inducible transcript is constitutively expressed. Incubation in rhG-CSF does not alter synthetase expression, although exogenous B-1 IFN doubles the cellular content of synthetase mRNA.

CONCLUSIONS

With three methods available for the analysis of gene expression, as defined by the mRNA content of cells, molecular biologists are in an enviable position. When technically feasible, conventional Northern analysis offers a wealth of information; by resolving the mRNA population according to transcript length, polymorphic mRNAs have been revealed, processing intermediates identified, and mRNA related through moderate sequence homology discovered. For the scientist new to this technology, as well as those working with a probe of unknown sequence, a single band on a Northern blot is good evidence that the appropriate specificity of detection has been attained.

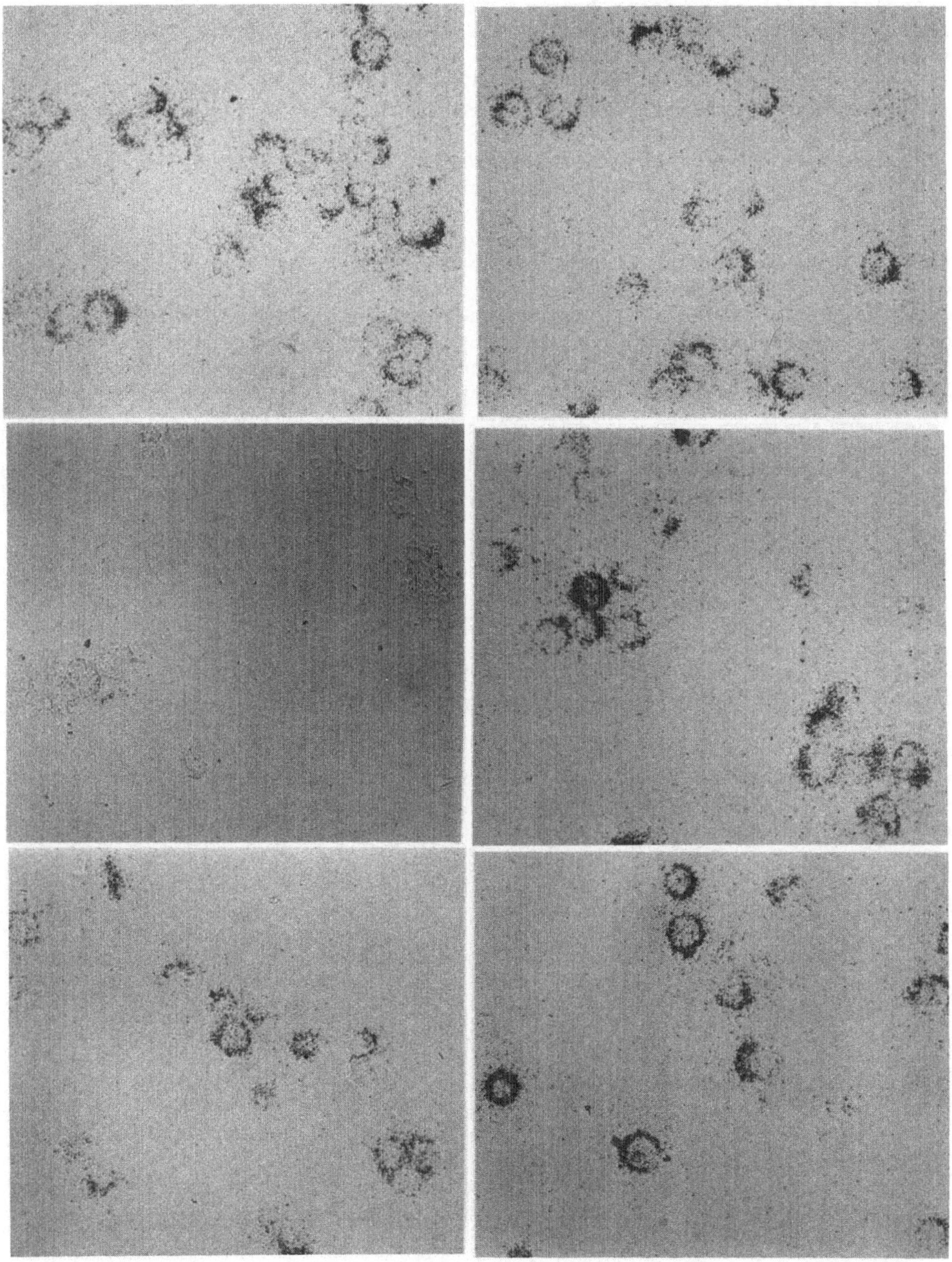

Fig. 4. *In situ* hybridization of HL-60 human promyelocytic leukemia cells. Top row: untreated samples. Middle row: cells exposed to 10,000 U/ml rhG-CSF for 7 days. Bottom row: cells exposed to 500 U/ml beta-1 interferon (kind gift of M. Revel) for 7 days. The left column shows the results obtained with a probe for c-myc transcripts; the right column shows the signal obtained with a probe for the interferon-inducible 2'5'-oglio(A) synthetase gene. Magnification: 400X. Exposure: 9 days.

However, as noted earlier, sample heterogeneity and size can limit the applicability of this technique. These can be resolved in the hematopoietic system by clonal expansion of progenitors in semi-solid medium, and analysis of the RNA content of the resulting colonies. In addition to the examples shown here, the RNA colony blot technique has proven useful in the analysis of retrovirally-infected bone marrow before and after engraftment in murine models (G. Keller, personal communication) and is currently being used to assess success of ADA gene replacement therapy in humans at the Sloan-Kettering Institute.

Unfortunately, this technique only allows one to examine the colonies resulting from maturation of precursors affected by in vitro culture conditions. A complete picture of gene expression during developmental hematopoiesis will require the identification of mRNA in selected normal progenitors, both exposed to a variety of cytokines and grown in contact with the microenvironment. It is in this context, as well as in diagnosis and evaluation of therapeutic modalities using patient material in vitro, that the power of in situ hybridization will become evident. With the development of more sensitive techniques, rapid screening will become a reality, and our understanding of nature's secrets advanced at a hitherto unattainable pace. An exciting time is ahead for us all.

ACKNOWLEDGEMENTS

I would like to take this opportunity to recognize the efforts of Dr. Oliver Ottmann during the development of the in situ technique, and thank him for his encouragement. I would also like to thank Dr. Janice Gabrilove for her advice and unshakeable belief in the future of this project, and thank Oliver, Janice, Meri Firpo, Michel Revel, Larry Souza, Karl-Walter Sykora, Chris Paige, Norman Iscove, and Malcolm A.S. Moore for helpful discussions, materials and support. This work was supported by a post-doctoral fellowship from the Cancer Research Institute and the Gar Reichman Foundation.

REFERENCES

1. Paige, C.J., G. Wu, and C.L. Castiglia. 1986. Detection of RNA transcripts in normal lymphoid and myeloid colonies. J Imm Methods 93: 37.
2. Chirgwin, J.M., A.E. Przybyla, R.J. MacDonald, and W.J. Rutter. 1979. Isolation of biologically active ribonucleic acid from sources enriched in ribonuclease. Biochem 18: 5394.
3. Aviv, H., and P. Leder. 1972. Purification of biologically active globin messenger RNA by chromotography on oligothymidylic acid-cellulose. Proc Natl Acad Sci. 69: 1408.
4. Carmichael, G.G., and G.K. McMaster. 1980. The analysis of nucleic acids in gels using glyoxal and acridine orange. Methods in Enz. 65: 380.
5. Feinberg, A.P., and B. Vogelstein. 1984. A technique for radiolabeling DNA restriction endonuclease fragments to high specific activity. Anal Biochem. 137: 266.
6. Holmes, D.S., and J. Bonner. 1973. Preparation, molecular weight, base composition, and secondary structure of giant nuclear ribonucleic acid. Biochem. 12: 2330.
7. Gallagher, R., S. Collins, J. Trujillo, K. McCredie, M. Ahearn, S. Tsai, R. Metzgar, G. Aulakh, R. Tine, F. Ruscetti, and R. Gallo. 1979. Characterization of the continuous, differentiating myeloid cell line (HL-60) from a patient with acute promyelocytic leukemia. Blood 54: 713.

8. Platzer, E., K. Welte, J.L. Gabrilove, L. Lu, P. Harris, R. Mertelzmann, and M.A.S. Moore. 1985. Biological activities of a human pluripotent hemopoietic colony stimulating factor on normal and leukemic cells. J. Exp. Med. 162: 1788.
9. Harris, P.E., P. Ralph, J.L. Gabrilove, K. Welte, R. Karmali, and M.A.S. Moore. 1985. Distinct differentiation-inducing activities of gamma-interferon and cytokine factors acting on the human promyelocytic leukemia cell line HL-60. Cancer Res. 45: 3090.

HAEMOPOIETIC REGULATION AND THE ROLE OF THE MACROPHAGE IN ERYTHROPOIETIC GENE EXPRESSION

Ivan N. Rich

Department of Transfusion Medicine
University of Ulm
Ulm, F.R.G.

SUMMARY

The macrophage is considered as an "active" component of the haemopoietic cellular microenvironment with respect to erythropoietin (epo) production during embryonic, foetal and adult erythropoiesis. Emphasis is placed on steady-state rather than pathophysiological conditions. In addition, the signals capable of affecting the functional capacity of the macrophage with regard to colony stimulating factor and epo production are also taken into account. Evidence is given demonstrating that a subpopulation of resident macrophages in vitro and in the mouse bone marrow, under normal conditions, can express the epo gene. These results indicate that erythropoiesis can be regulated by short-range or cell-to-cell interactions within the bone marrow.

HAEMATOPOIETIC ONTOGENY, MICROENVIRONMENT AND THE COMMON DENOMINATOR

Mammalian haemopoietic development follows a sequential and orderly pattern of events. Haemopoiesis is initiated in the yolk sac and transferred to the fetal liver, spleen (not important in man) and finally the bone marrow. It may be assumed that during ontogeny, rapid growth is one of the primary reasons for the expansion requirements for blood cell formation, so that by the end of the neonatal period, haematopoiesis does not occur in single organs but in the combined marrow of the bones.

The transition in haemopoietic sites is not only accompanied by a change in organ, but also by a change in the predominant cell type being produced. Thus, whereas the yolk sac and early foetal liver are predominantly erythropoietic organs, the bone marrow is myelopoietic. This may be one of the reasons why in long-term bone marrow cultures myelopoiesis is more predominant than erythroid progenitor cells[1].

Several investigations suggest that the migration and seeding of haemopoietic stem cells from the yolk sac are responsible for the ontological haemopoietic transitions[2,3]. This has recently been taken a step further by Wong et al.[4,5] showing that two primitive populations are present in the murine yolk sac, one of which may be capable of colonising the fetal liver.

Exactly how and why hemopoiesis is initiated in the yolk sac is still unclear. The first erythroblasts appear as blood islands in the yolk sac by apparent intimate contact between the mesodermal cells and the underlying endoderm. Whether the haemopoietic initiation in this organ is due to the proliferation and differentiation of totipotent stem cells[6] whose genome has, in some way, been previously programmed to specifically produce haemopoietic cells is unknown. On the other hand, the appearance of blood islands similar to those described by Bessis and Breton-Gorius in 1962[7], would imply that hemopoietic cells other than erythroblasts must be present at this time during ontogeny and are presumably a prerequisite for erythropoiesis.

Another controversy associated with erythropoietic ontogeny is whether erythropoietic precursor cell growth is independent of erythropoietin (epo). By subjecting pregnant rodents to polycythaemic conditions, it was found that foetal erythroid production continued in a normal manner[8,9]. The interpretation of these results was that adult and embryonic/foetal erythropoiesis were controlled by different mechanisms and that the embryonic/foetal erythropoietic system was autonomous[8,9]. However, Bleiberg and Feldman[10] indicated that very low levels of endogenous epo could be present in polycythaemic animals. This in turn implied that adult bone marrow erythropoietic precursor cells could be less sensitive to epo than fetal liver cells.

Using suspension cultures, Cole and Paul[11] demonstrated for the first time that mouse foetal liver, but not yolk sac cells could be stimulated by epo. However, loss in erythropoietic stimulation potential was attributed to loss in epo sensitivity[11,2]. The development of the erythroid colony-forming technique in methyl cellulose [13] allowed a reappraisal of epo sensitivity during ontogeny. Not only were the late erythropoietic progenitor cells (CFU-E) in the foetal liver sensitive to epo throughout the hepatic stage of gestation[14], but the epo sensitivity of the CFU-E population was found to decrease with age[15]. Although Cole and Paul[11] could not detect stimulation of mouse yolk sac cells by exogenous epo in culture, the addition of epo to yolk sacs in organ culture allowed the haemoglobin switch to occur[16,17]. This, together with the change in epo sensitivity from the foetal liver to the adult stage of erythropoietic development, implies that yolk sac erythropoietic precursor cells may not only be sensitive to epo, but possibly more sensitive than foetal liver precursor cells.

In vitro investigations demonstrated that the foetal liver macrophage is probably the source of epo during hepatic erythropoiesis[18]. Morphologically, the macrophage is one of the first, if not the first cell to be recognized in the foetal liver rudiments[19]. The question is of course, why? It is because the macrophage provides a cellular microenvironment capable of supporting hepatic erythropoiesis? Is it because, for erythropoiesis to occur in the foetal liver, a source of epo has first to be produced? Is the macrophage responsible for both of these functional capacities? And are these functional capacities one and the same? The answers to these questions are still outstanding. However, they apply not only to hepatic erythropoiesis, but to all the haemopoietic organs, including the yolk sac.

Another overlooked, frequently observed phenomenon is that haemopoiesis can, in part, retrace its path of ontogeny. This simply means that under certain perturbations, haemopoiesis can be reinitiated in the adult spleen and even in the liver. Examples of this occur during haemolytic anaemia caused by phenylhydrazine[20], bleeding[21], irradiation[22], Friend and Rauscher virus infection[23,24,25] and after bone marrow transplantation[26]. However, probably the most interesting example of this phenomenon is observed during pregnancy[27,14], since this represents a naturally perturbated situation.

For haemopoiesis to be capable of retracing its path of ontogeny, a cellular microenvironment must exist that is common to only the bone marrow, spleen and liver. Furthermore, this same cellular microenvironment, or one very similar to it, must exist in the yolk sac and foetal liver during ontogeny. Indeed, the erythropoietic supporting, and epo-producing capacities of the foetal liver appear to be carried over to the adult. This is observed not only during extramedullary haematopoiesis, but also as part of the extra-renal epo production system[28,29].

The haemopoietic cellular microenvironment consists of two discrete biological systems. The first is the reticuloendothelial system comprising mostly of fibroblasts, endothelial and reticular cells and adipocytes. The second is the mononuclear phagocyte system consisting of monocytes, macrophages and macrophage-like cells present in all areas of the body. There is no doubt that the haemopoietic cellular microenvironment plays an extremely important role in haematopoietic regulation, presumbably by its ability to release regulatory molecules affecting the haemopoietic target cells. However, to date, the study of the haemopoietic cellular microenvironment has been primarily qualitative and implicative. But, the quantitative effect of the haemopoietic cellular microenvironment and the unequivocal demonstration of haemopoietic regulator molecule production by specific members of the cellular microenvironment has yet to be performed.

It is clear from the above discussion that the common denominator associated with both the haemopoietic cellular microenvironment and erythropoiesis, is the macrophage.

THE MACROPHAGE AND STEADY-STATE CONDITIONS

In general, a steady-state condition may be regarded as that in which all biological activities remain within certain limits of an equilibrium or homeostasis. The haemopoietic system may be considered to be in steady -state and therefore in equilibrium when cell renewal equals cell death. For the macrophage, the term steady-state is generally taken to mean a "resting" or "non-activated" cellular state. The difference between this "resting", steady-state situation and an "activated", non-steady-state situation has been reviewed by Adams and Hamilton[30].

The ubiquitous presence of the macrophage in the body demonstrates the importance of this cell's functional capacity. Probably the most important capacity of the macrophage is to sense various external signals and respond to them by adapting, modifying or changing its functional capacity in order to maintain the homeostasis of other biological systems. In short, the macrophage acts as a surveyance cell.

The macrophage, together with the neutrophil, forms the first, non-immunologic line of defense against foreign particles. Unlike the neutrophil, however, it is an absolute requirement for the second, immunologic line of defense. Under inflammatory conditions for example, it is not only responsible for the interaction with the T-lymphocyte system, but also changes its metabolism, physiology and morphology, resulting in what is sometimes called an "angry" macrophage. In this condition, the macrophage is capable of regulating the production and release of a multitude of different substances suited to the situation. This could be described as a temporary non-steady-state situation.

Although activated macrophages are useful to study the regulation and expression of functional capacities, macrophage activation can be considered to be limited to relatively short periods of time in the life of a mammal. Predominant, are the macrophage functional capacities that occur during normal, steady-state conditions.

EXTERNAL SIGNALS AND THE ROLE OF THE MACROPHAGE IN HAEMOPOIETIC REGULATION. (Figure 1)

Unfortunately, the functional capacities of macrophages under normal, steady-state conditions are extremely difficult to study. Yet, in the day-to-day regulation of both myelopoiesis and erythropoiesis, the macrophage plays an extremely important role.

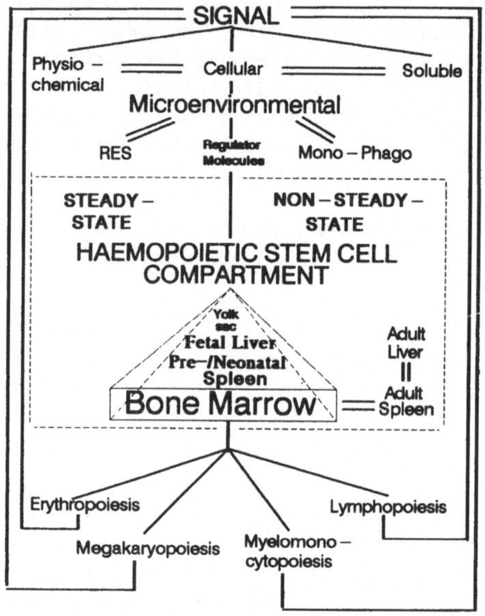

Fig. 1. Haemopoiesis: The role of biological signals regulating steady and non-steady-state haemopoiesis during ontogeny, adult and pathophysiological conditions. The mechanisms involved in the homeostatic production and death of functional haemopoietic end cells are fed back as biological signals of different types. These can be physio-chemical, as in the change in partial oxygen tension released by the erythrocytes in organs and tissues, or soluble as demonstrated by lactoferrin. In either case, these signals react with members of the haemopoietic cellular microenvironment, i.e. the reticuloendothelial system (RES) or the mononuclear phogocyte system (Mono-Phago). The cellular microenvironment is then responsible for the production of haemopoietic regulator molecules. The latter are "strictly" regulated within specific limits for both steady-state or naturally occuring perturbated (non-steady-state) system, seen during ontogeny and pregnancy. Under "activated" or pathophysiological conditions, the limits can be extended beyond the "point of no return". In this case, a temporary or a permanent non-steady-state condition occurs. This leads to disregulation of the homeostatic mechanisms.

(A) The Role of The Macrophage in Myelopoiesis.

It is not the purpose of this article to review myelopoietic regulation. Nevertheless, the role played by the macrophage in this haemopoietic cell lineage, of which it is itself a part, is unquestionable. Under normal conditions, the macrophage can release colony stimulating factor (CSF)[31]. Several CSFs have been described which act at various stages of differentiation. Some, such as Interleukin-3/Multi-CSF and granulocyte-macrophage CSF, demonstrate multifunctional action, while others, such as macrophage CSF and granulocyte CSF, are more or less cell lineage specific.

Although the physiological relevance of the CSFs still requires to be established, almost all CSFs have now been cloned so that hybridization studies should localize the production sites of these factors under normal and pathophysiological conditions.

If, however, CSF is responsible for the day-to-day production of granulocytes and macrophages, what is the signal(s) which maintains steady-state myelopoiesis. In a similar manner to the way macrophages are activated to study their functional capacities, so hemopoiesis is perturbated to study its regulatory mechanisms[32]. From a multitude of studies, it appears that substances produced peripherally represent the feedback signals which regulate granulopoiesis.

Some signals have been identified as prostaglandins, lactoferrin and acidic isoferritins[33]. Lactoferrin is interesting because, like transferrin, it is an iron-binding glycoprotein. Unlike transferrin, however, lactoferrin can act bacteriostatically by binding iron at the sites of infection so that bacteria are unable to proliferate[34]. In addition, it has been proposed that release of lactoferrin from the secondary granules of neutrophils results in the inhibition of CSF by macrophages[33]. This hypothesis has been extremely controversial because (a) in clinical disorders and experimental situations an accelerated production and maturation of granulocytes would be expected[32,34,35,36], and (b) in vitro and in vivo animal model studies have not been able to confirm the hypothesis[37,38,39]. In fact, either no effect of lactoferrin on CSF production by macrophages[38,39] or a stimulatory effect[37] has been observed. Nevertheless, from the estimated concentrations of lactoferrin produced under normal and bacterial-infection conditions[34], the specific effect of this molecule on macrophages and CSF production remains an interesting phenomenon.

(B) The Role Of The Macrophage In Erythropoiesis

The role of the macrophage during the hepatic phase of erythropoiesis and that of the Kupffer cell during extrarenal epo production, indicates that perhaps macrophages in other hemopoietic organs, i.e. the spleen and bone marrow, may also be involved in this function. This in turn would imply that the macrophage is an "active" common member of the haemopoietic cellular microenvironment in these organs. It would further imply that the macrophage probably also contained an oxygen sensor in order to respond to external variations in partial oxygen tension.

The problem of cellular oxygen sensing is one of the most difficult to study. The problem actually starts when molecular oxygen comes into contact with the plasma membrane. Taking the defense mechanisms against oxygen toxicity into account, no information exists regarding the intercytoplasmic signaling process. An enzymatic "domino effect" may occur leading to a change in the activity of an enzyme which is particularly sensitive to minute changes in oxygen tension. Even if this were the case, we do not know whether this causes a change in the degree of genetic expression of the epo gene or whether the effect is at a lower level, i.e. protein synthesis, glycosylation or secretion. Furthermore, no information exists as to whether cytoplasmic messengers are responsible for this process.

Nevertheless, it remains an undeniable fact that epo, regardless of where it is produced, is absolutely dependent on the prevailing oxygen tension.

In vitro studies indicated that bone marrow-derived macrophages incubated under physiological oxygen tensions respond by modulating epo release[40]. Furthermore, the cells respond to changing oxygen tensions in the 2% 5% range within 24 hr (unpublished results). Since production of epo mRNA can be initiated within 2-3 hr of erythropoietic stimulation[41,42], it would be expected that sensing and response mechanisms are extremely rapid.

Returning again to epo production sites, there is a large amount of evidence indicating that the kidney in the adult mammal produces this hormone. In fact, the information dates back to 1957[43]. To date, however, very little, if any, epo has been detected under normal, steady-state conditions in the kidney[44]. Recent evidence using Northern transfer analysis has not detected any epo mRNA in the kidney under these conditions[41,42,45]. Only when erythropoiesis is drastically stimulated can epo mRNA and the glycoprotein be detected in kidney extracts[41,42,45] and also in the liver[41]. Such results do not favour the kidney as being the primary site for the day-to-day production of epo. In fact, they imply that the kidney becomes and epo-producing organ under stress or emergency situations.

The cellular localization of epo production in the kidney after erythropoietic stimulation is a long-standing controversy. Both the tubular[42,46,47] and glomerular/mesangial cells[48,49,50,51] are candidates. However, until in situ hybridization studies are performed to definitively localize the cell responsible for epo production, this controversy will continue.

ERYTHROPOIETIN PRODUCTION UNDER STEADY-STATE CONDITIONS

We return once again to steady-state conditions. In the light of the above discussion, it is necessary to ask the question, where is epo produced under normal conditions if not in the kidney? It is clear that this question cannot be completely answered by in vitro techniques; in vivo studies have to be performed. In order to demonstrate that epo is being produced, both gene expression and the appearance of the mature protein have to be shown. Northern transfer analysis can be performed on the haemopoietic organs, but this does not provide the most important information of cellular localization within the organ and the cell type involved. Only in situ hybridization can provide these answers.

In order to study epo gene expression, the in situ hybridization technique was used and first tested on a 98% pure population of bone marrow-derived primary macrophages obtained after 14 days in culture[31]. Using both radioisotopic and non-isotopic labelling of a 1.2 kilobase, double-stranded, mouse epo DNA fragment subcloned into the PUC-19 plasmid (a generous gift from Dr. Eugene Goldwasser at the University of Chicago), in situ hybridization was performed[52,53,54]. The most interesting results have been obtained using biotin-labelled DNA, and a streptavidin-gold signal detection system[53,54]. Reflected light scatter is then observed using the reflection-contrast microscope, which also serves to amplify the signal. The latter appears as aggregates of silver-white light within the cell. Controls included cell lines such as the Interleukin-3/Multi-CSF - dependent FDCP-2 cell line derived from bone marrow long-term cultures and the L929 fibroblast cell line. In addition, control PUC-19 and lambda DNA were used as well as cell hybridized with labelled DNA, but which had been treated with RNase prior to the hybridization procedure. No specific hybridization was observed with any of the negative controls. As a positive control, dabbed preparations of kidney derived from severely anaemic mice were used.

Typical hybridization of biotin-labelled epo dNA to in vitro cultured macrophages is demonstrated in Figure 2. The picture shows a mature macrophage and a second smaller macrophage precursor cell with two nuclei. The larger cell has white aggregates in the cytoplasm demonstrating positive hybridization. Since the gold particles have a 5 nm diameter, the aggregates comprise of many streptavidin-gold molecules binding to the biotin-labelled DNA clumped together to produce total light scatter. A second label using a monoclonal antibody directed against the mouse, macrophage-specific F4/80 antigen conclusively demonstrated that the cells were in fact macrophages. Of the 98% macrophages present after 14 days in culture, 34% demonstrated epo gene expression.

Substantiating the Northern transfer studies using kidneys from mice with stimulated erythropoiesis, in situ hybridization demonstrated positive epo gene expression. However, no gene expression has been observed in kidneys from normal mice.

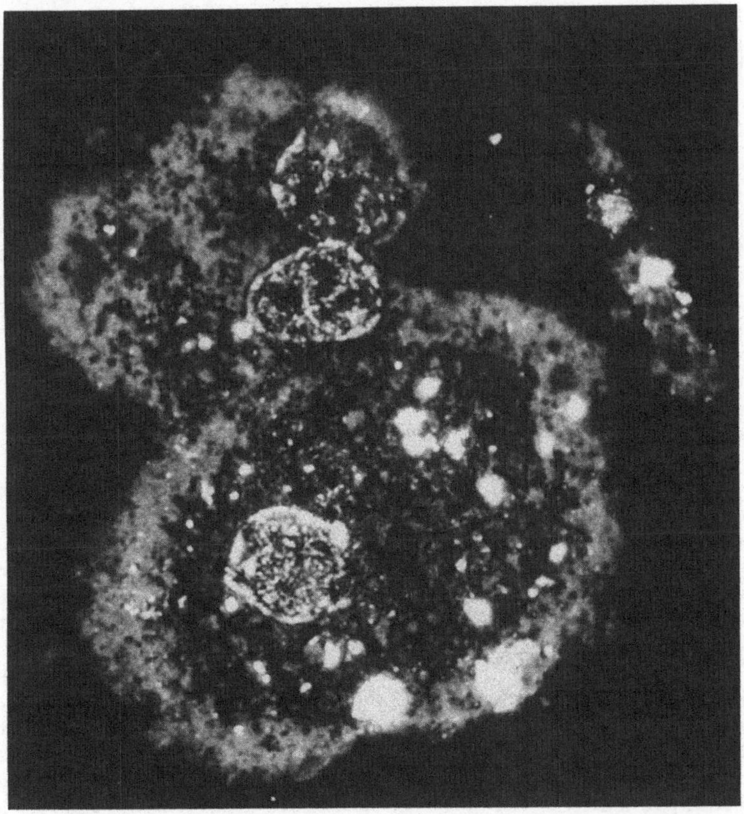

Fig. 2. In situ hybridisation demonstrating erythropoietin gene expression in 14 day cultured macrophages. The erythropoietin DNA probe was biotinylated and the signal detected using streptavidin-gold. The signal was observed and amblified by reflection-contrast microscopy. Magnification, x 360.

In situ hybridization was therefore performed on normal mouse bone marrow smears, with the result that of the estimated 10% F4/80 positive macrophages, 2-3% demonstrated simultaneous epo gene expression. Furthermore, these macrophages were always associated with surrounding cells which appeared to be in contact with the "central macrophage" as if forming a blood island[53,54].

IMPLICATIONS OF ERYTHROPOIETIN GENE EXPRESSION IN MACROPHAGES

These results prove, for the first time, that the macrophage is capable of expressing the epo gene. Whether it actually releases a mature glycoprotein still remains to be seen. Since the in vitro-derived macrophages are, to all intents and purposes, functionally and phenotypically "resident" in nature[52], they can, to a certain degree, be compared favorably with the bone marrow "resident" macrophages from normal mice. The proportion of in vitro cultured and bone marrow macrophages expressing the epo gene is approximately the same; about one-third. That only a proportion of the macrophages express the epo gene indicates that a subpopulation is responsible for this function, which appears to be similar both in vitro, under the culture conditions used, and in vivo. It therefore follows that those macrophages expressing the epo gene, represent an "active" subpopulation within the haemopoietic cellular microenvironment capable of short range or cell-to--cell interaction with the developing erythropoietic precursor cells. In short, epo could be produced at the sites of erythropoiesis under normal, steady-state conditions.

THE FUTURE

The impetus provided by the recent advances in the molecular biology of epo should be used as a tool to answer basic outstanding questions and provide answers to current controversies. There is no doubt that the in situ hybridization technique will be extremely valuable, especially as it provides simultaneous information at the molecular, cellular and morphological levels. It will be important to apply this technique to understand extramedullary erythropoiesis and, of course, to determine which cells in the kidney are responsible for epo production. In addition, it will be possible to determine if, when and where epo is produced in theyolk sac to obtain a better understanding of erythropoietic ontogeny.

Although the macrophage can be shown to express the epo gene and therefore is presumably capable of releasing the mature hormone, there is one exception to this working model; the anaemia of chronic renal failure. This situation is typically characterized by an epo titre that does not correspond to the degree of anaemia -- it is usually too low. If the macrophage is responsible for epo production, why does it not respond by increasing its output under these conditions? Why does extrarenal epo production not take over? Why does the epo level return to near normal when a successful renal transplant occurs? These are relevant, critical questions. Many investigations have implicated inhibitors of epo production. However, no specific inhibitor has yet been identified. It may therefore be possible that unspecific inhibitors are responsible and that these are removed after a successful transplantation (or even temporarily after dialysis). However, it is possible that other factors, which have not as yet been investigated, are responsible. Perhaps the ongoing clinical trials using recombinant human epo as a therapy for the anaemia of chronic renal failure may throw more light on the situation.

Finally, it remains to be seen whether bone marrow macrophages which express the epo gene are also capable of expressing CSF genes, or whether this is a function of a different macrophage subpopulation, or of other cell types under steady-state conditions. At present, a great deal of effort is being used to study the effect, either individually or in multiple combinations, of recombinant haemopoietic and other regulator factors on haemopoietic progenitor cells. This may provide important information as to how the progenitor cells could react under various pertubated situations, since many factors are not produced under steady-state conditions. On the other hand, it can also produce highly confusing information, especially when the data have to be interpreted from in vitro experiments and eventually extrapolated to the in vivo situation. It is therefore not only necessary to know the mechanism of action of these factors, but to determine where these molecules are produced and how they are regulated. Only when both sides of the coin are understood, will it be possible to understand haemopoietic regulation.

ACKNOWLEDGEMENT

This work was supported by the German Red Cross.

REFERENCES

1. Dexter, T.M. 1981. Self-renewing haemopoietic progenitor cells and the factors controlling proliferation and differentiation. In: Microenvironments in Haemopoietic and Lymphoid Differentiation. Ciba Foundation Symposium. Vol 84. 22-31.
2. Moore, M.A.S., and J.J.T. Owen. 1967. Stem cell migration in develop ing myeloid systems. Lancet 2: 658-659.
3. Johnson, G.R. and M.A.S. Moore. 1975. Role of stem cell migration of mouse fetal liver hemopoiesis. Nature 258: 726-728.
4. Wong, P.M.C., S-W. Chung, S.M. Reicheld, and D.H.K. Chui. 1986 Hemoglobin switching during murine embryonic development. Evidence for two populations of embryonic erythropoietic progenitor cells. Blood 67: 716-721.
5. Wong, P.M.C., S-W. Chung, D.H.K. Chui, and C.J. Eaves. 1986. Properties of the earliest clonogenic hemopoietic precursors to appear in the developing murine yolk sac. Proc. Natl. Acad. Sci. USA 38: 3851-3854.
6. Fleischman, R.A., R.P. Custer, and B. Mintz. 1982. Totipotent hemopoietic stem cells: Normal self-renewal and differentiation after transplantation between fetuses. Cell 30: 351-359.
7. Bessis, M.C. and J. Breton-Gorius. 1962. Iron metabolism in the bone marrow as seen by electron microscopy: a critical Review. Blood 19: 635-663.
8. Jacobson, L.O., E.K. Marks, and E.O. Gaston. 1959. Studies on erythropoiesis. II. The effect of transfusion-induced polycythemia in the mother on the fetus. Blood 14: 644-652.
9. Lucarelli, G., A. Porcellini, C. Carnevali, A. Carmena, and F. Stohlman, Jr. 1968. Fetal and neonatal erythropoiesis. Ann. N.Y. Acad. Sci. 149: 544-559.
10. Bleiberg, I. and M. Feldman. 1969. On the regulation of hemopoietic spleen colonies produced by embryonic and adult cells. Dev. Biol. 19: 566-580.
11. Cole, R.J. and J. Paul. 1966. The effects of erythropoietin on haemsynthesis in mouse yolk sac and cultured foetal liver cells. J. Embryol. Exp. Morph. 15: 245-260.
12. Cole, R.J., T. Regen, S.L. White, and E.M. Cheek. 1975. The relationship between erythropoietin-dependent cellular differentiation and colony-forming ability in prenatal haemopoietic tissue. J. Embryol. Exp. Morph. 34: 575-588.

13. Stevenson, J.R., A.A. Axelrad, D.L. McLeod, and M.M. Shreeve. 1971. Induction of colonies of haemoglobin-synthesizing cells by erythropoietin in vitro. Proc. Natl. Acad. Sci. USA 68: 1542-1546.
14. Rich, I.N. and B. Kubanek. 1979. The ontogeny of erythropoiesis in the mouse detected by the erythroid colony-forming technique. I. Hepatic and maternal erythropoiesis. J. Embryol. Exp. Morph. 50: 57-74.
15. Rich, I.N. and B. Kubanek. 1980. The ontogeny of erythropoiesis in the mouse detected by the erythroid colony-forming technique. II. Transition in erythropoietin sensitivity during development. J. Embryol. Exp. Morph. 58: 143-155.
16. Labastie, M-C., J-P. Thiery, and N.M. Le Douarin. 1984. Mouse yolk sac and intraembryonic tissues produce factors able to elicit differentiation of erythroid burst-forming units and colony-forming units respectively. Proc. Natl. Acad. Sci. USA 81: 1453-1456.
17. Dieterlan-Lievre, F. 1987. Respective roles of programme and differentiation factors during hemoglobin switching in the embryo. In: Molecular and Cellular Aspects of Erythropoietin and Erythropoiesis, Rich, I.N. editor. NATO ASI Series, Vol H8, Springer-Verlag Berlin Heidelberg, pp 127-145.
18. Gruber, D.F., J.R. Zucali, and E.A. Mirand. 1977. Identification of erythropoietin producing cells in fetal mouse liver cultures. Exp. Hemat. 5: 392-398.
19. Kelemen, E. and M. Janossa. 1980. Macrophages are the first differentiated blood cells formed in human embryonic liver. Exp. Hemat. 8: 996
20. Hara, H. and M. Ogawa. 1976. Erythropoietic precursors in mice with phenylhydrazine-induced anemia. Am. J. Hemat. 1: 453-458.
21. Boggs, D.R., A. Geist, and P.A. Chervenick. 1969. Contribution of the mouse spleen to post-hemorrhagoic erythropoiesis. Life Sci. 8: 587-599.
22. Crandall, T.L. and D.R. Boggs. 1980. Response of hepatic hematopoiesis to whole body irradiation. Exp. Hemat. 8: 25-31.
23. Koury, M.J., S.T. Sayer, and M.C. Bondurant. 1984. Splenic erythroblasts in anemia-inducing Friend disease. A source of cells for studies of erythropoietin-mediated differentiation. J. Cell Physiol. 121: 526-532.
24. Kreja, L. and H-J. Seidel. 1985. Role of the spleen in Friend virus (F-MULV-P) erythroleukemia. Exp. Hemat. 13: 623-628.
25. Opitz, U., H-J. Seidel, and I.N. Rich. 1977. Erythroid stem cells in Rauscher virus-infected mice. Blut 35: 35-44.
26. Arnold, R., W. Calvo, B. Heymer, T. Schmeiser, H. Heimpel, and B. Kubanek. 1985. Extramedullary haemopoiesis after bone marrow transplantation. Scand. J. Haemat. 34: 9-12.
27. Fruhman, G.J. 1968. Blood formation in the pregnant mouse. Blood 31: 242-248.
28. Peschle, C., I.A. Rappaport, G.P. Jori, M. Chiarello, and A.S. Gordon. 1976. Sustained erythropoietin productin in nephrectomized rats subjected to hypoxia. Brit. J. Haemat. 25: 187.
29. Schooley, J.C. and L.J. Mahlman. 1974. Extrarenal erythropoietin production by the liver in the weanling rat. Proc. J. Soc. Exptl. Biol. Med. 145: 1081-1083.
30. Adams, D.O. and T.A. Hamilton. 1984. The cell biology of macrophage activation. Ann. Rev. Immunol. 2: 283-318.
31. Rich, I.N. 1986. A role for the macrophage in normal hemopoiesis. I. Functional capacity of bone marrow-derived macrophages to release hemopoietic growth factors. Exp. Hemat. 14: 738-745.
32. Cronkite, E.P., H. Burlington, A.D. Chanana, and D.D. Joel. 1985. Regulation of granulopoiesis. In: Hematopoietic Stem Cell Physiology. Cronkite, E.P., Dainiak, N., McCaffrey, R.P., Palek, J. and Quenberry, P.J., editors. Alan R. Liss, Inc., New York. 129-144.

33. Broxmeyer, H.E. 1984. Negative regulators of hematopoiesis. In: Long-Term Bone Marrow Culture. Alan R. Liss, Inc., New York. 363-397.
34. Sawatzki, G. 1987. The role of iron-binding proteins in bacterial infection. In: Iron Transport in Microbes, Plants and Animals. Winkelman, G., von de Helm, D. and Neilands, J.B., editors. VCH Verlagsgesellschaft, Weinheim, F.R.G. 477-489.
35. Robinson, W.A. and A. Mangalik. 1975. The kinetics and regulation of granulopoiesis. Sem. Hemat. 12: 7-25.
36. Quesenberry, P., A. Morley, F. Stohlman, Jr., K. Richard, D. Howard, and M. Smith. 1972. Effect of endotoxin on granulopoiesis and colony-stimulating factor. New Engl. J. Med. 286: 227.
37. Rich, I.N. and G. Sawatzki. 1987. The role of lactoferrin in regulating colony stimulating factor production. In: The Inhibition of Hematopoiesis. Nayman, A., Guigon, M., Gorin, N-C., and Mary, J-Y., editors. John Libbey Eurotext Ltd. 162: 63-66.
38. Strickmans, P., A. Deforge, and R.B. Amson. 1986. Lactoferrin: No evidence for its role in regulation of CSA production by human lymphocytes and monocytes. Blood Cells 10: 369-395.
39. Winton, E.F., J.M. Kinkade, W.R. Vogler, M.B. Parker, and K.C. Barnes. 1981. In vitro studies of lactoferrin and murine granulopoiesis. Blood 57: 574-578.
40. Rich, I.N. 1986. A role for the macrophage in normal hemopoiesis. II. Effect of varying physiological oxygen tensions on the release of hemopoietic growth factors from bone marrow-derived macrophages in vitro. Exp. Hemat. 14: 746-751.
41. Bondurant, M. and M. Koury. 1986. Anemia induces accumulation of erythropoietin mRNA in the kidney and liver. Mol. Cell Biol. 6: 2731-2733.
42. Schuster, S.J., J.H. Wilson, A.J. Erselv, and J. Caro. 1987. Physiologic regulation and tissue localisation of renal erythropoietin messenger RNA. Blood. 70: 316-318.
43. Jacobson, L.O., E. Goldwasser, W. Fried, and L. Plazck. 1957. Role of the kidney in erythropoiesis. Nature. 179: 633-634.
44. Sherwood, J.B., and E. Goldwasser. 1978. Extraction of erythropoietin from normal kidneys. Endocrinol. 103: 866-870.
45. Goldwasser, E., J. McDonald, and N. Beru. 1987. The molecular biology of erythropoietin and the expression of its gene. In: Molecular and Cellular Aspects of Erythropoietin and Erythropoiesis. Rich, I.N., editor. NATO ASI Series, Vol. H8. Springer-Verlag, Heidelberg. 11-21.
46. Caro, J. and A.J. Erselv. 1984. Biologic and immunologic erythropoietin in extracts from hypoxic whole rat kidneys and in their glomerular and tubular factions. J. Lab. Clin. Med. 103: 922-931.
47. Caro, J., J. Hickey, and A.J. Erslev. 1984. Erythropoietin production by an established kidney proximal tubular cell line (LLCPK1). Exp. Hemat. 12: 357A.
48. Fisher, J.W., G. Taylor, and D. Porteous. 1965. Localisation of erythropoietin in glomeruli of sheep kidney by fluorescent antibody technique. Nature. 205-611.
49. Burlington, H., E.P. Cronkite, U. Reincke, and E. Zanjani. 1972. Erythropoietin production in cultures of goat renal glomeruli. Proc. Natl. Acad. Sci. UjSA. 69: 3547.
50. Kurtz, A., W. Jelman, F. Sinowatz, and C. Bauer. 1983. Renal mesangial cell cultures as a model for study of erythropoietin production. Proc. Natl. Acad. Sci. USA. 80: 4008.
51. Jelkman, W., A. Kurtz, and C. Bauer. 1983. Extraction of erythropoietin from isolated renal glomeruli of hypoxic rats. Exp. Hemat. 11: 581-588.
52. Rich, I.N. 1987. Erythropoietin production by macrophages: Cellular response to physiological oxygen tensions and detection of ery-

thropoietin gene expression by in situ hybridisation. In: Molecular and Cellular Aspects of Erythropoietin and Erythropoiesis. Rich, I.N. editor. NATO ASI series, Vol H8. Springer-Verlag, Heidelberg. 291-310.

53. Vogt, Ch., S. Pentz, and I.N. Rich. A role for the macrophage in normal hemopoiesis: III. In vitro and in vivo in situ hybridisation. Submitted for publication.

54. Rich, I.N., Ch. Vogt, and S. Pentz. 1988. Erythropoietin gene expression in vitro and in vivo detection by in situ hybridisation. Blood Cells. In Press.

MOLECULAR IMPLICATIONS OF Ph (+) MYELODYSPLASTIC SYNDROME

Keisuke, Toyama., Kazuma, Ohyashiki., and Junko, H. Ohyashiki

Department of Internal Medicine
Tokyo Medical College
6-7-1 Nishishinjuku
Shinjuku-ku
Tokyo 160, Japan

ABSTRACT

We report a case of 62-year-old Japanese male with a myelodysplastic syndrome (MDS) with a Philadelphia (Ph) chromosome. Cytogenetic analysis revealed the bone marrow cells to contain a Ph chromosome due to t(?;11;22)(?;q11;q11), as well as -5, -7, +8, -12 and an extra Ph, in addition to cells with a normal karyotype. Molecular analysis using breakpoint cluster region probes (5' bcr and 3' bcr) did not detect a rearrangement within the bcr DNA sequences, indicating that the breakpoint at 22q11 occurred outside the bcr. Furthermore, the bone marrow cells from this patient did not express an 8.5 kb c-abl mRNA. Thus, the Ph translocation in this case differs from that of Ph-positive chronic myelogenous leukemia.

INTRODUCTION

A Philadelphia (Ph) translocation is seen in more than 90% of cases with chronic myelogenous leukemia (CML) and in 10-25% of acute lymphoblastic leukemia (ALL)[1], though a Ph chromosome has been found very rarely in myelodysplastic syndromes (MDS), having been identified in only five cases to date[2-6]. We encountered a case of MDS with a variant type Ph translocation and performed molecular investigations in order to ascertain the genesis of the Ph in this case.

MATERIALS AND METHODS

Case:

A 62-year-old Japanese male was first seen in November because of pancytopenia and was admitted to the Tokyo Medical College Hospital on December 25, 1986. He had no lymphadenpathy or hepatosplenomegaly. His peripheral blood showed pancytopenia without blast. A bone marrow aspiration yielded a hypocellular marrow with 8.2% blasts in myeloid series. A biopsy specimen of iliac bone marrow revealed increased fatty tissue and hyperplastic marrow islets with increased numbers of neutrophils and eosinophils. His neutrophil alkaline phosphatase score was within normal limits. These findings led us to make the diagnosis of MDS, i.e., refractory anemia with an excess of blasts (RAEB). In the end of January, 1987, the patient was diagnosed as having RAEB in transformation, since blasts appeared in his peripheral blood and 20.8% of the blasts were found in his bone marrow. He had never been treated with any anti-leukemic chemotherapy, and died of pneumonia on February 13, 1987. Autopsy disclosed a hyperplastic marrow resembling CML and no evidence of Ewing's sarcoma.

Chromosomes and Molecular Studies:

Chromosome analysis was carried out by usual cytogenetic procedure and with Hoechst 33258 and quinacrine mustard double staining technique.

DNA and RNA were extracted by urea lysis/cesium chloride method. Southern or Northern blot were performed as described previously. The probes used in this study were a 1.1 kb HindIII/EcoRI 3' side of the bcr[7] and a 2.0 kb bgIII/HindIII fragment of 5' bcr (kindly provided by Dr. C.R. Bartram, Department of Pediatrics II, University of Ulm, Ulm, FRG). For Northern blot analysis, 1.8 kb EcoRI digested c-abl fragment from pc-abl 18.2 which contains a part of the kinase domain and some 3' abl sequences was used as a probe (provided by Dr. C.R. Bartram).

RESULTS

Cytogenetic analysis on December 25, 1986 revealed 2 of 20 cells had a normal male karyotype; the remaining 18 cells had a typical Ph chromosome. Q-banding revealed a representative karyotype to be 43,XY,-5,-7,+8,-11,-12,-13,-22,del(11)(q11),+der(11)t(?;11;22)(?;q11-q23;q11-qter), +del(22)(q11). Both chromosome 9 seemed to be normal. Eleven of the 18 metaphase cells with a Ph chromosome had an extra Ph chromosome. The deleted chromosomal material of a 22 (22q11-qter) was attached to a chromosome region 11q23. And this chromosome attached to unidentified chromosome material, producing a marker chromosome which can be described as a t(?;11;22)(?;11q11-q23;22q11-qter) (Fig. 1).

BgII, BamHI or HindIII digests DNAs from this case did not show aberrant fragments hybridized to the 1.1 kb 3' bcr probe. On the other hand, BgIII digested DNA from a Ph-positive CML case showed an aberrant 4.3 Kb fragment hybridized to the 3' bcr probe, in addition to the 5.0 Kb germ line fragment (Fig. 2). BglII digested DNA from

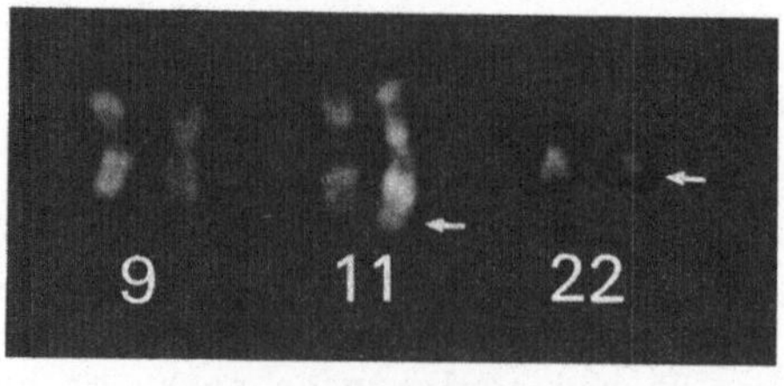

Fig. 1. Partial karyotype from a bone marrow cell, showing a translocation between chromosomes no. 11 and 22, producing a marker chromosome, i.e., t (?;11;22)(?::q11-q23::q11-qter;22pter-22q11::?).

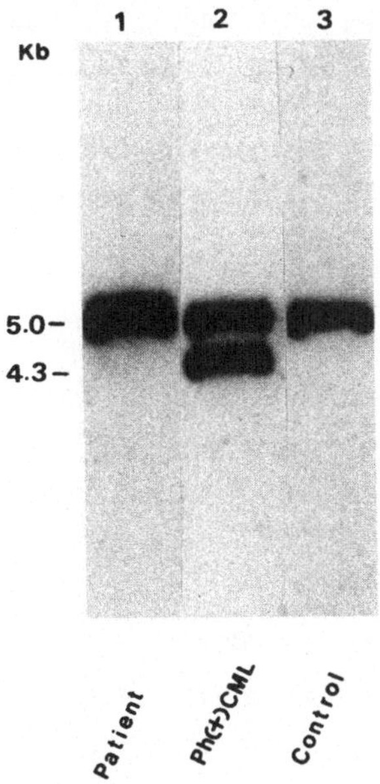

Fig. 2. Southern blot analysis of DNAs from this patient, a CML patient with a standard Ph translocation (Ph-positive CML, and Epstein-Barr virus transformed lymphoblastoid cells with a normal karyotype (Control). In Ph-positive CML cells, BglII digested DNA showed an abnormal fragment. No abnormal fragments were detected in the DNA from this patient.

the case did not show abnormal fragments hybridized to the 2.0 kb 5' probe. Thus, taken together with the data with the 3' bcr probe, DNA from this case did not show rearrangements within the 5.8 kb bcr.

In RNA from a Ph-positive CML an 8.5 kb aberrant fragment which hybridized to the c-abl probe was noted, in addition to the normal 6.0 and 7.0 kb fragments. K562 cells also expressed the aberrant message at a high level; however, in RNA obtained from the MDS case only normal band hybridized with the c-abl probe were detected.

DISCUSSION

In the literature, only five cases of MDS with Ph chromosome or Ph translocation have appeared[2-6] (Table 1), however, no molecular approach was made in these cases. On the other hand, it has been shown that a Ph translocation is a hallmark of CML, and juxtaposition of c-abl gene to aberrant bcr sequences producing a chimeric (bcr/c-abl) message[8].

The results presented here suggest that the neoplastic cells in our MDS case, even though they contained a Ph chromosome, were subject to a mechanism of neoplastic transformation different from that in CML with a Ph translocation. Thus, similar studies in MDS with a standard type Ph translocation might answer the question as to whether a Ph chromosome in MDS has a molecular background similar to that in CML.

Table 1. Reported Cases of Ph-Positive Myelodysplastic Syndrome

Age/Sex	Diagnosis	Karyotypes	References
1. 43/M	Preleukemia ↓ 6 years Acute leukemia	46,XY(90%)/46,XY,Ph (10%) ↓ 46,XY,Ph (100%)	[2]
2. 7/M	Preleukemia ↓ 3 years Acute leukmeia	46,XY(100%) ↓ 46,XY,Ph /47,XY,+G,Ph /45,XY,-C,Ph (100%)	[3]
3. 49/F	MDS(RA?) ↓ 3 years Acute leukemia	46,XX(62%)/46,XX,tPh (38%) ↓ 46,XX(50%)/46,XX,tPh (50%)	[4]
4. 42/F	Hodgkin's disease ↓ 5 years Preleukemia (secondary MDS: RAEB?)	46,XX,-7,+8,t(18;22)(p11;q11) (100%)	[5]
5. 69/M	MDS(phenylbutazone- ↓ 3 y. induced?) RAEB-T	ND 46,XY,tPh (100%)	[6]
6. 62/M*	RA ↓ RAEB ↓ RAEB-T	46,XY(10%)/ 43,XY,-5,-7,+8,-11,-12,-13 -22,del(11)(q11)+der(11) t(?;11;22)(?;q11-q23;q11-qter), +del(22)(q11) with others (90%)	present case

Ph : Philadelphia chromosome, tPh :t(9;22)(q34;q11),
* : bcr rearrangement studied only our case; no rearrangement

ACKNOWLEDGEMENTS

This work was supported in part by Grant-in-Aid for Cancer Research from the Ministry of Education, Science and Culture, and in part by Grant-in-Aid for Japanese Group study for Intractable Diseases from the Ministry of Health and Welfare, Japan (K.T.).

REFERENCES

1. Sandberg, A.A., 1980. The chromosomes in Human Cancer and Leukemia Elsevier/North-Holland Inc., New York.
2. Canellos, G.P., and J. Whang-Peng. 1972. Philadelphia-chromosome-positive preleukemic state. Lancet i. 1227-1228.
3. Gracy, M., Suharjono, Sunoto. 1980. Myelomonocytic leukaemia with a preleukemic syndrome and ph[1] chromosome in monozygotic twins. Arch. Dis. Child. 52: 72-80.
4. Roth, D.G., C.M. Tichman, and J.D. Rowley. 1980. Chronic myelodysplastic syndrome (preleukemia) with the Philadelphia chromosome Blood 56: 262-264.
5. Michalske, K.A., J.H. Miles, and M.C. Perry. 1982. Unusual Ph translocation in preleukemia. Cancer Genet Cytogenet. 6: 89-90.
6. Ohyashiki, K., J.H. Ohyashiki, A. Raza, H.D. Preisler, and A.A. Sandberg. 1987. Phenylbutazone-induced myelodysplastic syndrôme with Philadelphia translocation. Cancer Genet Cytogenet. 26: 213-216.

7. Ohyashiki, K., J.H. Ohyashiki, A.J. Kinniburgh, J. Rowe, K.B. Miller, A. Raza, H.D. Preiler, and A.A. Sandberg. 1987. Transposition of breakpoint cluster region (3'bcr) in CML cells with variant Philadelphia translocations. Cancer Genet Cytogenet. 26: 105-115.
8. Stam, K., N. Heisterkamp, G. Grosveld, A. de Klein, B.S. Verma, M.Cleman, H. Dosck, and J. Groffen. 1985. Evidence of a new chimeric bcr/c-abl mRNA in patients with chronic myelocytic leukemia and the Philadelphia chromosome. N. Engl. J. Med. 313: 1429-1433.

EVALUATION OF C-SIS mRNA EXPRESSION BY HUMAN MEGAKARYOCYTIC CELLS IN NORMALS AND PATIENTS WITH MYELOPROLIFERATIVE DISORDERS

L. Kanz, R. Mielke, A.A. Fauser and G.W. Löhr

Department of Medicine
Division of Hematology-Oncology
Albert-Ludwigs-Universität
Freiburg, West Germany

INTRODUCTION

Chronic myelogenous leukemia (CML), polycythemia vera, essential thrombocythemia, primary myelofibrosis (PMF) and acute myelofibrosis (acute megakaryoblastic leukemia) form a group of hematological diseases, called myeloproliferative disorders, with varying patterns of lineage involvement in each disorder. There is laboratory and clinical evidence that bone marrow fibrosis in these diseases is promoted, at least in part, by platelet-derived growth factor (PDGF) (for review s.[1,2]). PDGF displays growth promoting activity for stroma cells (for review s.[3-5]) and possibly, by indirect action for hemopoietic progenitor cells[6-8]. Its synthesis has been shown by a number of cells, including megakaryocytes, and the B-chain of PDGF is encoded by the c-sis protooncogene[3-5].

A link between PDGF originating from megakaryocytes and the occurrence of myelofibrosis is indicated by the finding of a large megakaryocytic cell compartment in most myeloproliferative diseases and of a topographical relationship between megakaryocytes and fibroblast proliferation in bone marrow[9]. Of pathogenetic significance might be a loss of PDGF - synthesized at a normal rate on a per cell basis - into the bone marrow interstitium during maturation, because of intrinsic abnormalities of alpha-granules in megakaryocytic cells of the neoplastic lone[10,11]. Support for this hypothesis is derived from studies of the inherited platelet disorder gray platelet syndrome, which indicate a relationship between the loss of alpha-granule content from megakaryocytes and the myelofibrosis found in this syndrome[13,14]. Alternatively/additionally c-sis might be activated in megakaryocytic cells, resulting in an enhanced PDGF production, responsible for the progressive fibrosis observed. Current knowledge argues against the possibility of an abnormal expression as a consequence of chromosomal translocations[12]; an increased expression, however, might be primed in an undetermined way in megakaryocytic cells in the course of myeloproliferative disorders. Interestingly, phorbol ester stimulation of chronic myelogenous leukemia cells and of K562 human CML blast crisis cells induces megakaryoblastic differentiation and a concomitant increase in c-sis mRNA and secretion of PDGF[15,16].

The aim of our study was to determine, whether an enhanced expression of c-sis mRNA in megakaryocytic cells can be observed in myeloproliferative disorders. Because of the rarity of megakaryocytic cells in human bone marrow and the low number of cells that can be picked from megakaryocytic colonies, we performed in situ hybridization, which allows detection of mRNA sequences at the single cell level.

METHODS

Experimental Procedure

The experimental procedure is outlined in Fig. 1: bone marrow samples were aspirated from the posterior iliac crest into Ca-Mg-free Hanks' solution, containing citrate and theophylline (according to CATCH-medium[17]) as well as DNAse (25U/ml), HEPES pH 7,0 and 1% HSA. This harvesting medium was used throughout the isolation procedure.

To isolate megakaryocytic cells, bone marrow cells (or peripheral blood cells) were separated by density centrifugation, using a 4 step Percoll gradient [density fractions of 1.015 g/ml, 1.030 g/ml, 1.050 g/ml and 1.077 g/ml (Pharmacia, Freiburg, FRG)]. Cells derived from the gradient layers of density <1.030 g/ml and <1.050 g/ml were stained with monoclonal antibodies directed against glycoprotein IIIa (C17) followed by FACS analysis and sorting; anti- GPIIIa positive cells were sorted directly onto coated areas of regular microscope glass slides[18]. Cells harvested after Percoll density centrifugation from the fraction 1.077 g/ml resp. mononuclear cells after Ficoll-separation were grown in methylcellulose cultures as previously described[19-21]. After 12-14 days of culture, individual megakaryocytic colonies were plucked by micropipette, washed in Hanks solution and transferred to the reaction areas of the slides[18].

In previous experiments we have demonstrated that progenitor cells which give rise to megakaryocytic colonies ($<$density 1.077 g/ml) do not express glycoprotein IIb/IIIa, whereas megakaryocytic progenies (density <1.050 g/ml and <1.030 g/ml) express this antigenic determinant (L. Kanz et al., Experimental Hematology, submitted).

In Situ Hybridization

In order to detect mRNA within single megakaryocytic cells-either isolated from bone marrow resp. peripheral blood cells, or aspirated from in vitro colonies - in situ hybridization was performed on regular

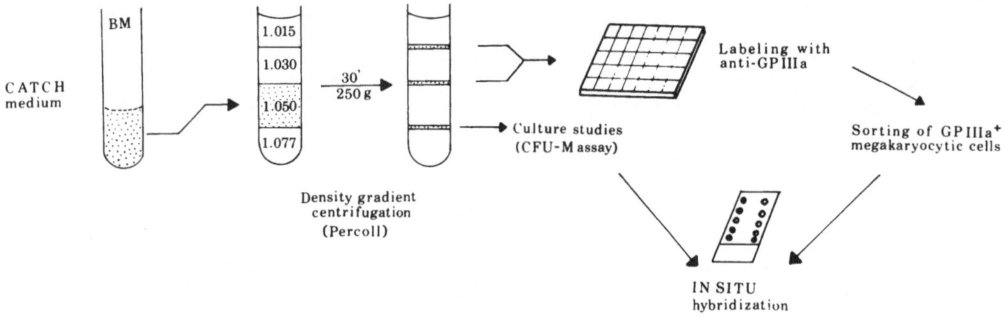

Fig. 1. Experimental procedure (for details, s. text).

microscope glass slides with small poly-L-lysine or fibronectin-coated reaction areas (1-5 mm in diameter) (18) as described[22]. In brief: cells were fixed with ethanol/acetic acid (3:1) for 15 min, followed by dehydration through increasing concentrations of ethanol. Subsequently the cells were treated with HCl (0,1 N) and proteinase K (1µg/ml). Acetylation was performed immediately before hybridization according to Hayashi et al[23]. All washing steps included the addition of vanadyl-ribonucleoside complex (10mM) to the washing solutions. Prehybridization (12h, 37°C) and hybridization (72h, 37°C; wet chamber) were performed with formamide (50%)-containing standard (pre-)hybridization solutions. Posthybridization treatment included washes at room temperature with formamide and incubation with 2 x SSC at 55°C. NTB2 emulsion (Kodak) and D19-developer (Kodak) were used for autoradiography. Finally the slides were stained with azur B and eosin Y[24].

Probe

V-sis specific DNA probe (983 bp; highly homologuous to c-sis transcript[25]; available from Oncor Inc., Gaithersburg, MD, USA) was labeled by standard nick translation reaction with ^{35}S-dGTP and/or ^{35}S-dCTP (1200 Ci/mmol; NEN, Dreieich, FRG).

Specific activity: 2-5x10^8dpm/µg (dGTP) and 5-8x10^8dpm/µg (dGTP plus dCTP).

RESULTS

In situ hybridization was performed with cell samples from normal volunteers and from patients with myeloproliferative disorders, as indicated in Table 1: glycprotein IIIa positive megakaryocytic cells (GPIIIa$^+$) were isolated by fluorescence activated cell sorting from bone marrow samples and from peripheral blood cells. In addition megakaryocytic cells were plucked from in vitro colonies (CFU-M), grown from bone marrow cells as well as from peripheral blood cells (PMF, CML); the cells derived from the patient with essential thrombocythemia showed spontaneous CFU-M colony formation, which was increased by PHA-LCM (phytohemagglutinin-stimulating, leukocyte-conditioned medium; source of megakaryocyte colony stimulated activity) to 70-90 CFU-M colonies per 10^5 cells cultured. Cytological analysis of megakaryo-

Table 1. Cell samples used for in situ hybridization to detect c-sis mRNA in megakaryocytic cells. (ET = essential trombocythemia; PMF = primary myelofibrosis; CML = chronic myelogenous leukemia)

	Controls (n=4)	E T (n=1)	O M F (n=1)	C M L (n=6)
GPIIIa$^+$ cells (bone marrow)	*	*	-	*
GPIIIa$^+$ cells (blood)	-	-	*	-
CFU-M colonies (bone marrow)	*	*	-	*
CFU-M colonies (blood)	-	-	*	*

cytic cells from the patients revealed - with the exception of the patient with essential thrombocythemia - atypical megakaryocytes, as described by Rabellino et al.[26].

The patient with essential thrombocythemia did not display myelofibrosis, whereas medullary fibrosis was moderate (grade II) in the patient with PMF and varied from normal to moderate in CML patients.

The precental distribution of FACS-isolated megakaryocytic cells from normal bone marrow with respect to differing levels of cellular maturation is shown in Table 2; all stages of cytoplasmic and nuclear maturation could be observed. Promegakaryoblasts, however, only numbered 2.5% of pure preparations of megakaryocytic cells from density fraction 1.050 g/ml, a finding which impaired mRNA analysis of these very immature megakaryocytic cells.

All hybridization experiments revealed that there was no positive reaction with any of the samples studied, e.g. only background levels of silver grains in the developed autoradiographs were observed with megakaryocytic cells sorted from bone marrow and picked from in vitro colonies. This held true for all the probes from normals as well as for all samples from the patients studied, independent of the degree of medullary fibrosis.

The sensitivity of the in situ hybridization procedure, using a ^{35}S labeled v-sis probe (983 bp; specific activity $4x10^{8}$dpm/μg) can be calculated as follows: 325 kd (hybridizing copy) x $0.168x10^{-23}$g = $5.5x10^{-12}$μg x $4x10^{8}$dpm/μg = 0.002 dpm, corresponding to 9 disintegrations after 72 hours of exposure; based on an efficiency of grain development for ^{35}S of 0.5 grains per disintegration[27], this results in 4-5 grains per cell per copy; as background values varied from 0-6 grains per cell surface area in our experiments, sensitivity might be calculated to 2 copies per cell. However, these theoretical considerations assume among other - that the probe is perfectly representative and exclude self reassociation of the complementary probes synthesized by nick translation. Calibration experiments, which we performed

Table 2. Percental distribution of FACS-isolated megakaryocytic cells from normal bone marrow with respect to cellular differentiation

	density fraction	
	1.030 g/ml	1.050 g/ml
Promegakaryoblast	-	2,5%
Megakaryoblast	5,4%	48,3%
Promegakaryocyte	7,3%	16,2%
Megakaryocyte,granul.	51,9%	16,4%
Megakaryocyte, mature	35,4%	16,4%
Number of cells examined	1102	450
Number of experiments	7	5

with PHA-stimulated and non-stimulated T4-lymphocytes demonstrated, that the sensitivity of the procedure which we used equals about 5 copies per cell (evaluated by in situ hybridization and solution hybridization using a ^{35}S-myc probe: the average number of grains per cell after in situ hybridization was plotted against the average number of c-myc mRNA copies per cell; copy numbers were quantified after dot blot hybridization of total cellular RNA, which was isolated by the guanidinium isothiocyanate method; results not shown).

Reactivity of the ^{35}S-labeled v-sis with sis mRNA was shown by cytoplasmic dot hybridization experiments[28] using SSV-transformed NIH-3T3 fibroblasts as well as A172 glioblastoma cells, and NIH 3T3 cells as controls (results not shown) (all cell lines were kindly provided by Dr. S. A. Aaronson, NIH).

DISCUSSION

PDGF has been implicated in the pathogenesis of medullar fibrosis during the course of myeloproliferative disorders[1,2]. Based on the conspicuous megakaryocytic hyperplasia in some of these disorders, associated with a probable reactive proliferation of bone marrow fibroblasts in the proximity of megakaryocytic cells, megakaryocytes are considered the most likely cells to be involved in PDGF-induced medullary fibrosis. Current evidence suggests a premature release of PDGF from alpha-granules of impaired, clonally transformed megacaryocytic cells[10,11].

We asked the question, whether additionally resp. alternatively c-sis, which encodes for the B-chain of PDGF, might be abnormally activated in megakaryocytic cells, leading to an enhanced PDGF synthesis on a per cell basis. Whether PDGF has a role in the neoplastic proliferation in myeloproliferative disorders itself was not addressed by these studies.

So far the c-sis oncogene has not been directly implicated in the pathogenesis of any hematological malignancy. C-sis expression, as studied by blotting techniques, was only seen in some patients with accelerated or blastic phase of CML[12]. These observations, however, do not exclude the possibility that c-sis might be expressed in megakaryocytic cells, because blotting techniques are limited if the mRNA to be hybridized is expressed only in a small subpopulation such as megakaryocytes in bone marrow.

In our experiments, therefore, c-sis mRNA expression in megakaryocytic cells was evaluated by in situ hybridization with a nick translated DNA probe.

Our results indicate that neither megakaryocytes from bone marrow nor cultured megakaryocytes from normals nor megakaryocytic cells derived from patients with myeloproliferative disorders display significantly increased hybridization signals compared to controls. These available data suggest that at least some of the patients with myeloproliferative disorders do not express c-sis in megakaryocytic cells above basal levels, as assessed by in situ hybridization - a procedure which in our experience displays a sensitivity of about 5 copies per cell, as studied by calibration experiments. To further evaluate the role of c-sis gene expression for bone marrow fibrosis, analysis should be performed in neoplastic megakaryocytic cells that phenotypically are comparable to the most immature normal megakaryocytic cells, that is promegakaryoblasts, which probably are candidate cells where normal sis transcription occurs. Sis expression by these cells - as recently demonstrated in a patient with acute megakaryoblastic leukemia[29] however, does not necessarily argue for an abnormal expression but may reflect, and correspond to normal sis RNA expression at this very early stage of cellular differentiation. It would be essential to directly compare normal and neoplastic megakaryocytic cells at corresponding stages of cellular differentiation. To date, however, we are not able to prove sis-mRNA expression in normal promegakaryoblasts, as on the one hand these cells comprise only about 2% of FACS isolated normal megakaryocytic cells, and one the other we consistently observe unspecific sticking of labeled probes to 0-1% of the cells analyzed. Refined techniques to further enrich for the most immature megakaryocytic cells are necessary.

On the balance, the evidence at present appears to favour a concept of ineffective megakaryocytopoesis with a normal rate of PDGF synthesis per cell but with PDGF release from intrinsically abnormal megakaryocytes as a possible mechanism of PDGF-mediated fibrosis. Further experiments should include analysis with specific antibodies against PDGF to assess, whether in megakaryocytic hyperplasia of myeloproliferative diseases that concurs with myelofibrosis, PDGF is lost at some stage of megakaryocytic maturation. The role of other potential mediators of bone marrow fibrosis - such as TGF-ß and PF4[30] - as well as the role of other effector cells-such as monocytes/macrophages and endothelial cells - remains to be defined.

ACKNOWLEDGEMENT

This work was supported by Deutsche Forschungsgemeinschaff (UA 585/1-3).

REFERENCES

1. Romano, M., and A. Poggi. 1986. Myelofibrosis: A role for platelet derived growth factor (PDGF)? Haematologica 71: 359.
2. McCarthy, D.M. 1985. Fibrosis of the bone marrow: content and causes. Br. J. Haematol. 59: 1.
3. Heldin, C.H., A. Wasteson, and B. Westermark. 1985. Platelet derived growth factor. Mol. and Cell Endocrinol. 39: 169.
4. Deuel, T.F., B.D. Tong, and J.S. Huang. 1985. Platelet derived growth factor: Structure, function, and roles in normal and transformed cells. In: Current topics in cellular regulation, Vol. 26: 51.
5. Ross, R. 1987. Platelet derived growth factor. An Rev. Med. 38: 71.
6. Michalevicz, R., G.E. Francis, G.M. Price, and A.V. Hoffbrand. 1985. The role of platelet-derived growth factor in human pluripotent progenitor (CFU-GEMM) growth in vitro. Leukemia Res. 9: 399.
7. Rosenfeld, M., A. Keating, D.F. Bowen-Pope, J.W. Singer, and R. Ross. 1985. Responsiveness of the in vitro hematopoietic microenvironment to platelet-derived growth factor. Leukemia Res. 9: 427.
8. Delwiche, F., E. Raines, J. Powell, R. Ross, and J. Adamson. 1985. Platelet derived growth factor enhances in vitro erythropoiesis via stimulation of mesenchymal cells. J. Clin. Invest. 76: 137.
9. Breton-Gorius, J., M. Bizet, F. Reyez, E. Dupuy, C. Meat, J.P. Vannier, and P. Tron. 1982. Myelofibrosis and acute megakaryoblastic leukemia in a child: topographic relationship

between fibroblasts and megakaryocytes with an alpha-granule defect. Leukemia Res. 6: 97.
10. Groopman, J.E. 1980. The pathogenesis of myelofibrosis in myeloproliferative disorders. Ann. Int. Med. 92: 857.
11. Castro-Malaspena, H., E.M. Rabellino, A. Yen, R.L. Nachman, and M.A.S. Moore. 1981. Human megakaryocyte stimulation of proliferation of bone marrow fibroblasts. Blood. 57: 781.
12. Romero, P., M. Blick, M. Talpaz, E. Murphy, J. Hester, and J. Gutterman. 1986. C-sis and c-abl expression in chronic myelogenous leukemia and other hematologic malignancies. Blood. 67: 839.
13. Breton-Gorius, J., W. Vainchenker, A.T. Nurden, S. Levy-Tolendano, and J.P. Caen. 1981. Defective alpha-granule production in megakaryocytes from gray-platelet syndrome. Ultrastructural studies of bone marrow cells and megakaryocytes growing in culture from blood precursors. Am. J. Pathol. 102: 10.
14. Caen, J.P., J.F. Deschamps, E. Bodevin, M.C. Bryckaert, E. Dupuy, and A. Wasteson. 1987. Megakaryocytes and myelofibrosis in gray platelet syndrome. Nouv Rev. Fr. Hematol. 29: 109.
15. Lopez, A.R., and A. Deisseroth. 1986. Is secretion of PDGF CML cells accompanied by changes in expression of the transforming growth factor beta gene? Blood. Suppl 1, 170 a.
16. Colamonici, O.R., J.B. Trepel, C.A. Vidal, and L.M. Neckers. 1986. Phorbol ester induces c-sis gene transcription in stem cell line K-562. Mol. and Cell. Biol. 6: 1847.
17. Levine, R.F., and M.E. Fedorko. 1976. Isolation of intact megakaryocytes from guinea pig femoral marrow. J. Cell Biol. 69: 159.
18. L. Kanz, K.J. Bross, R. Mielke, G.W. Löhr, and A.A. Fauser. 1986. Fluorescence activated sorting of individual cells onto poly-L-lysine coated slide areas. Cytometry 7: 491.
19. Fauser, A.A. and H.A. Messner. 1979. Identification of megakaryocytes, macrophages and eosinophils in colonies of human bone marrow containing neutrophilic granulocytes and erythroblasts. Blood 53: 1023.
20. Kanz, L., G. Straub, K.J. Bross, and A.A. Fauser. 1982. Identification of human megakaryocytes derived from pure megakaryocytic colonies (CFU-M), megakaryocytic-erythroid colonies (CFU-M/E), and mixed hemopoietic colonies (CFU-GEMM) by antibodies against platelet associated antigens. Blut. 45: 267.
21. Kanz, L., G.W. Löhr, and A.A. Fauser. 1986. Lymphokine(s) from isolated T-lymphocyte subpopulations support multilineage hemopoietic colony (CFU-GEMM) and megakaryocytic colony (CFU-M) formation. Blood. 68: 991.
22. Kanz, L., R. Mielke, A.A. Fauser, and G.W. Lohr. 1988. Detection of messenge RNAs within single hemopoietic cells by in situ hybridization on small slide areas. Exp. Hematol. in press.
23. Hayashi, S., I.C. Gillam, A.D. Delaney, and G.M. Tener. 1978. Acetylation of chromosome squashes of Drosophila Melanogaster decreases the background in autoradiographs from hybridization with 125I-labeled RNA. J. Histochem. Cytochem. 26: 677.
24. Wittekind, D.H., V. Kretschmer, and I. sohmer. 1982. Azure B - eosin Y stain as the standard Romanowski-Giemsa stain. Br. J. Haematol. 51: 391.
25. Durga Rao, C., H. Igarashi, I.M. Chiu, K.C. Robbins, and S.A. Aaronson. 1986. Structure and sequence of the human c-sis/ platelet-derived growth factor 2 (SIS/PDGF2) transcriptional unit. Proc. Natl. Acad. Sci. 83: 2392.
26. Rabellino, E.M., R.B. Levene, R.L. Nachman, and L.K.L. Leung. 1984. Human megakaryocytes III: Characterization in myelo-

proliferative disorders. Blood. 63: 615.
27. Haase, A., M. Brahic, L. Stowring, and H. Blum. 1984. Detection of viral nucleic acids by in situ hybridization. Methods in Virology, Vol 7: 189.
28. White, B.A., and F.C. Bancroft. 1982. Cytoplasmic dot hybridization. J. Biol. Chem. 257: 8569.
29. Sunami, S., A. Fuse, B. Simizu, M. Eguchi, Y. Hayashi, K. Sugita S. Nakazawa, Y. Okimoto, T. Sato, and H. Nakajima. 1987. The c-sis gene expression in cells from a patient with acute megakaryoblastic leukemia and Down's syndrome. Blood. 70: 368.
30. Burstein, S.A., T.W. Malpass, E. Yee, M. Kadin, M. Brigden, J.W. Adamson, and L.A. Harker. 1984. Platelet factor-4 excretion in myeloproliferative disease: implications for the aetiology of myelofibrosis. Br. J. Haematol. 57: 383.

EXPRESSION OF SPECIFIC ISOFORMS OF PROTEIN 4.1

IN ERYTHROID AND NON-ERYTHROID TISSUES

Tang K. Tang, Thomas L. Leto, Vincent T. Marchesi
and Edward J. Benz, Jr.

Departments of Internal Medicine and Human Genetics
Department of Pathology
Yale University School of Medicine
New Haven, CT 06510

ABSTRACT

Protein 4.1 in red cells is an important submembrane linking protein that binds to spectrin actin complexes at one end of its structure and to transmembrane proteins, such as glycophorin, at the other. Protein 4.1 thus contributes to the strength and flexibility of the erythrocyte membrane, a fact dramatically exemplified by the appearance of hereditary hemolytic anemias in patients with absent or abnormal protein 4.1. Recently, protein 4.1 forms have been discovered in many non-erythroid tissues. Their intracellular locations raise the possibility that these isoforms might have different functions. We have thus conducted comparative analysis of erythroid and non-erythroid protein 4.1 forms by cloning and sequencing erythroid and lymphoid protein 4.1 cDNAs. The lymphoid protein 4.1 isoforms exhibit at least five nucleotide sequence motifs that appear to be either inserted or deleted relative to the erythroid mRNA sequence by alternative splicing of a common mRNA precursor. One of these motifs, located within the spectrin-actin binding domain, is found only in erythroid cells and is specifically produced during erythroid cell maturation. The selective expression of this alternatively spliced mRNA during erythroid maturation implies the existence of a lineage specific splicing mechanism whose activity is triggered by terminal maturation. Two motifs alter the 5' untranslated region of the "prototypical" erythroid mRNA in such a way as to permit synthesis of a novel larger isoform. This form appears to localize preferentially in the nucleus. We thus conclude that a single gene gives rise to multiple protein 4.1 isoforms with potentially diverse locations and functions.

INTRODUCTION

The strength and flexibility of circulating erythrocytes is due in large part to the presence of a supporting network of structural proteins called the membrane cytoskeleton. The cytoskeleton contains several proteins, notably spectrin, actin, ankyrin, protein 4.9 and protein 4.1[1,2]. Recently, non-erythroid homologues of these proteins have been described in many tissues[3-10]. These isoforms exhibit considerable structural and functional diversity. Delineation of their structures and genetic regulation is currently a subject of considerable interest. For example, selective accumulation of the distinctive erythroid forms of these proteins is a major prerequisite for erythroid cell maturation.

Protein 4.1 is an essential component of the cytoskeleton, whose structure and function have been analyzed in detail in erythrocytes (for review, see Reference 11). As shown in Figure 1, protein 4.1 is an 80 kDa phosphoprotein that contains several domains definable by partial digestion and functional analysis[12]. Of particular note are the amino terminal 30 kDa

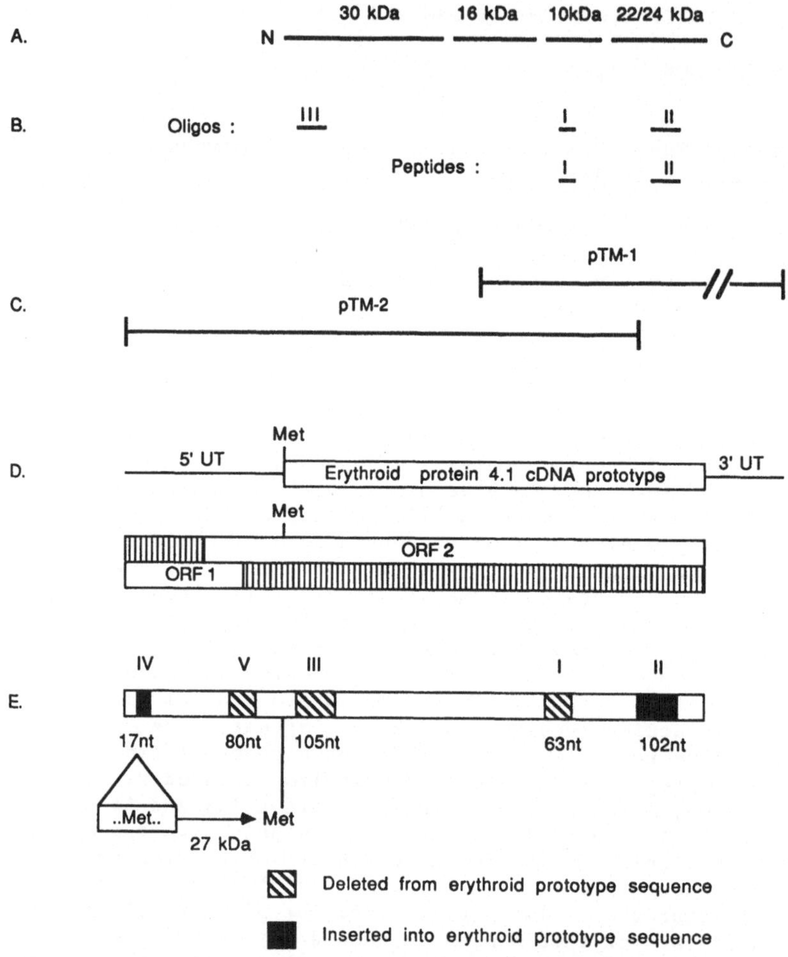

Fig. 1. Structure model of protein 4.1 and alignment of synthetic oligonucleotide and peptide probes against erythroid protein 4.1 domains. (A) Structural domains of human erythroid protein 4.1 deduced by proteolytic studies[12]. (B) Synthetic oligonucleotides and peptides used for gene probes and site-specific antibodies. (C) cDNA clones (pTM-1, pTM-2) derived from MOLT 4 lymphocytes. (D) Erythroid protein 4.1 cDNA clone reported by Conboy et al.[17]. Two open reading frames (ORF 1 and 2) are found in the erythroid cDNA. (E) Variable motifs (I to V) defined by comparison of erythroid[17] and lymphoid cDNA sequences.

domain, thought to be important for binding of an integral membrane protein, glycophorin[11,13], and the 10 kDa (8 kDa) domain that appears to contain sites for spectrin-actin binding[14,15]. Protein 4.1 is thought to serve a bridging function, binding cytoskeletal elements, such as spectrin and actin[14,15], on the one hand, and transmembrane proteins, such as glycophorin[13] and band 3[16], on the other. In this manner protein 4.1 attaches the cytoskeleton to the membrane.

Very little is known about the structure, function, or number of the non-erythroid isoforms of protein 4.1 Immunochemically reactive isoforms of protein 4.1 have been found in fibroblasts[3], platelets[4,5], endothelial cells[6], lens[7,8], neurons[9,10], and granulocytes[4]. In many tissues multiple cross reactive protein bands have been demonstrated[3-10]. The functions of these proteins remain obscure. These protein 4.1 isoforms are not strictly confined to the plasma membrane skeleton. For example, in fibroblasts protein 4.1 is distributed uniformly along stress fibers[3]. In aortic endothelial cells protein 4.1 seems to localize to the perinuclear region and areas of cell - cell contact[6]. Thus, non-erythroid protein 4.1 isoforms may serve functions distinct from those of the erythroid cytoskeletal protein 4.1.

Further understanding of this complex group of proteins will require more detailed understanding of the exact structures of different isoforms. Recently, Conboy and coworkers[17] have isolated cDNA clones encoding the complete amino acid sequence of at least one erythroid protein 4.1 form synthesized in human reticulocytes. This group has also presented evidence that only one copy of the protein 4.1 gene exists in the human genome. In this communication, we present new information concerning the regulated expression of isoforms of protein 4.1 in erythroid cells and in the T-cell leukemia line. This new isoform, identified by molecular cloning, exhibits structural features distinct from the erythroid isoform, at least one of which appears to represnet a general difference between erythroid and non--erythroid homologues. In an inducible erythroleukemia cell line, the isoform containing the erythroid structural motif is expressed only when erythroid maturation is induced. Our results also suggest that these isoforms most likely arise from a single protein 4.1 gene by alternative splicing of a common mRNA precursor. Two of these splicing events generate a potentially larger protein 4.1 isoform that may have novel properties, including intranuclear localization.

Identification and Isolation of Protein 4.1 cDNA Clones

The current structural model for erythrocyte protein 4.1, derived by limited α chymotryptic digestion, provides for four major structural domains (30 kDa, 16 kDa, 10 kDa, 22/24 kDa) (Figure 1A, reference 12). In order to isolate cDNA clones for protein 4.1, we used amino acid sequence which has been determined for a fragment, located in the central domain (16 kDa region) of the molecule. We deduced a synthetic oligonucleotide probe 63 bases long from this region[12,17] and screened a cDNA library that was prepared from human bone marrow. Several positive recombinant DNA clones were obtained. One of these clones, γA2.1, containing a 0.7 kb DNA insert, was studied in more detail. DNA isolated from the γA2.1 clone was subcloned into the plasmid vector SP65 (renamed pA2.1) and M13 bacteriophage (mp 19) for DNA sequencing. Identification of the γA2.1 cDNA clone was verified by comparing the amino acid sequence deduced from the nucleotide sequence with known peptide sequence of the spectrin actin binding domain (10 kDa region) of erythrocyte protein 4.1[15].

Using the pA2.1 clone as a probe, we then isolated cDNA clones (pTM-1, pTM-2), encoding lymphoid protein 4.1 from a MOLT 4 cDNA library[18]. Nucleotide sequence data from MOLT 4 cDNA clones allowed alignment of the clones as shown in Figure 1C.

Nucleotide Sequence Differences Between Erythroid and Lymphoid Protein 4.1

Conboy, et al.[17] have recently reported the cDNA sequence for the protein coding region of an erythroid reticulocyte protein 4.1 cDNA. Comparative analysis of our nucleotide sequence data with their published sequence revealed that the lymphoid protein 4.1 isoforms exhibit at least five nucleotide sequence "motifs" that appear to be inserted or deleted by alternative splicing of a common mRNA (Figure 1E, 2). The first MOTIF (MOTIF I) is 63 nucleotides long. It encodes a 21 amino acid segment within the spectrin - actin binding domain (10 kDa region) that is present in the erythroid protein 4.1, but absent from the MOLT 4 sequence. The second MOTIF (MOTIF II) encodes 34 amino acids near the carboxyl end of the 22/24 kDa domain. This sequence is present in MOLT 4 cDNA but absent from the erythroid cDNA sequence determined by Conboy et al.[17]. The third MOTIF (MOTIF III), encoding 35 amino acids near the amino end of the 30 kDa domain, is located in the erythroid cDNA but absent in the lymphoid form. MOTIFs IV and V are discussed later.

Expression of Sequence MOTIFs I-III in Erythroid and Lymphoid Cells

In order to investigate further whether these motifs are representative of tissue specific expression patterns, we developed molecular hybridization (synthetic oligonucleotide) and immunochemical (polyclonal antisera) probes (Fig. 1B, 2). Oligonucleotide I and peptide I are derived from sequences present within MOTIF I. Oligonucleotide II and peptide II are derived from regions within MOTIF II. Oligonucleotide III is complementary to the MOTIF III region of the mRNA. The synthetic oligonucleotides were used directly for Northern blot analysis of poly A+ RNA from various tissues. Rabbit antisera were prepared from each peptide after coupling to KLH (keyhole limpet hemocyanin); the antisera were then affinity purified by Sepharose - CL 4B columns bearing immobilized synthetic peptides as ligands[15]. The affinity purified antisera were then used directly to search for protein 4.1 forms containing either MOTIF I or MOTIF II in various tissues.

In order to assess the patterns of expression of mRNA's encoding these motifs, we performed both RNAse protection assays and Northern blot analysis. As shown in Figure 3B, Human reticulocyte RNA contains at least two Protein 4.1 mRNA species detectable by RNAse protection of an anti-sense ^{32}P labeled RNA probe spanning MOTIF I. The 333 base protected fragment indicates presence of an mRNA containing MOTIF I, and the 267 nucleotide band fragment indicates expression of an mRNA lacking MOTIF I. In contrast, the mRNA specimens from several other cell lines contained only the latter form.

For Northern blot analysis, poly A+ RNA isolated from reticulocytes and MOLT 4 cells was hybridized with the synthetic oligonucleotide probes, as well as with a 1 kb MOLT 4 cDNA probe that spans MOTIF I and II and also extends into the 3' untranslated region. Oligonucleotide I was 40 nucleotides long, whereas oligonuleotides II and III were 54 nucleotides long (Fig. 2). Using these probes, as shown in Figure 4 (A,C,D) we detected one major mRNA species in MOLT 4 cells, about 6.6 kb in length.

Both the cDNA and oligonucleotide II also detected another minor mRNA species (4.4 kb on heavily loaded blots). However, one of these mRNA species hybridized to the probe specific for MOTIF I (Fig. 4B). Figure 4 also shows that a 6.5 kb mRNA species encoding protein 4.1 can be detected in Poly A+ mRNA from sheep reticulocytes by the cDNA probe. This band is equally well detected by all oligonucleotide probes.

These results, which are typical of mRNA preparations obtained from a variety of hematopoietic and non-hematopoietic tissues, suggest strongly that the mRNA encoded in MOTIF I is present only in reticulocyte mRNA preparations. The findings we obtained with the oligonucleotide II and III probes were surprising. Both reticulocyte and MOLT 4 mRNAs were found to contain sequences complementary to oligonucleotide II and III even though MOTIF II is absent from the previously reported human reticulocyte cDNA sequence[17], whereas MOTIF III is absent from the MOLT 4 cDNA described in this communication. These results imply that both erythroid and lymphoid protein 4.1 isoforms are heterogeneous.

The above results were further confirmed by immunoblotting studies. Fig. 5A shows that both erythroid and lymphoid (MOLT 4) cells contained an 80 kDa protein reacting strongly with antisera raised against intact protein 4.1. The MOLT 4 cells also contained significant amounts of an additional larger reactive band. Similarly, both red cells and MOLT 4 cells reacted strongly with antipeptide II antibody (Fig. 5C), demonstrating that at least some forms containing MOTIF II are present in both erythroid and non-erythroid tissues. In contrast, antipeptide I antibody reacted only with the red cells proteins (Fig. 5B), implying that the 21 amino acid MOTIF I is produced exclusively in the red cells.

Protein 4.1 in red cells migrates as a doublet (4.1a, 4.1b), differing by 2,000 daltons[12,19]. Immunoperoxidase staining of human erythroid ghost proteins enabled greater resolution of these two erythroid protein 4.1 peptides. Both 4.1 a and b (80 & 78 kDa) were detected by immunoperoxidase staining, with anti protein 4.1 as well as both of the peptide sequence specific antibody reagents (Fig. 5D). Thus the apparent size difference observed between these 4.1a and 4.1b is not due to the presence or absence of either MOTIF I or MOTIF II.

<u>Selective Expression of Sequence MOTIF I During Erythroid Cell Differentiation</u>

Further studies demonstrated that expression of MOTIF I is induced when erythroid cells undergo terminal maturation. MELC are mouse erythroleukemia cells that undergo terminal erythroid maturation when exposed to compounds such as DMSO[20]. As shown in Figure 6, uninduced MELC produce significant levels of protein 4.1 mRNA. This mRNA contains both MOTIF II and III (Fig. 6D, 6E), but not MOTIF I (Fig. 6C). Interestingly, protein 4.1 mRNA in mouse cells was found to be smaller (5.6 kb) than in sheep (6.5 kb) or human lymphoid cells (6.6) kb. After DMSO induction, two changes occurred. First, the total amount of protein 4.1 mRNA increased. Second, MOTIF I was detected in a significant fraction, but not all, of the mRNA molecules.

Fig. 2. (on next page) Nucleotide sequence differences between erythroid and lymphoid protein 4.1. The nucleotide sequence of reticulocyte[17] and MOLT 4 cDNA clones are compared. The first base of lymphoid sequence corresponds to the 63rd base of reticulocyte sequence previously reported by Conboy et al.[17]. The cDNA inserts from the MOLT 4 clones were subcloned into M13mp19 vector to generate a sequencing library by the deletion method[25]. Single-strand DNA was prepared and sequenced by the dideoxynucleotide chain termination method[26] modified for the use of [^{35}S] dATP and using 6% buffer gradient gels[27]. The sequence regions marked MOTIF I to V are described in the results section. The synthetic oligonucleotide and peptide probes are underlined. The possible splicing sites of MOTIF I to V are shown in the open boxes. The amino acid sequence is deduced from the MOLT 4 cDNA sequence. The nucleotide differences between lymphoid and erythroid cDNA's are documented by ".". Two potential translation initiation codons are double underlined.

```
RETIC.                   AGAACGCGGTCGGCCCGGTCCCCGCCGCACCCAGCCCAG--- MOTIF IV ----AGAAGAGTTTAGTGACTGAGGCCGAAAATTCACAGCACCAACAGAAGGA
MOLT 4                   AGAACGCGGTCGGCCCGGTCCCCGCCGCACCCAGCCCAGCAACATCATGACAACAGAGAAGAGTTTAGTGACTGAGGCCGAAAATTCACAGCACCAACAGAAGGA
                                                                           MetThrThrGluLysSerLeuValThrGluAlaGluAsnSerGlnHisGlnGlnLysGl

                                                                                                                               226
RETIC.  AGAGGGTGAGGAAGCCATAAACTCAGGCCAACAAGAACCTCAGCAGGAGGAATCTTGTCAAACAGCAGCTGAAGGAGATAATTGGTGTGAACACAAGGTGAAAGCTTCTAATGGAGACAC
MOLT 4  AGAGGGTGAGGAAGCCATAAACTCAGGCCAACAAGAACCTCAGCAGGAGGAATCTTGTCAAACAGCAGCTGAAGGAGATAATTGGTGTGAACAGAAGCTGAAAGCTTCTAATGGAGACAC
        uGluGlyGluGluAlaIleAsnSerGlyGlnGlnGluProGlnGlnGluGluSerCysGlnThrAlaAlaGluGlyAspAsnTrpCysGluGlnLysLeuLysAlaSerAsnGlyAspTh

                                                                                                                               346
RETIC.  TCCTACACATGAAGACTTGACCAAGAACAAGGAGCGGACATCAGAAAGCAGAGGACTTTCACGACTATTCTCCTCGTTTCTCAAAAGGCCCAAATCTCAGGTGTCCGAGGAAGAAGGCAA
MOLT 4  TCCTACACATGAAGACTTGACCAAGAACAAGGAGCGGACATCAGAAAGCAGAGGACTTTCACGACTATTCTCCTCGTTTCTCAAAAGGCCCAAATCTCAGGTGTCCGAGGAAGAAGGCAA
        rProThrHisGluAspLeuThrLysAsnLysGluArgThrSerGluSerArgGlyLeuSerArgLeuPheSerSerPheLeuLysArgProLysSerGlnValSerGluGluGluGlyLy

                                                                                                                               466
RETIC.  AGAAGTAGAGTCAGATAAAGAAAAAGGTGAAGGAGGTCAGAAAGAGATAGAATTTGGAACCAGTCTTGATGAAGAGATCATTTTAAAGGCCCCAATTGCAGCTCCTGAACCGGAACTCAA
MOLT 4  AGAAGTAGAGTCAGATAAAGAAAAAGGTGAAGGAGGTCAGAAAGAGATAGAATTTGGAACCAGTCTTGATGAAGAGATCATTTTAAAGGCCCCAATTGCAGCTCCTGAACCGGAACTCAA
        sGluValGluSerAspLysGluLysGlyGluGlyGlyGlnLysGluIleGluPheGlyThrSerLeuAspGluGluIleIleLeuLysAlaProIleAlaAlaProGluProGluLeuLy

                                                                                                                               586
RETIC.  AACAGACCCATCTTTGGATCTTCATTCATTAAGCAGTGCAGAAACACAGCCTCACCATTACAATTAAGAATTATTTTTAGAGTCTTCTTATTTCTGAAGCATGTGAATATTATCTCGATC
MOLT 4  AACAGACCCATCTTTGGATCTTCATTCATTAAGCAGTGCAGAAACACAG----------------------  M O T I F V  ------------------------------
        sThrAspProSerLeuAspLeuHisSerLeuSerSerAlaGluThrGln

                                                                                                                               706
RETIC.  GTTAAAAGTCCTGCTCAGGAAGAACTCAGAGAAGATCCAGATTCTGAAATTAAGGAAGGAGAAGGACTTGAAGAGTGCTCCAAAATAGAAGTAAAAGAAGAAAGCCCTCAATCAAAAGCA
MOLT 4  ---------CCTGCTCAGGAAGAACTCAGAGAAGATCCAGATTTTGAAATTAAGGAAGGAGAAGGACTTGAAGAGTGCTCCAAAATAGAAGTAAAAGAAGAAAGCCCTCAATCAAAAGCA
                 ProAlaGlnGluGluLeuArgGluAspProAspPheGluIleLysGluGlyGluGlyLeuGluGluCysSerLysIleGluValLysGluGluSerProGlnSerLysAla

                                                                                                                               826
RETIC.  GAAACAGAATTAAAAGCTTCCCAAAAACCAATCAGAAAACACAGGAACATGCACTGCAAGGTTTCTTTGTTGGATGACACAGTTTATGAATGTGTTGTGGAGAAACATGCTAAGGGACAA
MOLT 4  GAAACAGAATTAAAAGCTTCCCAAAAACCAATCAGAAAACACAGGAACATGCACTGCAAGGTTTCTTTGTTGGATGACACAGTTTATGAATGTGTTGTGGAG---------------------
        GluThrGluLeuLysAlaSerGlnLysProIleArgLysHisArgAsnMetHisCysLysValSerLeuLeuAspAspThrValTyrGluCysValValGluLysHisAlaLysGlyGln

                             Oligo III                                                                                         946
RETIC.  GATTTGCTTAAACGAGTATGTGAGCATCTCAATCTTTTGGAAGAAGACTATTTTGGTCTAGCCATTTGGGATAACGCAACCTCTAAGACATGGCTGGATTCCGCCAAAGAAATAAAAAAG
MOLT 4  ---------- ----------------------  M O T I F III  -----------------------------------------ACATGGCTGGATTCCGCCAAAGAAATAAAAAAG
        AspLeuLet .yr. rgValCysGluHisLeuAsnLeuLeuGluGluAspTyrPheGlyLeuAlaIleTrpAspAsnAlaThrSerLysThrTrpLeuAspSerAlaLysGluIleLysLys

                                                                                                                               1066
RETIC.  CAGGTTCGTGGTGTCCCTTGGAATTTTACATTTAATGTAAAGTTTTATCCACCTGACCCAGCACAGTTAACAGAAGACATAACAAGATATTATTTATGTCTTCAGCTTCGGCAGGACATA
MOLT 4  CAGGTTCGTGGTGTCCCTTGGAATTTTACATTTAATGTAAAGTTTTATCCACCTGACCCAGCACAGTTAACAGAAGACATAACAAGATATTATTTATGTCTTCAGCTTCGGCAGGACATA
        GlnValArgGlyValProTrpAsnPheThrPheAsnValLysPheTyrProProAspProAlaGlnLeuThrGluAspIleThrArgTyrTyrLeuCysLeuGlnLeuArgGlnAspIle

                                                                                                                               1186
RETIC.  GTTGCAGGACGTCTGCCCTGTTCCTTTGCAACCTTAGCATTATTAGGTTCTTACACCATCCAGTCTGAACTGGGAGACTACGACCCAGAACTCCATGGCGTGGATTATGTTAGTGATTTT
MOLT 4  GTTGCAGGACGTCTGCCCTGTTCCTTTGCAACCTTAGCATTATTAGGTTCTTACACCATCCAGTCTGAACTGGGAGACTACGACCCAGAACTCCATGGCGTGGATTATGTTAGTGATTTT
        ValAlaGlyArgLeuProCysSerPheAlaThrLeuAlaLeuLeuGlySerTyrThrIleGlnSerGluLeuGlyAspTyrAspProGluLeuHisGlyValAspTyrValSerAspPhe

                                                                                                                               1306
RETIC.  AAACTGGCCCCGAATCAGACCAAGGAACTTGAAGAGAAGGTCATGGAACTGCATAAGTCATACAGGTCCATGACTCCAGCTCAGGCTGACTTGGAGTTTCTTGAGAATGCCAAAAAGTTG
MOLT 4  AAACTGGCCCCGAATCAGACCAAGGAACTTGAAGAGAAGGTCATGGAACTGCATAAGTCATACAGGTCCATGACTCCAGCTCAGGCTGACTTGGAGTTTCTTGAGAATGCCAAAAAGTTG
        LysLeuAlaProAsnGlnThrLysGluLeuGluGluLysValMetGluLeuHisLysSerTyrArgSerMetThrProAlaGlnAlaAspLeuGluPheLeuGluAsnAlaLysLysLeu

                                                                                                                               1426
RETIC.  TCTATGTATGGAGTTGATCTTCATAAAGCAAAGGACTTGGAAGGAGTAGATATCATCCTAGGTGTCTGCTCTAGTGGCCTTCTGGTTTACAAAGATAAGCTGAGAATTAACCGCTTCCCT
MOLT 4  TCTATGTATGGAGTTGATCTTCATAAAGCAAAGGACTTGGAAGGAGTAGATATCATCCTAGGTGTCTGCTCTAGTGGCCTTCTGGTTTACAAAGATAAGCTGAGAATTAACCGCTTCCCT
        SerMetTyrGlyValAspLeuHisLysAlaLysAspLeuGluGlyValAspIleIleLeuGlyValCysSerSerGlyLeuLeuValTyrLysAspLysLeuArgIleAsnArgPhePro
```

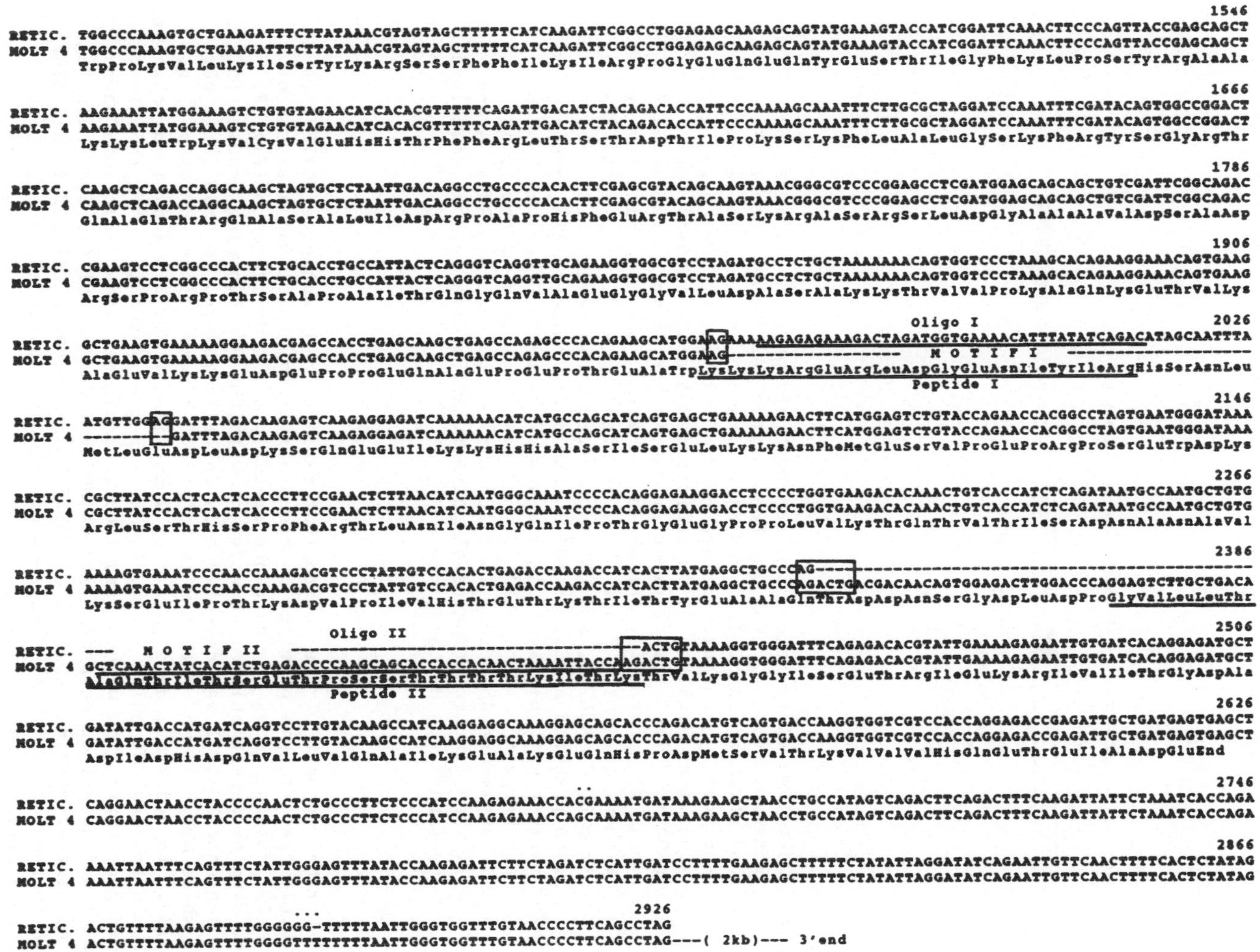

Fig. 2. (see page 85)

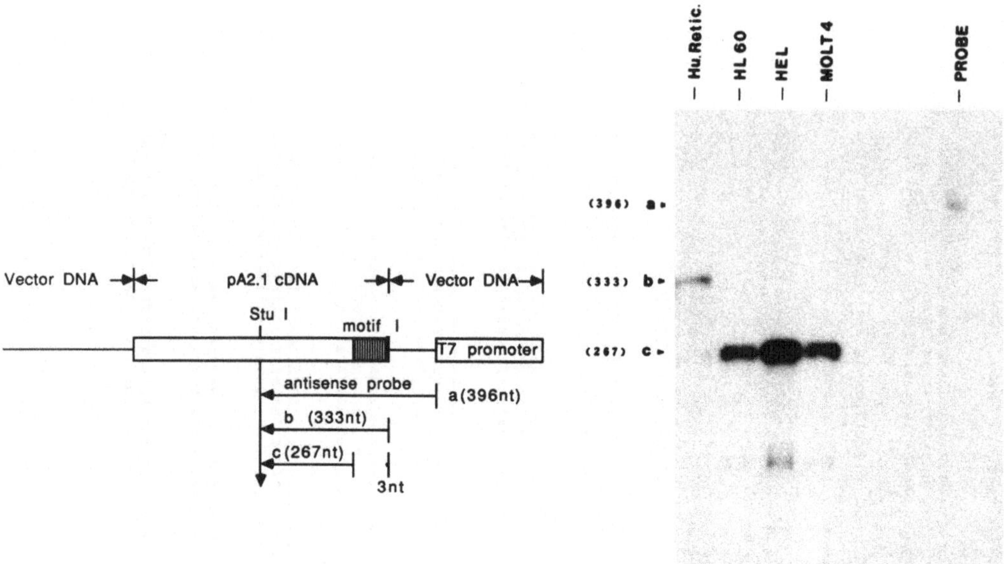

A B

Fig. 3. Detection of erythroid and nonerythroid protein 4.1 mRNAs by RNAse protection. (A) The bone marrow pA2.1 cDNA insert (722 bp) which spans MOTIF I and also extend 3 nucleotides beyond the motif was subcloned into pGEM-4 blue vector (Promega), and linearized with Stu I: An antisense RNA probe (396nt) was generated from the T7 promoter within the vector. Protected fragment lengths for the whole probe (a), the mRNA with (b) or without (c) MOTIF I are shown in the diagram A. (B) Total RNA from HL60, MOLT 4, and HEL (human erythroleukemia) cell lines were prepared as described[28]. Human reticulocyte RNA from a patient with hemolytic anemia was prepared as described by Benz and Forget[29]. Total RNA (30 ug) was hybridized to the 396nt T7 antisense RNA probe (5 x 10^5 cpm), digested with RNase A and T1, and fractionation of 5% polyacrylamide/urea gels as described by Zinn[30] and Melton[31].

Further support of the notion that expression of sequence MOTIF I is erythroid specific was provided by comparative immunoblotting analysis of MEL cell proteins using the sequence specific antibody reagents (Fig. 7). These studies were completely consistent with the RNA blotting results in that sequence MOTIF I was detected only in cells induced by DMSO (Fig. 6C, 7A), while MOTIF II was detected in both induced and uninduced cells (Fig. 6D, 7B). In addition, no significant induction of MOTIF I and II was observed in hemin treated MEL cells (Fig. 7A, 7B). Hemin, which is known to stimulate globin gene expression without any induction of most other erythroid phenotypes[21] had no effect on production of MOTIF I in MEL cells. These findings are consistent with the absence of MOTIF I in HEL cells (Fig. 3B), and K562 cells (not shown) which also express globin gene in the absence of complete erythroid differentiation[22].

Translation of Novel Protein 4.1 Isoforms From mRNA's Containing MOTIF IV But Lacking MOTIF V

As indicated in Figure 1D, band 4.1 from erythroid cells mRNA contains

2 long open reading frames that could potentially encode protein. The first, ORF I, occurs in the 5' untranslated region. In actuality it is unable to encode protein because it lacks an initiator methionine codon. ORF I is also "closed" within a few hundred bases by several in frame termination codons. ORF II, is shifted one codon "out of phase" from ORF I. It is fully functional, encoding the 80 kDa protein 4.1 forms in both lymphoid and erythroid tissues.

Lymphoid protein 4.1 mRNA exhibits critical differences from its erythroid counterpart in the 5' extremity of the sequence (MOTIF IV/V rearrangement) (Fig. 2E). First, the insertion of MOTIF IV (17 bases), introduces an initiator methionine codon, surrounded by a functional consensus initiation sequence ("Kozak" box, Table I) in the same reading frame as ORF I. Second, the deletion of sequences encompassed by MOTIF V not only removes the termination codons from ORF I but further shifts the reading frame so that ORF

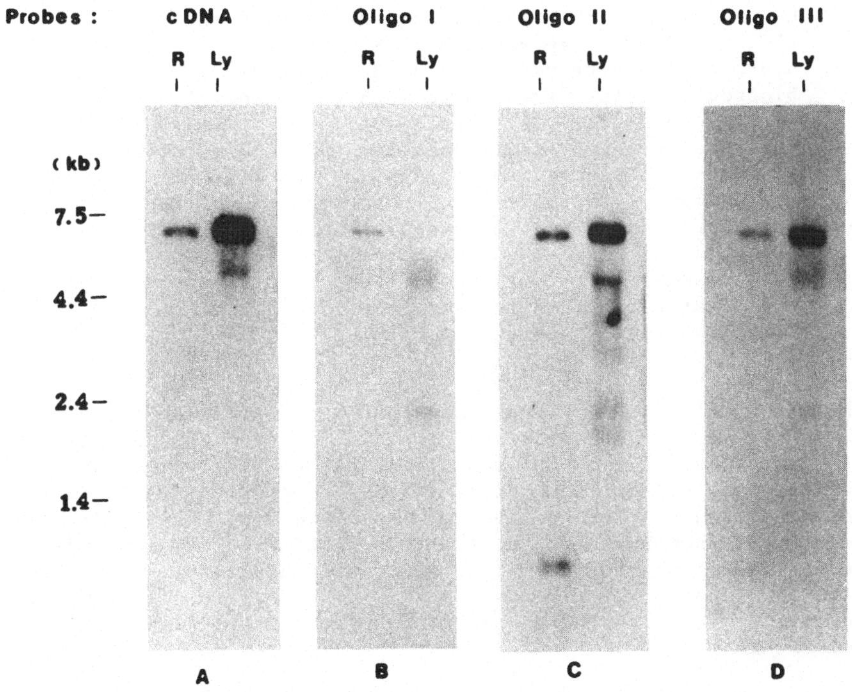

Fig. 4. Differential expression of sequence MOTIF I, II and III in protein 4.1 mRNAs. Total RNA from cultured T-cell leukemia cells (MOLT 4), was prepared as described in Fig. 3. Sheep reticulocyte poly(A+) RNA was a kind gift from Dr. Philip Dunham. Poly(A+) RNA from MOLT 4 was selected by oligo dT-cellulose chromatograph[33]. 5 ug of poly(A+) RNA was fractionated by electrophoresis through a 1.4% agarose formaldehyde gel, transferred to Nytran filters (Schleicher & Schuell) and hybridized to: (A), a probe generated from the 1 kb EcoRI fragment derived from the 5' end of the pTM-1 clone; (B), and B oligo I probe; and (C), the oligo II probe. (D) Blot shown in 4B rehybridized with the oligo III probe. Conditions for hybridization and washing by using cDNA probe were described by Thomas[32]. The synthetic oligonucleotide probes was 5' end labeled with polynucleotide kinase[33]. Conditions for hybridization and washing by using oligo probes were as previously described[34], except that hybridization & washing temperatures were 45°C for the oligo I probe, 56°C for oligo II probe.

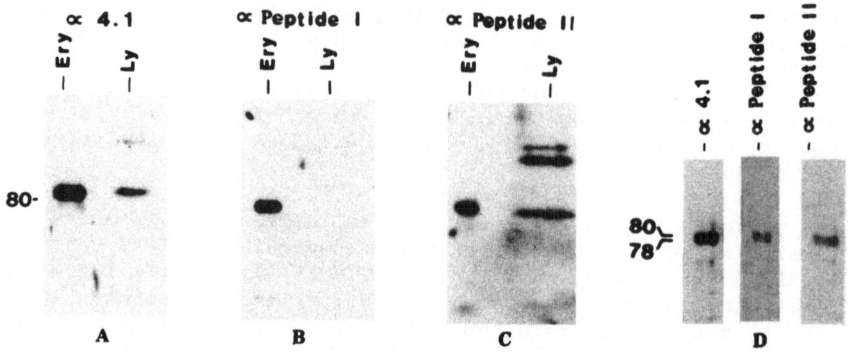

Fig. 5. Differential expression of sequence MOTIF I and II in protein 4.1 isoforms detected by immunoblotting analysis. Solubilized proteins from human red cell ghosts (5ug;Ery) and MOLT 4 cells (160ug;Ly) were separated by SDS-PAGE and blotted onto nitrocellulose as described[6,12]. Blotted proteins were probed with polyclonal anti-protein 4.1. (A), anti-peptide I antibody (B), and anti-peptide II antibody (C), and detected by I^{125}-staphylococcus protein A as described[15]. (D) Primary antibodies bound to protein 4.1 isoforms from human red blood cell ghosts (5ug) were detected with a goat anti-rabbit IgG-peroxidase conjugated as previously described[35]. Apparent molecular weights of reactive peptides are given in kilodaltons on left.

I and ORF II now form a single long, continuous open reading frame that potentially encoded a 27 kDa peptide. Translation from the new upstream start site generates a 122 kDa protein identical to the 80 kDa form except for the addition of the amino terminal "42 kDa" "headpiece". It is interesting to note that the size of "headpiece" estimated from the SDS-PAGE (42 kDa) is different from that deduced from the nucleotide sequence (27 kDa). This discrepancy may be due in part to aberrant migration in SDS-PAGE. Translation from the "original" downstream site generates the prototypical 80 kDa form. Two protein classes can thus be potentially synthesized from one mRNA molecule.

We synthesized functional synthetic mRNA templates from cDNA clones that we had engineered to contain or lack the MOTIF IV/V rearrangement (Fig. 2E). The synthetic mRNA was prepared by subcloning into "Gemini" vectors containing both SP6 and T7 RNA polymerase promoters, followed by "CAPping" to add the 5' CAP sequence needed for full activity. These mRNA's were then translated in a rabbit reticulocyte lysate system. The ^{35}S methionine labeled proteins were fractionated by SDS-PAGE.

As shown in Figure 8, the mRNA corresponding to the prototypical erythroid form promoted synthesis only of the 80 kDa protein (Fig. 8, lane 3). However, the single lymphoid mRNA containing the aforementioned MOTIF IV/V rearrangement, encoded 2 proteins, as predicted: the 80 kDa form and the 122 kDa form (Fig. 8, lane 2). Therefore, one protein 4.1 mRNA can give rise to additional isoform heterogeneity by encoding two distinctive isoforms.

The function of the 122 kDa isoform is currently under investigation. We have prepared antisera against synthetic peptide derived from the "new" upstream amino acid sequence and used the sera as immunochemistry probes

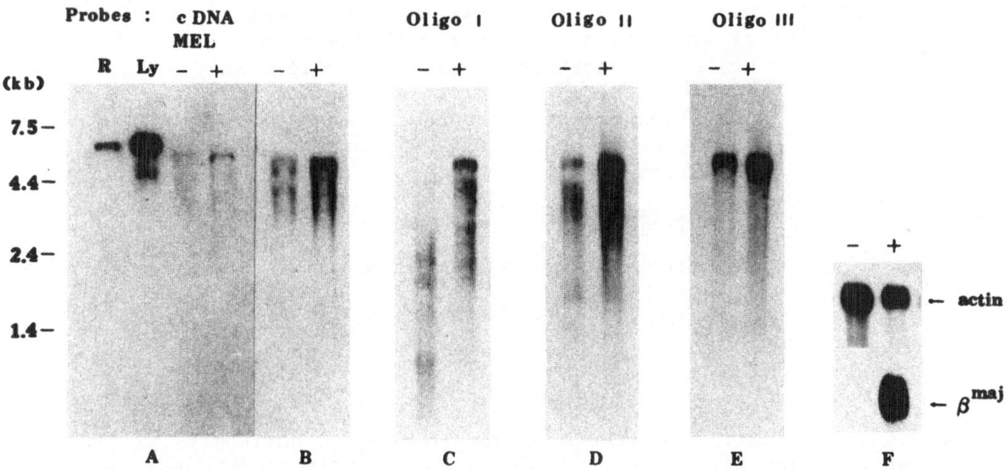

Fig. 6. Introduction of sequence MOTIF I in MELC detected by northern blot analysis. MEL cells (strain 745A) were maintained in Dulbecco's modified Eagle's medium (DMEM) containing 10% fetal calf serum. MEL cells treated (+) or not treated (-) with 1.8% DMSO (Dimethylsulfoxide) were harvested 96hr later. (A) 5 ug poly(A+) RNA from sheep reticulocyte (R), MOLT 4 (ly), uninduced (- lane), and DMSO induced (+ lane) MEL cells were hybridized to a 1 kb cDNA probe. 10ug poly(A+)RNA from uninduced (- lane) and induced (+ lane) MELC were hybridized to: (B), a 1kb cDNA probe; (C) the oligo I probe; (D), the oligo II probe; and (E), the oligo III probe. Fig. 6F, the upper panel: blot shown in 6C rehybridized with actin gene probe. Lower panel: blot shown in 6 D rehybridized with mouse β major globin gene probe. Fig. 6F indicates that induction of globin gene expression by DMSO, and equal loading of uninduced & induced MELC mRNAs.

in Western blot and intact cell immunofluorescence surveys. Our preliminary data suggest that the 122 kDa form is widespread in tissues and localized largely to the nucleus (manuscript in preparation).

DISCUSSION

The results presented in this communication define a structural feature of protein 4.1 isoforms that is differentially expressed in lymphoid and erythroid tissues. This difference alters the 10 kDa domain which has been shown by Correas et al.[14,15] to be important for spectrin - actin binding. Erythroid protein 4.1 includes at least some isoforms that contain 21 amino acids not present in any lymphoid isoforms detectable by Northern blot analysis of mRNA or by immunoblot analysis of proteins.

Previous studies by Correas et al.[14,15] have shown that an antibody raised against a synthetic peptide contained within MOTIF I efficiently inhibited spectrin actin binding by purified erythroid protein 4.1. It is thus reasonable to speculate that MOTIF I represents an important region that may mediate fundamental functional differences between erythroid and non-erythroid isoforms of protein 4.1, at least with respect to the ability of the individual isoforms to interact with other submembranous cytoskeletal elements. In this context, it is important to note that antibodies directed against MOTIF II, as well as an antiserum raised against the total protein 4.1 molecule, detect at least 2 isoforms of protein 4.1 in lymphoid cells

Table I. Comparison of the first two AUG codons of lymphoid 4.1 isoform with the Kozak consensus sequence[36] for eukaryotic translational initiation site. The exact positions of the first and second AUG codons are shown in Fig. 2 (double underline).

(Fig. 5A, 5C); yet, neither of these appear to contain MOTIF I (Fig. 5B). We have examined many tissues for the presence of MOTIF I with the antipeptide I antiserum, using both immunofluorescent and immunochemical approaches. In no case have we identified any proteins containing MOTIF I (unpublished observation). Moreover, we have demonstrated specific "induction" of the expression of MOTIF I upon induction of MELC to an erythroid phenotype by DMSO (Fig. 6,7). We thus conclude that this sequence region is a unique feature of erythroid protein 4.1.

Our data also demonstrate an unexpected heterogeneity of protein 4.1 forms in red cells as well as in lymphoid cells. Conboy and Coworkers[17] have isolated a human red cell protein 4.1 cDNA that lacks sequence MOTIF II. However, as indicated by Figures 4C and 5C, our results demonstrate that MOTIF II is present not only in multiple lymphoid isoforms but also

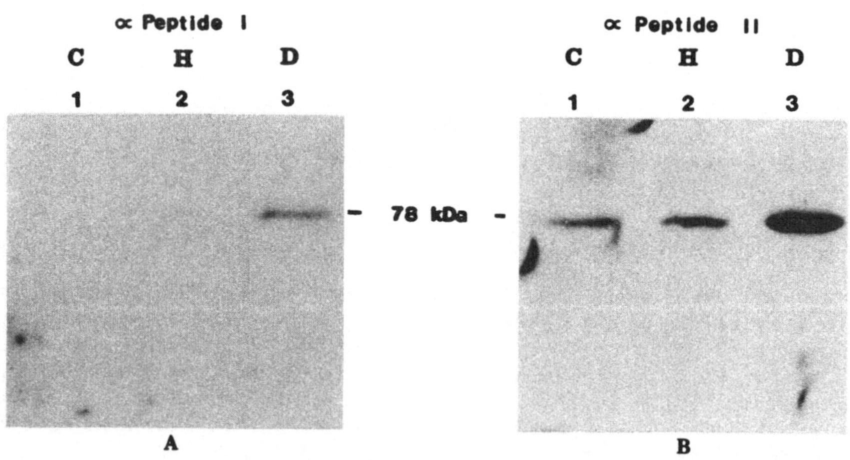

Fig. 7. Induction of sequence MOTIF I & II in MEL cells detected by immunoblot analysis. Polyclonal antipeptide I & II were used as described in Fig. 5. Solubilized proteins (160ug) from uninduced (C) or DMSO induced (D) MEL cells were described in Fig. 5. Lanes marked H contained proteins (160ug) isolated from MEL cells treated with 50 uM hemin for 96 hrs. Signal intensities were normalized with each antibodies by including a parallel lane (not shown) containing equal quantities of protein 4.1 from red cell ghosts and adjusting exposure to give equal intensity of this band on each autoradiography.

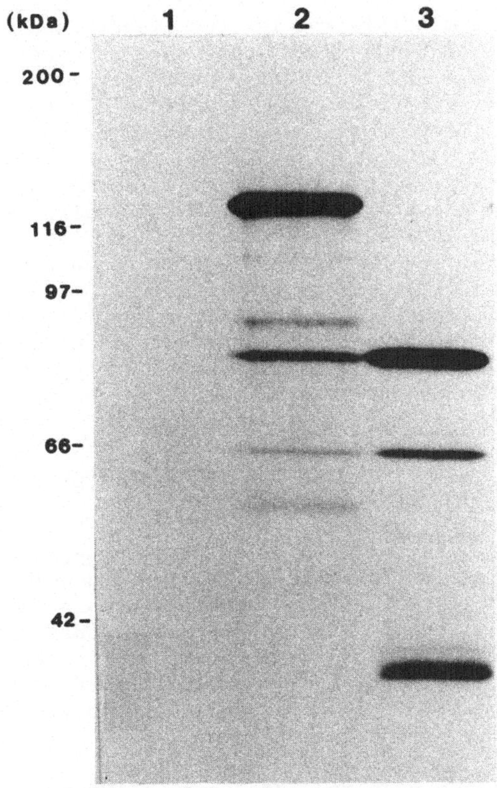

Fig. 8. Translation of synthetic mRNA's in vitro using the rabbit reticulocyte lysate. Lymphoid clone (pLym), which contains the first two translation initiation codons (ATG), and a truncated cDNA (pEry), which deleted MOTIF's IV and V that causes removal of the first ATG codon, were constructed and subcloned into a transcription vector (pGEM^{TM} 4-blue). Sense mRNA transcripts, made from SP6 or T7 RNA polymerase, were translated in vitro by adding rabbit reticulocyte lysate (Promega). The translated proteins were labeled by ^{35}S-methionine and separated by SDS-PAGE. Lane 1, no mRNA; lane 3, pLym sense mRNA, and lane 3, pEry sense mRNA.

in mRNA and protein derived from erythroid sources of human and sheep. Upon examination of multiple tissues, we have found that every tissue we analyzed produced at least some isoforms containing MOTIF II (data not shown). In addition, the RNAse protection results (Fig. 3) indicate that at least two different protein 4.1 mRNA's classes are present in human reticulocyte cells: one containing MOTIF I (major form) and one lacking this sequence (minor form). Therefore, we conclude that there must be several structurally distinct protein 4.1 isoforms present in red cells. Heterogeneity of protein 4.1 isoforms was also encountered in lymphoid cells. Lymphoid mRNA containing MOTIF III was detected by oligo III probe (Fig. 4D), even though this motif is not present in the MOLT 4 cDNA (Fig. 2). These results imply that lymphoid protein 4.1 mRNAs contain at least two forms, one containing and one lacking MOTIF III. It is interesting to note that RNA blot analysis of Protein 4.1 mRNA's in multiple species and tissues consistently yields rather diffuse patterns of bands (Fig. 4, 6). It is possible that most cells produce multiple 4.1 mRNA's differing slightly in molecular weight and giving a broad banding pattern. The functional importance of these different isoforms in different tissues remains to be investigated.

In this communication, we provide direct evidence not only for the existence of multiple protein 4.1 isoforms that are differentially produced in human tissues (Fig. 3,4,5) but also for the induction of an erythroid specific protein 4.1 isoform when erythroid cells undergo terminal differentiation (Fig. 6,7). Our findings are concordant with those of Lazarides and coworkers, who have identified multiple protein 4.1 isoforms in different tissues of avian species[8,23], and suggested that the heterogeneity of avian erythroid isoforms was generated by differential RNA processing[24]. The genetic regulatory mechanism determining the relative amounts of each isoform produced in specific tissues must clearly be complex. It is very likely that only a single copy of the gene for protein 4.1 exists in both human[17] and avian genomes[24]. Comparison of the nucleotide sequence of our cDNA clones described with that reported by Conboy et al.[17] is consistent with this view, since we have encountered almost "complete" nucleotide sequence homology between lymphoid and erythroid isoforms, except for the five regions that we have designated as motifs. This pattern of homology interrupted by interstitial en bloc insertions and deletions within one mRNA relative to the other is most consistent with the hypothesis that the mRNAs encoding these isoforms arise from a common mRNA precursor by alternative splicing. In addition, the selective expression of MOTIF I in erythroid cells raises the possibility that induction of the erythroid phenotype may involve appearance of a lineage specific splicing mechanism.

The discovery of a new open reading frame in lymphoid protein 4.1 mRNA's has allowed us to ask whether each of the protein 4.1 isoforms mentioned above might also have a "larger" counterpart produced by use of an upstream start site and opening of the reading frame by the MOTIF IV/V rearrangement, as noted earlier. We have shown that a larger 4.1 isoform of this type can be translated, at least in vitro. It is interesting to speculate about the function of these larger forms. Since the C-terminal 80 kDa are identical to "typical" 4.1 forms, the additional 27 kDa at the amino terminus could simply target large protein 4.1 isoforms to new locations without altering functions. Our preliminary data suggest that these large 4.1 forms are, indeed, localized largely to the nucleus (manuscript in preparation).

In summary, protein 4.1 has proven to be a complex group of closely related proteins, all of which are probably encoded by a single gene. The tissue specific regulation of expression of different forms suggest that some isoforms are adapted to specific functions, especially in the erythroid lineage. Elucidation of the structures of these forms should facilitate analysis of the functions of this fascinating protein family.

REFERENCES

1. Marchesi, V.T.. 1985. Ann.Rev.Cell Biol. 1: 531.
2. Bennett, V. 1985. Ann.Rev.Biochem. 54: 273.
3. Cohen, C.M., S.F. Foley, and C. Korsgren. 1982. Nature (London) 299: 648.
4. Spiegel, J.E., D.S. Beardsley, F.S. Southwick, and S.E. Lux. 1984. J. Cell Biol. 99: 886.
5. Davis, G.E., and C.M. Cohen. 1985. Blood 65: 52.
6. Leto, T.L., B.M. Pratt, and J.A. Madri. 1986. J.Cell Physiol. 127: 423.
7. Aster, J.C., M.J. Welsh, G.J. Brewer, and H. Maisel. 1984. Biochem.

Biophys. Res. Commun. 119: 726.
8. Granger, B.L., and E. Lazarides. 1984. Cell 37: 595.
9. Goodman, S.R., L.A. Casoria, D.B. Coleman, and I.S. Zagon. 1984. Science 224: 1433.
10. Baines, A.J., and V. Bennett. 1985. Nature 315: 410.
11. Leto, T.L., I. Correas, T. Tobe, R.A. Anderson, and W.C. Horne. 1986. In: Membrane Skeletons and Cytoskeleton-Membrane Association, V. Bennett et al., Eds. (Alan R. Liss, Inc., New York.) pp 201-209.
12. Leto, T.L., and V.T. Marchesi. 1984. J. Biol. Chem. 259: 4603.
13. Anderson, R.A., and R.E. Lovrien. 1984. Nature (London) 307: 655.
14. Correas, I., T.L. Leto, D.W. Speicher, and V.T. Marchesi. 1986. J. Biol. Chem. 261: 3310.
15. Correas, I., D.W. Speicher, and V.T. Marchesi. 1986. J. Biol. Chem. 261: 13362.
16. Pasternack, G.R., R.A. Anderson, T.L. Leto, and V.T. Marchesi. 1985. J. Biol. Chem. 260: 3676.
17. Conboy, J., Y.W. Kan, S.B. Shohet, and N. Monhandas. 1986. Proc. Natl. Acad. Sci. U.S.A. 83: 9512.
18. Alonso, M.A., and S.M. Weismann. 1987. Proc.Natl.Acad.Sci.U.S.A. 84: 1997.
19. Goodman, S.R., J.Yu, C.F. Whitfield, E.N. Culp, and E.J. Posnak. 1982. J. Biol. Chem. 257: 4564.
20. marks, P.A., and R.A. Rifkind. 1978. Ann.Rev.Biol. 47: 419.
21. Gusella, J.F., et al. 1980. Blood 56: 481.
22. Martin, P., and T. Papayannopoulou. 1982. Science 216:1233.
23. Granger, B.L. and E. Lazarides. 1985. Nature 313: 238.
24. Ngai, J., J.H. Stack, R.T. Moon, and E. Lazarides. 1987. Proc. Natl. Acad. Sci. U.S.A. 84: 4432.
25. Dale, R.M.K., B.A. McClure, and J.P. Houchins. 1985. Plasmid 13: 31.
26. Sanger, F., S. Nicklen, and A.R. Coulson. 1977. Proc. Natl. Acad. Sci. USA. 74: 5463.
27. Biggin, M.D., T.J. Gibson, and G.F. Hong. 1983. Proc. Natl. Acad. Sci. USA. 80: 3963.
28. Chirgwin, J.M., A.E. Przybyla, R.J. MacDonald, and W.J. Rutter. 1979. Biochemistry 18: 5294.
29. Benz, E.J., and B.G. Forget. 1971. J.Clin.Invest. 50: 2755.
30. Zinn, K., D. Dimaio, and T. Maniatis. 1982. Cell 34: 865.
31. Melton, D.A., et al. 1984. Nucl. Acids Res. 12: 7035.
32. Thomas, P.S. 1980. Proc. Natl. Acad. Sci. U.S.A. 77: 5201.
33. Maniatis, T., E.F. Fritsch, and J. Sambrook. Molecular Cloning: Laboratory Manual (Cold Spring Harbor Laboratory, Cold Spring Harbor, NY, 1982) pp 122-123, 197-198.
34. Linnenbach, A.J., D.W. Speicher, V.T. Marchesi, and B.G. Forget. 1986. Proc. Natl. Acad. Sci. USA 83: 2397.
35. D.A. Johnson, J.W. Gautsch, J.R. Sportsman, and J.H. Elder. 1984. Gene Anal. Tech. 1: 3.
36. Kozak, M. 1984. Nucl. Acids Res. 12: 857.

EXPRESSION OF HEME OXYGENASE IN HEMOPOIESIS

Nader G. Abraham, Steve M. Mitrione, W. John B. Hodgson, Richard D. Levere and Shigeki Shibahara*

Department of Medicine
New York Medical College
Valhalla, NY 10595

*Department of Applied Physiology
Tohoku University School of Medicine
Sendei, Miy agi, Japan

ABSTRACT

Heme oxygenase has been purified to electrophoretic homogeneity from detergent solubilized adult human liver microsomes. Treatment of microsomes with Triton X-100, sodium cholate and subsequent batchwise DEAE-cellulose, 2', 5' ADP-sepharose 4B, Sepharose CLB and hydroxylapatite column resulted in 17% yield of the purified heme oxygenase. The reconsituted system of heme oxygenase, composed of heme oxygenase, NADPH cytochrome c (P450) reductase and biliverdin reductase was equiactive with 1 mM NADPH and 4 nM NADH and showed complete dependence on added heme for catalytic activity. The Km values for NADPH and NADH were .046 and .526 mM, respectively. While NADPH concentration was held constant, the Km value for heme was 1.01 μM with a specific activity of 583 unit/mg protein. The activity of the reconstituted heme oxygenase system was not affected by preincubation with heavy metals despite their inhibitory effect of NADPH cytochrome c (P450) reductase and biliverdin reductase. However, the metalloporphyrins of these heavy metals were found to be strong inhibitors of the reconsituted system with Ki values of 0.015, 0.6, 2.3 and 5 μM for Sn-, Co-, Zn- and Mg- protoporphyrins, respectively. Similarly, the sulfhydryl inactivating reagents, $HgCl_2$, iodoacetamide and p-chloromercurylbenzoate, inhibited the reconstituted heme oxygenase activity.

Rabbits were immunized with purified human liver heme oxygenase and the resulting antibody preparation was used to examine the species specificity of the enzyme. Microsomal protein with a molecular weight of 32,000 from rat and human liver as well as $HepG_2$ cells were identified on dot and Western blots by their reaction with the anti-heme oxygenase similar to the purified enzyme protein. Anti-heme oxygenase precipitated quantitatively, the entire heme oxygenase of rat liver microsomes obtained from animals maintained on standard diet. The human bone marrow microsomal heme oxygenase activity was also quantitatively precipitated by this antibody. Antibody inhibition of rat and human heme xoygenase demonstrated a degree of conservation of both enzyme proteins between the species. As judged by Western blotting, the anti-heme oxygenase recognized only a single protein in spleen, liver, kidney, brain, heart, bone marrow, intestine and corneal epithelium.

The human heme oxygenase cDNA was isolated by screening a cDNA library in the Okayama-Berg vector with a rat liver cDNA and was subjected to nucleotide sequence analysis. The deduced human heme oxygenase is also composed of 288 amino acids with a molecular mass of 32,800 Da. Following hemin treatment of human leukemic cell line K562 there was an increase in the amount of heme oxygenase protein and mRNA.

INTRODUCTION

The microsomal heme oxygenase [EC 1.14.99.3] plays an essential role in the physiological degradation of heme to biliverdin IX[1]. In mammals, the microsomal heme oxygenase system consists of three individual enzymes; heme oxygenase, NADPH cytochrome c (P450) reductase [EC 1.6.2.4] and biliverdin reductase [EC 1.3.1.24] and requires NADPH and molecular oxygen for bilirubin formation[2-5].

The involvement of NADPH cytochrome c(P450) reductase in the heme oxygenase reaction has been previously demonstrated[6]. The NADPH cytochrome c(P450) reductase is responsible for the reduction of the oxygenated heme-heme oxygenase complex to form hydroxy-heme[7]. However, Maines, et al. have shown that NADH catalyzed heme degradation was accomplished by heme oxygenase[3]. This finding was confirmed in rat bone marrow[8] and in a purified preparation of bovine spleen heme oxygenase[2]. Heme oxygenase has been shown to be induced by heavy metals, bromobenzene, benzene and hemin[9-12]. In contrast to induction by certain metals, metalloporphyrins have been shown to inhibit heme oxygenase activity and/or block induction both _in vitro_ and _in vivo_, respectively[2,13]. Certain hormones and endogenous factors seem to play a significant role in perturbation of heme oxygenase levels. Thyroid hormone, glucagon, epinephrine and adrenalectomy have been shown to increase hepatic heme oxygenase activity[14-17]. However, much less has been published about heme oxygenase in man[18,19].

The interrelationship between the regulation of erythroid heme synthesis and catabolism has never been fully elucidated, although previous studies demonstrate reciprocal oscillations in hepatic ALA synthase and heme oxygenase _in vivo_ following intravenous heme administration. This is explained as an initial repression of ALA synthase mRNA by hemin, then succeeding induction of heme oxygenase, followed by decreased cellular heme and depression of ALA synthase. In hemopoietic cells, heme also has been shown to be required for erythroid cell development, differentiation, and for other cell function (Figure 1) including the accumulation of globin mRNA[20-22].

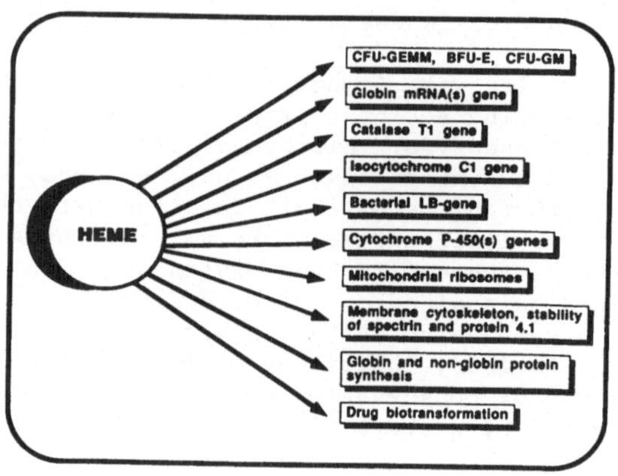

Fig. 1. Schematic presentation for utilization of heme for various important functions in cells.

Erythroid heme metabolism is finely coordinated to ensure a balance of both heme and globin synthesis. Therefore, expression of specific heme enzymes such as ALA synthase, ALA dehydratase and PBG deaminase has been shown to be a prerequisite for accumulation of heme levels needed for globin and non-globin protein synthesis, and for normal cell growth and differentiation[18,23,24]. However, when cellular heme levels exceed that need, an induction of heme oxygenase may occur to ensure the coordination of heme and globin biosynthesis. Neither the heme concentration needed for the essential process of normal cell development and differentiation, nor the mechanism of heme oxygenase induction by heme is known. Therefore, purification of human heme oxygenase and antibody preparation will enable us in the identification of heme oxygenase in a small number of hemopoietic analysis of its role in human disease associated with disturbances of heme metabolism. Furthermore, our study will facilitate analysis of the structural and functional constraints of the heme oxygenase molecule in hemopoiesis.

METHODS

Purification of Human Liver Heme Oxygenase

Human liver and other tissues were obtained from the Laboratories and Research Department, Westchester County Medical Center, Valhalla, New York, by autopsy, a few hours after death. Approximately 200 to 300 gm of liver were suspended in 100 ml of 100 mM Tris-HCl buffer, pH 7.6, containing 150 mM sodium chloride and 250 mM sucrose and further homogenized as described previously[22]. Briefly, livers were sliced and homogenized in a Waring blender with 2.5 volumes of Tris-HCl buffer, pH 7.6, containing 250 mM sucrose and 0.4 mM PMSF. Homogenate was successively centrifuged at 800 x g for 10 min, at 13,500 x g for 20 min to pellet the mitochondria. These mitochondria were washed twice with the same buffer. The supernatant was centrifuged at 105,000 x g for 90 min to isolate the microsomal fraction. The microsomal pellet was washed with 100 mM sodium pyrophosphate, Ph 7.5, to eliminate the hemoglobin[23]. The microsomes were then suspended in 20 mM Tris-HCl buffer, pH 7.7, containing 0.5 mM EDTA and 0.4 mM PMSF to final concentration of 20 mg/ml. Triton X-100 and sodium cholate were added to the microsomal suspension to a final concentration of 2.0% and 1.0%, respectively. Cholate was re-crystallized in 50% ethanol using activated charcoal and celite in the first step, dried in vacuum over silica, and accurately made up to a 20% solution (w/v) and adjusted to pH 7.5 with NaOH and filtered through Whatman paper. The suspension was gently stirred for 30 min at 4°C and then centrifuged at 105,000 x g for 90 min.

DEAE-Cellulose Chromatography

The soluble heme oxygenase preparation was loaded onto a column (4.4 x 60 cm) of DEAE cellulose equilibrated with 20 mM Tris-HCl buffer, pH 7.7, containing 20% glycerol, 0.5 mM EDTA, 0.5% Triton X-100 and 0.2% sodium cholate. The DEAE cellulose was washed with 2 leters of the same buffer. The enzyme was then eluted with 2 liters of similar buffer in which the KCl concentration was increased by a concave gradient (R1:R2=1:3) from 0. to 0.5 M KCl at a flow rate of 140 ml/hr. Fractions were collected in 11 ml aliquots; heme oxygenase activity eluted at approximate concentrations of 0.08=0.14 mM KCl (Fraction A). The latter fractions, between 0.2-0.5 M, had highest activity of NADPH cytochrome c (P450) reductase (Fraction B). Fraction A, which had the activity not only of heme oxygenase, but also the activity of NADPH cytochrome c (P450) reductase, was pooled and pH adjusted to 6.8 and then diluted 2-fold with 10 mM potassium phosphate, pH 7.7, containing 20% glycerol, 0.05 mM EDTA, 0.1 mM DTT, 0.2% Triton X-100, and 0.2% sodium cholate (Buffer A). This fraction was concentrated overnight using a PM-10 filter in an Amicon ultrafiltration system.

2'5' ADP Sepharose 4B Column

The concentrated fraction was applied directly to a 2'5' ADP-sepharose 4B column (1.7 x 13 cm) equilibrated with 10 mM potassium phosphate buffer, pH 7.7, 0.2% Triton X-100, 0.1 mM EDTA, and 0.1 mM DTT (Buffer B), and subsequently washed with 250 ml of buffer B containing 0.3 M potassium phosphate and 0.1 mM 2' AMP. Heme oxygenase activity was eluted using buffer B; fractions containing its activity were concentrated on Amicon ultrafiltration system with a PM10 filter to a final volume of 50 ml and dialyzed overnight against buffer B. The other steps involved in the purification and production of antibodies was essentially as that described[25].

Heme Oxygenase Assay

The reaction mixture contained the following components in a final volume of 1 ml: 1 mM NADPH or its generating system (2 units of glucose 6-phosphate dehydrogenase, 1 mM NADP and 2 mM glucose-6-phosphate); 3 mM $MgCl_2$; 30 µM hemin; 100 mM potassium phosphate buffer, pH 7.7, 10 units of biliverdin reductase, and various concentrations of microsomes. The reaction was started by the addition of hemin to the incubation mixture which was kept in the dark in a shaking waterbath at 37°C for 20 min. The reaction was terminated by either placing the tubes on ice or by adding 0.1 ml of 10% $HgCl_2$. Bilirubin was extracted twice with 2 ml of chloroform and its formation was determined by using the difference in absorption from 460 to 530 mm and an extinction coefficient of 43.5 mM^{-1} cm^{-1}.[3] Metalloporphyrins were dissolved in 2 ml of 0.1 N NaOH. After addition of 10 mg of Tris-base, the pH was adjusted to 8.4 with a few drops of 0.1 M HCl and made up to a final volume with distilled water.

The reconstituted system of the purified heme oxygenase contained purified heme oxygenase (10 units), NADPH cytochrome c (P450) reductase (5 units), biliverdin reductase (15 units), 50 µg of dilauroylphosphatidylcholine (dispressed by sonication), 15 mM $MgCl_2$, 90 mM KCl and 100 mM potassium phosphate buffer, pH 7.7 in a final volume of 300 µl. This mixture was incubated for 15 min at 22°C and 1 mM NADPH or 4 mM NADH were added. Five minutes later, 30 µM hemin was added to initiate the reaction. The reaction was carried out for 20 min at 37°C and stopped by addition of 1 ml chloroform. Bilirubin was extracted and measured as described above.

Immunotitration of Human and Rat Heme Oxygenase

For immunotitration of the enzymatic activity, serial dilutions of preimmune and immune purified rabbit sera were prepared in 0.9% sodium chloride containing 0.02 M potassium phosphate buffer, pH 8.0, in a final volume of 100 µl. Each tube then received 40 µl of the enzyme-containing preparations to be titrated. The mixture was incubated for 25 min at 25°C; various amounts of rabbit antihuman IgG (Dako Corporation, Santa Barbara, CA) were added, and the tubes were stored overnight at 4°C. After centrifugation at 15,000 x g in a microfuge for 20 min, the supernatant fractions were assayed for enzymatic activity as described above. The results are expressed as the percentage of the control activity remaining in the supernatant fluid. In control vessels, the enzyme activity did not change significantly during the pre-incubation and the preimmune serum did not precipitate detectable quantities of heme oxygenase enzyme protein.

Western Blotting and Dot Blot Identification of Heme Oxygenase

Purified human heme oxygenase, and rat and human microsomes were subjected to SDS gel electrophoresis and separated proteins blotted[26] electrophoretically onto nitrocellulose filters[26] and as previously described[25].

Further Purification and Assay of NADPH Cytochrome c (P450) Reductase

All steps were carried out at 4°C in the presence of 20% glycerol, and 0.4 mM PMSF. The fractions containing the reductase activity, i.e., Fraction B of the DEAE cellulose, were pooled and concentrated in an Amicon concentrator and dialyzed against 20 volumes of 10 mM potassium phosphate buffer pH 7.7 containing 0.2% Triton X-100, 0.2% sodium cholate, 0.05 mM EDTA and 0.1 mM DTT (Buffer C) for 10 hours and used for further purification as described previously[27]. The concentrated fraction was then applied to a 2'5' ADP Sepharose 4B column (1.7 x 13 cm) previously equilibrated with Buffer C. The column was washed with 250 ml of 200 mM potassium phosphate buffer pH 7.7 containing 20% glycerol, 0.1% sodium cholate, 0.05 mM EDTA, 20% glycerol, and 0.1% sodium cholate. The column was then eluted with 50 ml of Buffer C without cholate but containing 5 mM 2' AMP. Those fractions having reductase activity were pooled and concentrated on an Aminco ultrafiltration cell. The concentrated fraction was passed through a Sephadex G100 column (2.2 x 40 cm) preequilibrated with Buffer C to remove unbound 2' AMP and stored in small aliquots at -70°C[27]. Cytochrome c reductase was assayed in a final volume of 1.0 ml, 300 mM potassium phosphate buffer, pH 7.7, 0.1 mM NADPH, an appropriate amount of the purified NADPH cytochrome C (P450) reductase and the acceptor, cytochrome c. The reaction was carried out at 30°C following the reduction of cytochrome c at 550 nm using an extinction coefficient of 21 mM^{-1} cm^{-1}(28) on SLM/Aminco DW-2C dual beam spectrophotometer.

Purification and Measurement of Bilirubin Reductase

The 105,000 x g supernatant of the microsomal fraction of human-liver was used for purification of biliverdin reductase as described[38,29]. The final assay mixture of biliverdin reductase contained 0.1 mM potassium phosphate buffer, pH 7.6, 10 μM biliverdin; 1 mg bovine serum albumin; 100μM NADPH and appropriate amount of enzyme protein in a final volume of 2.0 ml. The complete reaction mixture was preincubated for 10 min at 37°C. The reaction was started by the addition of NADPH. Formation of bilirubin was calculated from the increase in optical density at 460 nm, utilizing SLM/Aminco DW-2C spectrophotometer and a extinction coefficient of 46.0 mM^{-1} cm^{-1} (30). One unit of enzyme activity was defined as that amount of enzyme which resulted in the formation of 1 nmol bilirubin per hr. The preparation obtained had a specific activity of 2800 units per mg protein.

Cell Lines

K562 cells were maintained at 37°C in a humidified atmosphere in RPM1 1640 medium, supplemented with 10% fetal calf serum as described[31]. Hep G_2 were maintained as described[32,33]. Induction of heme oxygenase was accomplished by adding hemin to a final concentration of 10μM. Hemin was prepared as described[18] and sterilized by filtration prior to addition to the cultures. Cell growth was continued for 8 hours after addition of hemin. Exposure to hemin under these conditions did not reduce cell viability as determined by trypan blue exclusion. Aliquots of cells were removed after 1.5, 3, 4.5, 6 and 7.5 hours of hemin exposure. After incubation, the 10,000 RPM supernatant from cell homogenate was prepared and used for the determination of heme oxygenase activity[18].

Preparation of RNA and RNA Blot Hybridization Analysis

Total RNA was prepared[34] from control and treated K562 cells. Cells were incubated with 100 μM hemin at 37°C for 5 hr. The enzyme activity was also determined. RNA blots were prepared as described[35], and hybridized with $\alpha^{32}P$-labelled XhI -AccI fragment (-64 to 230) derived[36] from pH Hol (See Figures 8,9). All DNA probes used in this study were labeled with [$\alpha^{32}P$] dCTP by the method of Feinberg and Vogelstein[37].

Cloning of Full-length cDNAs From Human Heme Oxygenase

Full-length cDNA from human heme oxygenase was prepared according to Yoshida et al.[36]. Briefly, poly(A)rich RNA was isolated by oligo (dT)-cellulose chromatography[38] from human macrophages U 937 treated with 10 µM hemin for 3 hr. Using 6 µg poly(A)rich RNA, a cDNA library was constructed in the Okayama-Berg vector. The library was screened with a ^{32}P-labelled XhoI-Gam HI fragment (nucleotide residues -59 to 297) derived from the rat heme oxygenase CDNA, pRHo1[35]. Nucleotide sequences were determined by the method of Maxam and Gilbert[40].

MATERIALS

Restriction endonuclease and E. coli DNA polymerase (Klenow fragment) were obtained from Boehringer Mannheim Biochemicals, Indianapolis, IN, oligo (dT)-cellulose was from Bethesda Research Laboratory, Gaithersburg, MD. NAD, NADH, glucose-6-phosphate, PMSF, hemin, bovine serum albumin, biliverdin, bilirubin, protein markers for molecular weight determination and cholic acid were purchased from Sigma Chemicals, Co., St. Louis, MO. Hydroxylapatite gel (Bio-Gel HA), acrylamide, SDS, bis-acrylamide, Coomassie brilliant blue R-250, Bio-Beads SM-2, glycine, N,N-dimethyl-bis-acrylamide were obtained from Biorad. Lab., Richmond, CA. Glucose-6-phosphate dehydrogenase was obtained from Boehringer Biochemicals. DEAE-cellulose (DE 52) was obtained from Whatman. Sepharose Cl-6B was purchased from Pharmacia Fine Chemicals, Piscataway, NJ. 2'AMP, 2'5'ADP Sepharose 4B, affinity gel NADP agarose were purchased from P.L. Biochemicals. Zinc-, tin-, cobalt-, magnesium-, and iron-protoporphyrins were purchased from Porphyrin Products, UT, (specific activity 3000 Ci/mmol), [α-^{32}P] dCTP multiprime DNA labelling system, and nitrocellulose were from Amersham Corporation, Arlington Heights, IL.

RESULTS AND DISCUSSION

Purification of Human Adult Heme Oxygenase

During the purification of human heme oxygenase, we were faced with several problems. First, since the spleen has the highest activity of heme oxygenase, we spent a great deal of time purifying the enzyme from human spleen. However, as has been shown with bovine spleen, pufication was with success, due to a lack of stability. During the initial steps of the purification, the human-spleen enzyme lost more than 50% of its activity. It was assumed that this was due to the presence of proteases which are not inhibited by PMSF. Secondly, we found that human hepatic microsomes, isolated at the 105,000 x g pellet, are highly contaminated by hemoglobin. Therefore, washing the microsomes with sodium pyrophosphate not only removed almost all of the contaminating hemoglobin but also preserved the content and activity of heme oxygenase.

A summary of the purification of heme oxygenase from human liver is outlined in Table I. Several features of the purification procedure are noteworthy. First, 2'5' ADP Sepharose, which has been used previously for the purification of NADPH cytochrome c (P450) reductase[27,28], proved to be useful in the separation of heme oxygenase from NADPH cytochrome (P450) reductase. Therefore, the DEAE cellulose column was washed with 500 ml of buffer B. The eluant from the loading and washing steps did

Table I. Purification of Human Liver Heme Oxygenase

Fractions	Protein (mg)	Total Activity (Units)	Specific Activity (units/mg)	Yield %	Purification Fold
1) Microsomes	11637	15128	1.3	100	1.0
2) Solubilized Microsomes	6886	14522	2.3	95.9	1.8
3) DEAE-Cellulose	315	11036	35.0	72.9	27.9
4) 2',5' ADP Sepharose and Hydroxylapatite Column	46	5849	127	38.7	98
5) Sepharose Cl-6B	6.8	3843	491	25.4	434
6) Hydroxylapatite and Bio-Beads	3.4	2575	757	17.0	582

The reaction mixtures contained 0.1mm01 of NADP, 30 μmol hemin in a final volume of 1.0 ml 0.1M of potassium phosphate buffer (pH 7.7). The rate of reaction were measured at 37°C in 0.1 M potassium phosphate (pH 7.7) as described under "Experimental Procedures". This purification procedure resulted in spectrally pure heme oxygenase as seen in Figure 3.

not contain significant amounts of either NADPH cytochrome c (P450) reductase or heme oxygenase, and was discarded. A linear 0 to 0.5 M KCl gradient was then used to elute the column (Figure 2). The gradient mixing flask contained 1000 ml of buffer B and the second flask contained 100 ml of buffer B to which 0.5 M KCl had been added. The resulting fractions obtained from this column were analyzed for heme oxygenase and NADPH cytochrome c (450) reductase. As seen in Figure 2A, heme oxygenase activity was eluted at KCl concentrations between 0.08-0.14M, whereas NADPH cytochrome c (P450) reductase activity eluted between 0.2-0.5M. The yield of heme oxygenase activity from this column was about 70% (Table 1).

The isolated heme oxygenase was then passed through 2'5' ADP Sepharose column as described in Methods to separate heme oxygenase from NADPH cytochrome c (P450) reductase followed by a hydroxylapatite column to further purify heme oxygenase (Figure 2B). Secondly, successful isolation of heme oxygenase, which is present in small quantities, was achieved using a minimum number of chromatography columns with high yield steps. The overall yield of the described procedure was 30.3%. Thirdly, the procedure is specifically designed so that the product of one chromatographic step can, in the majority of cases, be applied directly to the next column without an intervening process. Using this five-step purification procedure, heme oxygenase activity was purified by 600-fold with a high specific activity of 757 unit/mg protein (Table I). The Km and Vmax values of the purified heme oxygenase were determined for the two electron donors, NADPH and NADH, in the presence of 30μM hemin. The Km values as calculated from the Lineweaver-Burke plat for the reduced nucleotides were 0.046 and 0.526 mM for NADPH and NADH, respectively. This value for NADPH agrees with that found for the rat liver enzyme[34]. The activity of the heme oxygenase system reached 470 units/mg protein in the presence of 4 mM NADH wich was 80% of that observed with 1 mM NADPH, 588 units/mg protein. This result indicated that NADH can serve as an electron donor for heme oxidation in vitro[2,8]. The Km value for heme was 1.01 μM measured in the presence of 1 mM NADPH only, the Vmax value was 583 unit/mg protein.

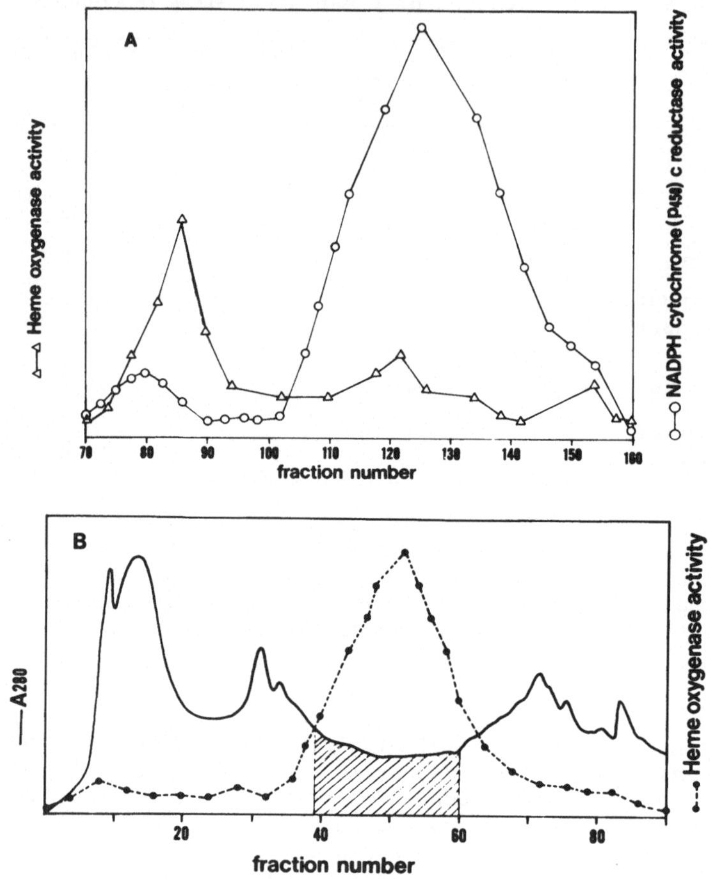

Fig. 2. Elution profiles of heme oxygenases: A) Chromatography of the Triton X-100 and sodium cholate solubilized human-liver microsomes on DEAE cellulose column. An amount of human liver microsomes containing 1,600 mg protein was washed with sodium pyrophosphate and solubilized as described in Methods. B) Chromatography on hydroxylapatite column. Fractions (11 ml each) were collected after 500 ml of buffer B had passed through the column. The activity of heme oxygenase and cytochrome c (P450) reductase was assayed as described in Methods.

Effects of Heavy Metals and Sulfhydryl Agents

We investigated the in vitro effects of some of the heavy metals on the reconstituted heme oxygenase system and its component enzymes. NADPH cytochrome c (P450) reductase, biliverdin reductase and heme oxygenase were pre-treated with the heavy metals and sulfhydryl inactivating agents and passed through a Sephadex G100 column (2.5 x 10 cm) to remove the excess of these reagents. The enzyme activities were then determined individually and in the reconstituted system. Table II summarizes the effect of the heavy metals and the sulfhydryl inactivating reagents on heme oxygenase, NADPH cytochrome c (P450) reductase, beliverdin reductase and the reconstituted system of heme oxygenase. Cobalt and copper were the only metals that slightly inhibited the heme oxygenase system due to their strong effect

Table II. Effects of Heavy Metals and Sulfhydryl Agents on the Heme Oxygenase Reconstituted System

		% inhibition			
Addition	Concentration (mM)	Reconstituted Heme Oxygenase System	Heme Oxygenase	Biliverdin Reductase	NADPH Cytochrome c (P450) Reductase
Co^{++}	0.1	13.8	0	31.1	0
Cu^{++}	0.1	14.0	0	6.2	62
Fe^{++}	0.1	0	0	0	0
Sn^{++}	0.1	0	10.1	0	10.1
Pb^{++}	0.1	2.8	0	5.1	8.5
P-chloromercuril-benzoate	0.1	35	12.3	85	79
Iodoacetamide	0.1	18.5	0	60	69
$HgCl_2$	0.1	100	10.5	100	100

The reconstituted heme oxygenase system contained heme oxygenase (30 units), NADPH cytochrome c (P450) reductase (2 units) and biliverdin reductase (10 units). The enzymes of heme oxygenase systems were pretreated with metal or SH reagent and passed through a column of Sephadex G100 to remove the non-bound agent. The enzyme activities were then determined as described in Methods.

Table III. Ki of Various Metalloporphyrins for the Purified Human-Liver Heme Oxygenase in a Reconsituted System.

Metalloporphyrin	Ki (μM)
SN-protoporphyrin	0.015
Co-protoporphyrin	0.6
Zn-protoporphyrin	2.3
Mg-protoporphyrin	5.0

The metalloporphyrin were dissolved in 2 ml of 0.1N NaOH and 10 mg of Tris-base. The pH was adjusted to 8.4 with 0.1M HCl and made up to a final volume with distilled water. The reconstituted heme oxygenase system was preincubated with different concentrations of the metalloporphyrin for 10 min. Heme oxygenase activity was measured as described in Methods. Ki were determined using double reciprocal plots of substrate concentration vs. velocity of heme oxygenase reconstituted system in the presence of the appropriate metalloporphyrin and by Dixon plots.

on biliverdin reductase and NADPH cytochrome c (P450) reductase and biliverdin reductase. These results are in agreement with those reported for rat liver and bovine liver heme oxygenase[2,41]. Hg was the strongest sulfhydryl inactivating reagent inhibitor of the heme oxygenase reconstituted system, i.e., 0.1 mM inhibited 100% of the activity. This potency _in vitro_ appears to be related to the sulfhydryl binding capacity of these reagents to NADPH cytochrome c (P450) reductase or biliverdin reductase in the reconstituted system of heme oxygenase.

Effect of Metalloporphyrins

Synthetic metalloporphyrins of Fe protoporphyrin are able to suppress neonatal jaundice, by appropriately regulating the rate of heme oxidation _in vivo_. The potential application of these metalloporphyrin treatments in humans is under examination. We tested the effect of various synthetic metalloporphyrins on the purified reconstituted system of heme oxygenase. Table III summarizes the Ki values of the synthetic metalloporphyrins tested.

The Ki values were determined using the double reciprocal plots of substrate concentrations vs. velocity of heme oxygenase reconstituted system in the presence of the appropriate metalloporphyrin and by Dixon plots. Tin-protoporphyrin was the strongest inhibitor of purified human heme oxygenase activity in a reconstituted system with a Ki of 0.015 μM, whereas cobalt-protoporphyrin at the same concentration had a moderate effect on the enzyme activity and a Ki value of 0.6 μM. However, on the other hand, zinc- and magnesium-protoporphyrins were less inhibitory of enzyme activity and demonstrated Ki values of 2.3 μM, 5 μM, respectively.

Immunoprecipitation of Heme Oxygenase Activity from Human and Rat Liver Microsomes

In order to determine the immunological identity of heme oxygenase activity in the microsome of rat liver, and their relation to human bone marrow heme oxygenase, we carried out the following immunotitration. Human and rat solubilized microsomes, as well as purified heme oxygenase, were reacted with antibodies prepared against the purified human heme oxygenase, and the results of the immunotitration with the antibodies are shown in Figure 3. The antibodies inhibited both rat and human bone marrow microsomal enzyme, whereas the preimmune IgG had no effect. Similar results were

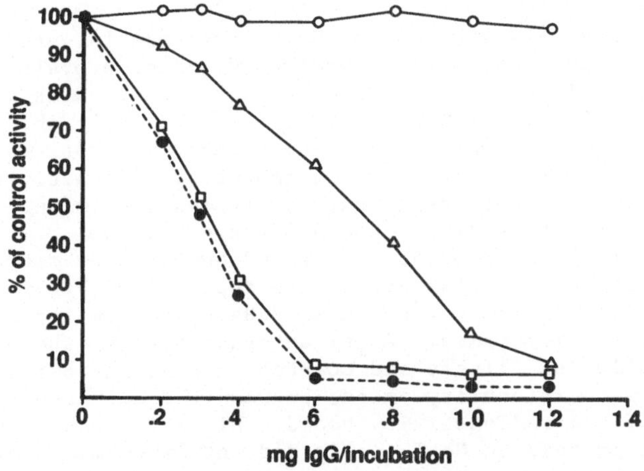

Fig. 3. Inhibition of bilirubin formation in rat (Δ-Δ) and human (-) xolubilized bone marrow microsomes and in the reconstituted system of the purified human-liver heme oxygenase (o---o) by anti-heme oxygenase. The preimmune IgG-treated human-bone marrow microsomes (0——0). The activity was measured as described in Methods. The antibodies (in indicated amounts) were added to microsomes or purified enzymes and incubated overnight at 4°C; the samples were centrifuged at 15,000 x g and assayed for enzymatic activity in a total volume of 300 µl. The initial activity of human-bone marrow microsomes and rat-liver microsomes were 9.8 units and 9.2 units per mg. The reconstituted system of the purified heme oxygenase contained purified heme oxygenase (10 units), NADPH cytochrome c (P450) reductase (5.0 units), dilauroylphosphatidylcholine (10 µg) and biliverdin reductase (15 units) as described in Methods. Values represent the mean of duplicate determination for 2 separate experiments.

obtained when human liver solubilized microsomes were used instead of human bone marrow microsomes. In a similar experiment with purified human liver heme oxygenase, using the same ratio of antibody to enzyme, these antibodies inhibited the activity to 95% of the control value of 0.6 mg IgG per incubation mixture. Thus, human and rat liver enzymes cross react efficiently since the heme oxygenase activity in rat liver microsomes was gradually precipitated by the antibody to the human heme oxygenase as a function of antibody concentration. However, more antibody was required to precipitate the enzyme from the rat microsomes, although both microsomes had virtually identical initial activities. This finding suggests that the rat liver microsomes contain a larger quantity of enzyme protein with a lower specific activity. Thus it appears that rat and human heme oxygenases are quite similar antigenically, despite the rather striking differences in catalytic properties.

Dot Blot Titration of Heme Oxygenase

The following experiments were carried out in order to confirm the specificity and feasibility of using the antibodies to study the quantitative and qualitative distribution of heme oxygenase in various tissues. Various amounts of purified heme oxygenase (0.5-5 ng protein) were spotted onto nitrocellulose filters and treated with the antibodies as described in Methods. The diluted antibodies (1:100) were clearly able to detect 0.5 ng of heme oxygenase. After autoradiography, the dots were excised from the nitrocellulose filters and the amount of radioactivity bound was counted. There was a linear relationship between the amount of purified enzyme used and the radioactivity bound (data not shown). We further tested the specificity of the antibodies in rat and human liver microsomes and in the human hepatoma cell line $HepG_2$. As depicted in Figure 4, the antiserum at 1:500 dilution cross reacted with $HepG_2$ tissue lysates and with rat and human liver microsomes. After autoradiography, the dots were excised from the nitrocellulose filters and counted in a gamma counter. Figure 5 indicates the proportionality exists between the amount of tissue used and the amount of the radioactivity bound. Linearity was present between 0.1 and 12 μg of protein in $HepG_2$ cells and rat liver microsomes and between 0.2 and 20 μg in human microsomes.

Western Blot of Human Heme Oxygenase

To further identify the precipitate protein in rat and human microsomes and in $HepG_2$ cells, the molecular size of the precipitated protein was measured by the Western blotting technique. Microsomes from rat and human liver, $HepG_2$ homogenate and the purified enzyme were submitted on 10% SDS polyacrylamide gel electrophoresis as described in Methods. The proteins were transformed electrophoretically onto nitrocellulose filters

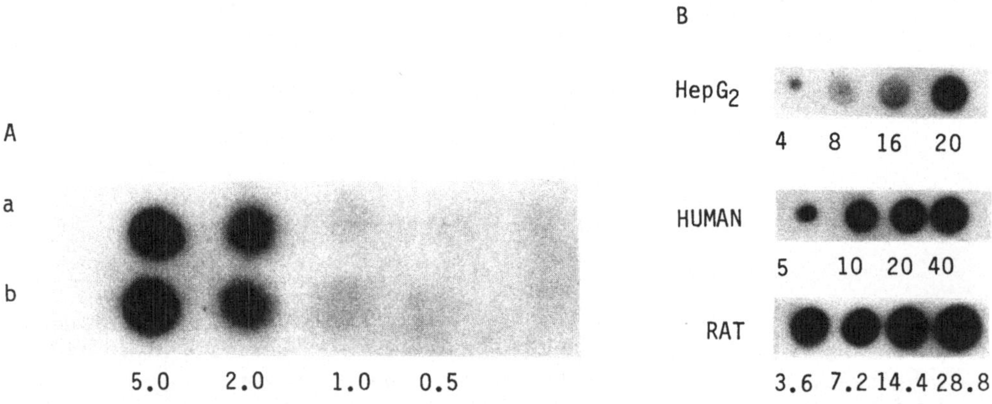

Fig. 4. Autoradiogram of dot-blot titration of the (A) purified heme oxygenase in duplicate and (B) rat and human microsomes and $HepG_2$ cells. Various amounts (0.5-5 ng protein) of purified heme oxygenase, rat and human microsomes (3.6-40 μg) and $HepG_2$ cells (4-20 μg) were spotted onto nitrocellulose filters and treated with a diluted antibody preparation (1:100). Dot blot analysis was performed using 1:500 antibody dilution as described in Methods.

and Western blot analysis was performed. Incubation of the blot with ^{125}I-protein A resulted in labeled protein bands corresponding to "Mr" of 32,000 (Figure 6). The Mr values of the precipitated protein in each preparation were identical to that obtained for the purified heme oxygenase on SDS polyacrylamide gel electrophoresis, i.e., 32,000. Western blot analysis using the preimmune IgG did not reveal any labeled proteins.

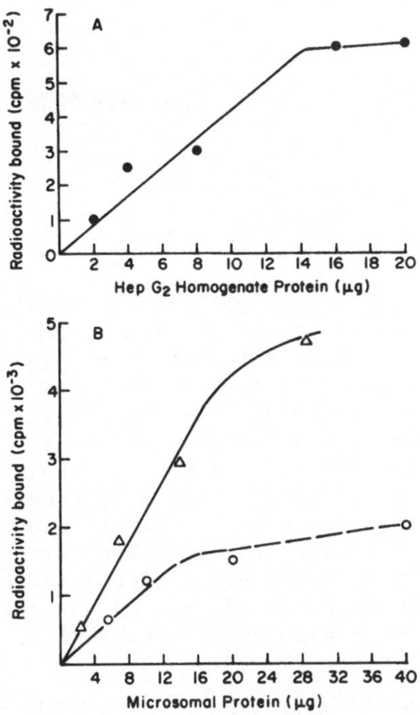

Fig. 5. Dot blot titration of human heme oxygenase; quantitations of the radioactivity bound in each blot. Various amounts of rat and human-liver microsomes and $HepG_2$ homogenate were applied onto nitrocellulose filters. Dot blot analysis was performed as described in Methods using a 1:500 antibody dilution. The radioactivity in each dot was counted in a gamma counter, (A) $HepG_2$ anti (B) human-(Δ-Δ) and rat-(O——O) liver microsomes.

Tissue Distribution of Heme Oxygenase

We investigated the presence and distribution of heme oxygenase in hepatic and extra-hepatic human tissues such as spleen, kidney, brain, bone marrow, heart, intestine and cornea. Microsomes from all tissues were prepared by the same method used for human liver microsomes. The Western blot analysis using microsomal protein from various tissues, shows a single band increasing in intensity with the increase in heme oxygenase levels in various amples with the least activity in the cornea followed

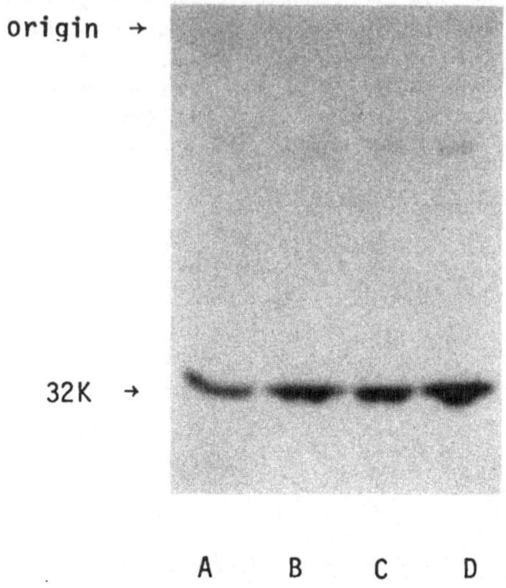

Fig. 6. Western blot analysis of heme oxygenase. Autoradiogram of the immunoblot after SDS-PAGE. The nitrocellulose filter was treated as described in Methods using a 1:1000 antibody dilution. Duplicate samples were also tested with control sera and no specific bands were detectable. Lanes: A) human-liver microsomes (10 μg protein); B) rat-liver microsomes (20 μg protein); C) $HepG_2$ cell lysates (30 μg protein); and D) purified-heme oxygenase (0.1 μg protein).

by the heart and intestine (Figure 7). Although the activity of heme oxygenase varied between tissues (data not shown), all samples gave a single band at the molecular weight of 32,000 indicating the presence of one single enzyme protein.

cDNA Cloning of Heme Oxygenase

In the early experiment, we screened human spleen λ gt 11 library with human heme oxygenase antibody raised in rabbits as described above. The screening protocol was as that described by Young, et al., 1983[43,44] during the screening several positive clones were detected. However, one of us, principally Dr. Shibahara, used rat heme oxygenase cDNA for the screening utilizing macrophages treated with hemin. Hemin treatment increases the amount of heme oxygenase in human macrophages[36], we used poly-(A)rich RNA, prepared from hemin-treated macrophages, as a template to construct a cDNA library. From about 5×10^4 transformants, we isolated eight cDNA clones which hybridized with the rat heme oxygenase cDNA probe. Of these clones, we chose two independent cDNA clones for sequence analysis, both of which apparently carry the full-length insert[36]. These two clones, PHH01 and pHH02, have the inserts of about 1700 nucleotides, including the dGtail and the poly(dA)tail (Figure 8). The size of human heme oxygenase mRNA was estimated by blot hybridization analysis of total RNA prepared from human macrophages treated with hemin[36]. The probe was hybridized to a discrete mRNA of about 185, which was similar to the size of rat heme oxygenase mRNA[35,36].

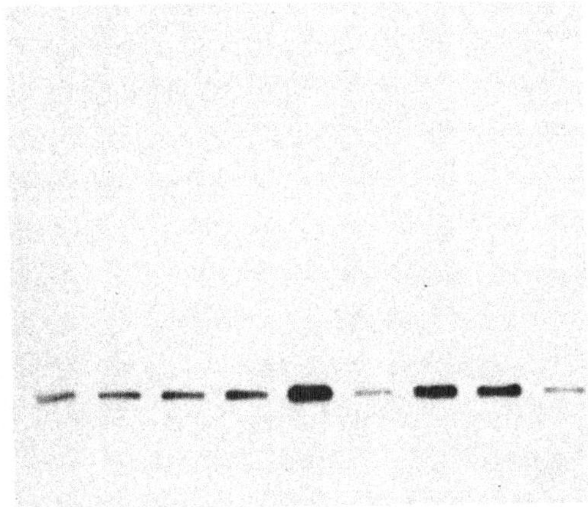

Fig. 7. Western blot analysis of heme oxygenase in human tissues. An SDS gel electrophoresis was run with the following samples: A. purified human-liver heme oxygenase (0.1 μg protein), B. spleen (10 μg protein), C. liver (10 μg protein), D. kidney (30 μg protein), E. brain (50 μg), F. bone marrow (30 μg protein), G. heart (50 μg protein), H. intestine (50 μg protein) and I. corneal epithelium cell (50 μg protein). Samples were applied to the gel. Proteins were transferred to nitrocellulose sheets and incubated with antibodies and ^{125}I-labeled protein A as described in Methods. The nitrocellulose sheets were autoradiographed at -80°C. A single immunoreactive band is seen for all the microsomal preparations studied as well as for the purified heme oxygenase.

Nucleotide Sequence of the Human Heme Oxygenase cDNA

The nucleotide sequence of pHHO1 was determined in both directions (Figure 9). The pHHO1 contains 1550 nucleotides excluding the dG tail of nine residues and the poly(dA) tract of about 45 residues (Figure 8). Using the reading frame of the rat heme oxygenase mRNA[35], we assign the ATG Codon at the position 1-3 as the translation-initiation site. The sequence flanking the assigned initiating methionine is consistent with the favored sequence for eukaryotic translation initiation sites[42]. The translation termination Codon TGA is present in the assigned reading frame at position 865-867 and is followed by 603 nucleotides of the 3'-untranslated region. A polyadenylation signal, AATAA is located 13 nucleotides upstream from the poly(dA) tail. We investigated the induction of human heme oxygenase in the erythroid leukemic cell line K562. As seen in Figure 10, hemin treatment increased the amount of functional heme oxygenase mRNA (Fig. 10. lane B as compared to untreated lane A). This confirms our previous finding of induction of heme oxygenase protein in the human leukemic cell line K562.

In this paper, we have been able to demonstrate with immunological techniques the presence of a single heme oxygenase protein and mRNA in

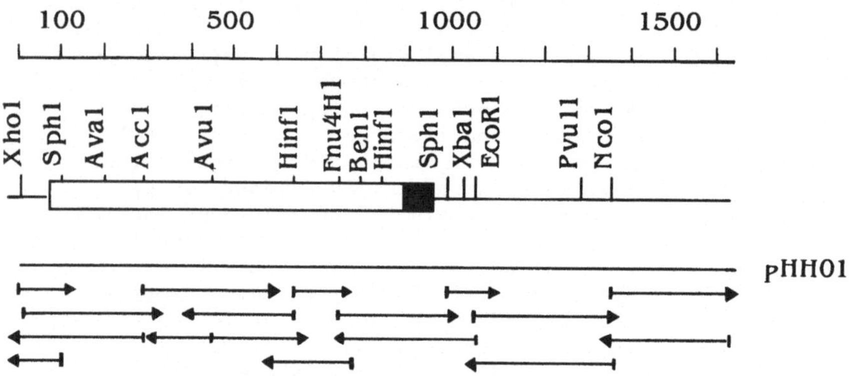

Fig. 8. Restriction map and sequencing strategy of cloned cDNA encoding human heme oxygenase. The restriction map shows only relevant sites. The poly(dA) tail and poly(dG) tail are not included in the restriction map. The protein-coding region is indicated by an open box, and the putative membrane segment is indicated by a closed box. The arrows indicate the direction and extent of sequence determinations. The short vertical lines and the slash marks at the end of arrows indicate the site of 5'-end labelling located in the cDNA and the vector DNA, respectively.

```
 -1·1           TCAACGCCTGCCTCCCCTCGAGCGTCCTCAGCGCAGCCGCCCGCGGAGCCAGCACGAACGAGCCCAGC   -11
ACCGGCCGGATGGAGCGTCCGCAACCCGACAGCATGCCCAGGATTTGTCAGAGGCCCTGAAGGAGGCCACCAAGGAGGTGCACACCCAG   81
           1                        10                       20
GCAGAGAATGCTGAGTTCATGAGGAACTTTCAGAAGGGCCAGGTGACCCGAGACGGCTTCAAGCTGGTGATGGCCTCCCTGTACCACATC  171
    30                        40                       50
TATGTGGCCCTGGAGGAGGAGATTGAGCGCAACAAGGAGAGCCCAGTCTTCGCCCCTGTCTACTTCCCAGAAGAGCTGCACCGCAAGGCT  261
    60                        70                       80
GCCCTGGAGCAGGACCTGGCCTTCTGGTACGGGCCCCGCTGGCAGGAGGTCATCCCCTACACACCAGCCATGCAGCGCTATGTGAAGCGG  351
    90                       100                      110
CTCCACGAGGTGGGGCGCACAGAGCCCGAGCTGCTGGTGGCCCACGCCTACACCCGCTACCTGGGTGACCTGTCTGGGGGCCAGGTGCTC  441
   120                       130                      140
AAAAAGATTGCCCAGAAAGCCCTGGACCTGCCCAGCTCTGGCGAGGGCCTGGCCTTCTTCACCTTCCCCAACATTGCCAGTGCCACCAAG  531
   150                       160                      170
TTCAAGCAGCTCTACCGCTCCCGCATGAACTCCCTGGAGATGACTCCCGCAGTCAGGCAGAGGGTGATAGAAGAGGCCAAGACTGCGTTC  621
   180                       190                      200
CTGCTCAACATCCAGCTCTTTGAGGAGTTGCAGGAGCTGCTGACCCATGACACCAAGGACCAGAGCCCCTCACGGGCACCAGGGCTTCGC  711
   210                       220                      230
CAGCGGGCCAGCAACAAAGTGCAAGATTCTGCCCCCGTGGAGACTCCCAGAGGGAAGCCCCCACTCAACACCCGCTCCCAGGCTCCGCTT  801
   240                       250                      260
CTCCGATGGGTCCTTACACTCAGCTTTCTGGTGGCGACAGTTGCTGTAGGGCTTTATGCCATGTGAATGCAGGCATGCTGGCTCCCAGGG  891
_____________________________________________________________________
   270                       280
CCATGAACTTTGTCCGGTGGAAGGCCTTCTTTCTAGAGAGGGAATTCTCTTGGCTGGCTTCCTTACCGTGGGCACTGAAGGCTTTCAGGG  981

CCTCCAGCCCTCTCACTGTGTCCCTCTCTCTGGAAAGGAGGAAGGAGCCTATGGCATCTTCCCCAACGAAAAGCACATCCAGGCAATGGC 1071

CTAAACTTCAGAGGGGGCGAAGGGGTCAGCCCTGCCCTTCAGCATCCTCAGTTCCTGCAGCAGAGCCTGGAAGACACCCTAATGTGGCAG 1161

CTGTCTCAAACCTCCAAAAGCCCTGAGTTTCAAGTATCCTTGTTGACACGGCCATGACCACTTTCCCCGTGGGCCATGGCAATTTTTACA 1251

CAAACCTGAAAAGATGTTGTGTCTTGTGTTTTTGTCTTATTTTTGTTGGAGCCACTCTGTTCCTGGCCCAGCCTCAAATGCAGTATTTTT 1341

GTTGTGTTCTGTTGTTTTTATAGCAGGGTTGGGGTGCTTTTTGAGCCATGCGTGGGTGGGGAGGGAGGTGTTTAACGGCACTGTGGCCTT 1431

GGTCTAACTTTTGTGTGAAATAATAAACAACATTGTCTG                                                   1470
```

Fig. 9. Nucleotide sequence of cDNA encoding human heme oxygenase and its deduced amino acid sequence. The nucleotide sequence of the message strand is shown. Nucleotides are numbered in the 5' to 3' direction and nucleotide residue 1 is the A of the initiating methionine codon ATG, and the nucleotides on the 5' side of residue 1 are indicated by negative numbers. The deduced amino acids are shown below the nucleotide sequence and are numbered beginning with the initiating methionine. The TC dinucleotides at the 5' end of the pHH01, cDNA (underlined) are the AA dinucleotides in pHH02. The putative membrane segment and the polyadenylation signal are also underlined. The poly(dG) and poly(dA) tails are not included.

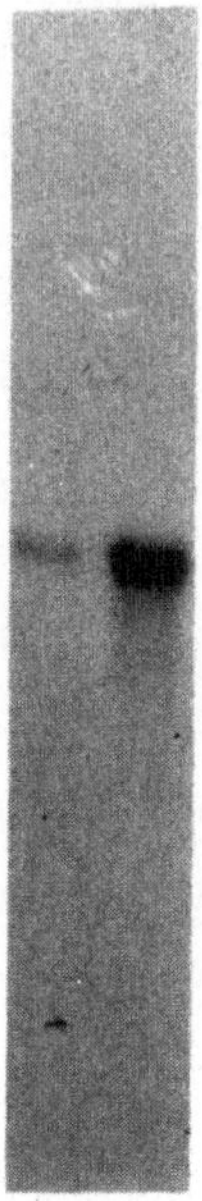

A B

Fig. 10. Autoradiogram of blot hybridization analysis of human macrophage RNA. Each lane contained 8 μg total RNA prepared from K562, untreated (A), incubated for 5 hr with hemin (B). The hybridization probe was the ^{32}P-labelled XhoI-AccI fragment (nucleotide residues - 64 to 230) derived from PHH01. The size markers were human rRNA and bacterial rRNA.

various human tissues studied including bone marrow. Further investigation is needed to characterize the molecular events that regulate the expression of the human heme oxygenase gene by hemin and various other environmental-hormonal factors.

ACKNOWLEDGEMENTS

We wish to thank Joyce Eshet for preparation and typing of the manuscript. These studies were supported by NIH grant #AM 29742 and in part by the Eye Bank for Sight Restoration. N.G. Abraham is a recipient of a Research Career Development Award #AM 00781.

REFERENCES

1. Schmid, R. 1977. Trans. Assoc. Amer. Physicians 89: 64-76.
2. Yoshinaga, T., S. Sassa, and A. Kappas. 1982. J. Biol. Chem. 257: 7786-7793.
3. Maines, M.D., N.G. Abraham, and A. Kappas. 1977. J. Biol. Chem. 252:5900-5903.

4. Docherty, J., B.A. Schacter, G.D. Firneisz, and S.B. Brown. 1984. J. Biol. Chem. 259: 13066-13069.
5. Yoshinaga, T., S. Sassa, and A. Kappas. 1982. J. Biol. Chem. 258: 7778-7785.
6. Schacter, B.A., E.B. Nelson, H.S. Marver and B.S.S. Masters. 1972. J. Biol. Chem. 247: 3601-3607.
7. Yoshida, T., M. Noguchi, G. Kikuchi. 1980. J. Biol. Chem. 255: 4418-4420.
8. Abraham, N.G., and R.D. Levere. 1980. Life Science 21: 4487-4491.
9. Pimstone, N.R., P. Engel, R. Tenhunen, P.T. Seitz, H.S. Marver, and R. Schmid. 1971. J. Clin. Inv. 50: 2042-2050.
10. Maines, M.D., and A. Kappas. 1977. Science 198: 1215-1221.
11. Guzelian, P.S. and N.A. Elshourbagy. 1979. Arch. Biochem. Biophys. 196: 178-185.
12. Abraham, N.G., J.D. Lutton, M.L. Freedman, and R.D. Levere. 1986. Am. J. Med. Sci. 29: 81-86.
13. Kappas, A. and G.S. Drummond. 1986. J. Clin. Invest. 77: 335-339.
14. Sardana, M.K., S. Sassa, and A. Kappas. 1980. J. Biol. Chem. 255: 11320-11323,
15. Sardana, M.K., S. Sassa, and A. Kappas. 1985. Biochem. Pharmacol. 34: 2937-2955.
16. Smith, T.J., G.S. Drummond, I.A. Kourides, and A. Kappas. 1982. Proc. Nat. Acad. Sci. 79: 7537-7541.
17. Bakken, A.F., M.M. Thaler, and R. Schmid. 1972. J. Clin. Invest. 51: 530-536.
18. Abraham, N.G., J.D. Lutton, R. Hoffman, and R.D. Levere. 1985. J. Lab. Clin. Med. 105: 593-600.
19. Schacter, B.A., B. Yoda, and L.G. Israels. 1979. J. Lab. Clin. Med. 93: 838-846.
20. Porter, P.N., R.H. Meints, and K. Mesner. 1979. Exp. Hematol. 7: 11-16.
21. Ross, J. and D. Sautner. 1976. Cell 8: 513-520.
22. Hoffman, R., M.J. Murnane, D. Burger, et al. 1981. In: Stamatogannopoulas G., Nienhuis, A.W., Eds., Hemoglobins in development and differentiation. Alan R. Liss, New York. pp. 487-506.
23. Sassa, S. 1980. In vivo and in vitro erythropoiesis: The friend system, G.B. Rossi, Ed., Elsevier/North Holland, Amsterdam, pp. 219-228.
24. Bern, N., K. Sahr, and E. Goldwasser. 1983. J. Cell Biochem. 21: 93-99.
25. Abraham, N.G., J.H. Lin, M.W. Dunn, and M.L. Schwartzman. 1987. Invest. Ophthalmol. Vis. Sci.
26. Towbin, H., T. Strachlin, and J. Gordon. 1979. Proc. Natl. Acad. Sci. 76: 4350-4354.
27. Schwartzman, M.L., P. Pagano, J.C. McGiff, and N.G. Abraham. 1986. Arch. Biochem. Biophys. 252: 635-645.
28. Yashukochi, Y., and B.S.S. Masters. 1976. J. Biol. Chem. 251: 5337-5344.
29. Kutty, R.K. and M.D. Maines. 1981. J. Biol. Chem. 256: 3956-3962.
30. Watt, J. and P. O'Carra. 1976. Biochem. Soc. Trans. 4: 866-868.
31. Rutherford, T.R., J.B. Clegg, and D. J. Weatherall. 1979. Nature 208: 164-165.
32. Knowles, B.B., C.C. Howe, and D.P. Aden. 1980. Science 209: 497-499.
33. Galbraith, R.A., S. Sassa, and A. Kappas. 1986. Biochem. J. 237: 597-600.
34. Chirgwin, J.M., A.E. Przybyla, R.J. MacDonald, and W.J. Rutter. 1979. Biochem. 18: 5290-5299.
35. Shibahara, S., R.M. Muller, H. Taguchi, and T. Yoshida. 1985. Proc. Natl. Acad. Sci. USA. 82: 7865-7869.
36. Yoshida, T., P. Biro, T. Cohen, R.M. Muller, and S. Shibahara. 1988.

Eur. J. Biochem., In Press.
37. Feinberg, A.P. and B. Vogelstein. 1983. Anal. Biochem. 132: 6-13.
38. Aviv, H., and P. Leda. 1972. Proc. Natl. Acad. Sci. USA 69: 1408-1412.
39. Okayama, H. and P. Berg. 1982. Mol. Cell. Biol. 2: 161-170.
40. Maxam, A.M. and W. Gilbert. 1980. Methods Enzymol. 65: 499-560.
41. Yoshida, T. and G. Kikuchi. 1979. J. Biol.Chem. 254: 4487-4491.
42. Kozak, M. 1981. Nucleic Acid Res. 9: 5233-5252.
43. Young, R.A., and R.W. Davis. 1983. Proc.Natl.Acad.Sci. 80: 1194-1198.
44. Young, R.A. and R.W. Davis. 1983. Science 222: 778-782.

REGULATION OF FETAL GLOBIN GENE EXPRESSION IN HUMAN ERYTHROLEUKEMIA (K562) CELLS

Maryann Donovan-Peluso, David O'Neill, Santina Acuto, and Arthur Bank

Columbia University
College of Physicians and Surgeons
Department of Genetics and Development
Department of Medicine
New York, NY

ABSTRACT

We have analyzed the transcription and induction of fusion globin genes comprised of portions of either gamma and beta globin sequences or gamma and neomycin resistance gene sequences. The analysis of gamma promoter beta and gamma-neo fusion genes indicates that 5' gamma flanking sequences are sufficient for tissue specific expression but not induction in K562 cells. A beta gene containing only the substitution of gamma IVS 2 for beta IVS 2 is expressed and induced when transcripts are analyzed with a 3' probe in contrast to the lack of expression seen with an intact beta gene. Thus, fusion globin genes containing gamma IVS 2 are both expressed and induced indicating that this region may be involved in the response to hemin stimulation, however, the mechanism is unclear. A gamma-neo fusion gene containing the gamma 5' region is expressed but not induced. When an EcoRI-Bgl II fragment containing the beta 3' enhancer is ligated downstream of the gamma-neo gene this gene is now inducible. Multiple genetic elements are involved in the regulated expression of gamma genes in fetal erythroid cells. These experiments begin to localize these sequences to specific regions within the gamma globin gene.

The mechanisms involved in the regulated expression of individual globin genes during development is a critical question in hematopoeisis that remains unsolved. The availability of a human erythroleukemia cell line (K562 cells) provides a system to explore the regulated expression of human globin genes in an embryonic-fetal erythroid environment[1-3]. K562 cells express the endogenous epsilon and gamma genes but do not express the endogenous beta genes[2-3]. When K562 cells are grown in the presence of hemin erythroid differentiation or "induction" occurs accompanied by the increased accumulation of globin mRNA transcripts and hemoglobin. Previous studies have shown that transfected epsilon genes are expressed and induced[4] whereas transfected beta genes are neither expressed nor induced in these cells[5]. Molecular cloning of a beta globin gene from K562 cells and its expression in a transient expression assay indicates that the lack of expression in K562 cells is not due to a structural defect within the beta globin gene in these cells[6,7]. The lack of beta expression in these cells results from either the absence of positive activators for beta gene expression or the presence of negative regulators that prevent its expression.

We have used K562 cells to begin to identify the important cis-acting elements that are involved in regulated globin gene expression. To do this, we have constructed hybrid gamma-beta fusion genes and subcloned these genes into pSV2neo, a plasmid that contains the neomycin resistance (neo) gene driven by and SV40 promoter[8]. Expression of this plasmid in eukayrotic cells confers resistance to G418, a neomycin analogue. These genes were transfected into K562 cells by a variety of techniques including calcium phosphate precipitation, protoplast fusion and most recently electroporation and clones containing the plasmid after stable integration were selected with G418. Resistant clones were grown in both the absence and presence of hemin and the transcripts produced by the fusion genes were analyzed by either S1 nuclease protection or Rnase protection[13].

Figure 1 shows a schematic representation of the various fusion genes that were analyzed. The prototype globin gene is indicated at the top and contains both 5' and 3' untranslated regions (U), three coding exons (E1,E2,E3) and two introns (I1,I2). The beta gamma fusion gene consists of beta 5' and coding sequences to the Eco RI site in exon 3 and gamma sequences 3'. Beta gamma 2 consists of a beta gene in which gamma IVS2 has been substituted for beta IVS 2. Gamma-beta consists of gamma 5' and coding sequence to the Eco RI site in exon 3 and beta sequences 3'. Gamma beta 2 consists of a gamma gene in which beta IVS2 has been substituted for gamma IVS 2. Gamma promoter beta consists of gamma 5' sequences fused at an Nco I site just 5' to the translational AUG to beta coding and 3' sequences.

Figure 1 also summarizes the expression and induction of the fusion genes that were analyzed and Table 1 examines the contribution of promoter, IVS 2 and the beta 3' enhancer on expression and induction. Neither

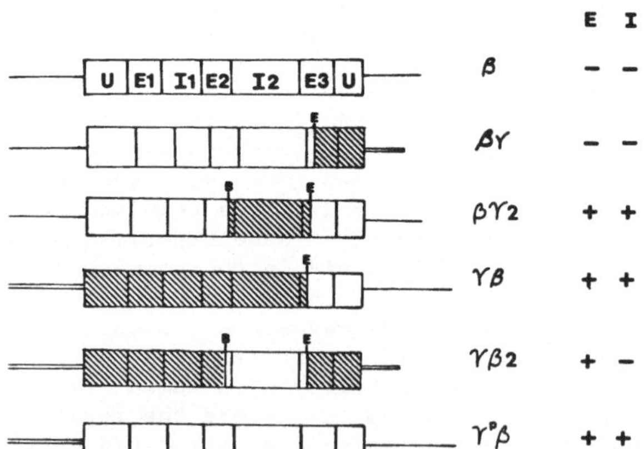

Fig. 1. Schematic representation of gamma-beta fusion genes analyzed. Beta gene sequences are indicated by single lines and open boxes, gamma gene sequence are indicated by double lines and hatched boxes. Relevant restriction sites are indicated by B (Bam HI) and E (Eco RI). At the right hand side of the figure the E column indicates expression and I column indicates induction.

Table 1. Analysis of the contribution of promoter, IVS 2 and beta 3' enhancer sequence on expression and induction of fusion genes.

	PROMOTER	IVS2	3' ENHANCER	EXP	IND
1. β	β	β	NO	-	-
2. βγ	β	β	NO	-	-
3. βγ2	β	γ	NO	+	+
4. γβ	γ	γ	YES	+	+
5. γβ2	γ	β	NO	+	-
6. $\gamma^{P}\beta$	γ	β	YES	+	+

the endogenous nor exogenous beta genes are transcribed in K562 cells. Similarly, the beta gamma fusion gene is not expressed suggesting that the information contained in the 3' portion of the gamma gene is insufficient to allow transcription to initiate from the beta promoter in these cells. In contrast, a beta gamma 2 fusion gene containing only the substitution of gamma IVS 2 for beta IVS 2 is transcribed and these transcripts accumulate in response to hemin induction when analyzed with a 3' fusion probe. However, when the 5' end of these transcripts is evaluated by both Rnase protection with a 5' beta specific probe and primer extension analysis using a uniformly labelled primer complementary to the 3' end of the mRNA to prime the fusion transcripts for extension with reverse transcriptase no discrete 5' end is detectable. The presence of gamma IVS 2 in the mRNA precursor seems to be able to stabilize aberrantly initiated transcripts, although the actual mechanism is not understood. It is clear, however, that the presence of a gamma IVS 2 in the context of a beta gene is not sufficient to activate the beta promoter and promote transcription from the canonical cap site.

The gamma beta fusion gene containing the gamma 5' region as well as the gamma structural gene to the Eco RI site in exon 3 is expressed in all clones containing an intact transcription unit. This gene also accumulates the fusion transcript in response to hemin induction in 50% of the lines. Thus, the gamma sequences present in this gamma beta fusion gene are sufficient for expression and induction in K562 cells. The transcription of the gamma beta fusion gene was evaluated using a fusion probe complementary to the exon 3 portion of the message. The gamma beta fusion gene is expressed in K562 cells at 10-40% of the endogenous gamma level and in many of the lines the expression of this gene increases in response to hemin induction. The expression and induction of this gene was evaluated in 17 lines that expressed the fusion gene. In 7 of the lines there was an increase in the transcript accumulation in response to hemin in 7 there was a decrease and in 2 lines there was no change. Transcripts from the gamma beta fusion gene were appropriately initiated as evidenced by primer extension analysis. The lack of induction in the other lines was investigated. All of the individual clones induced the endogenous epsilon and gamma genes providing evidence that there is no inherent change in the potential inducibility of the transfected cells containing the fusion gene that could account for the lack of induction of exogenous genes in individual clones. Differences in the ability of these fusion genes to be induced with hemin appears to be due to integration into different chromosomal positions although the details of this variable expression remain to be elucidated.

The role of gamma IVS 2 in gamma gene expression was further evaluated by a fusion gene in which beta IVS 2 was substituted for gamma IVS 2 (gamma beta 2). This gene was expressed but in most cases the transcripts either did not change or decreased in response to hemin induction. Thus, the substitution of beta IVS 2 for gamma IVS 2 in a gamma gene seems to affect the accumulation of this fusion gene transcript in K562 cells especially with induction. The lack of induction of fusion genes containing gamma 5' sequences and beta structural sequences including beta IVS 2 has been reported by others[9]. This provides further evidence for the idea that gamma IVS 2 may play a critical role in the regulated expression of gamma genes in a fetal erythroid environment.

To define the minimal sequence requirements for correct initiation of globin gene transcription in K562 cells a beta gene containing the gamma 5' flanking and promoter region as well as the gamma 5' untranslated region was constructed (gamma promoter beta). This gene was expressed and induced in both pools of clones as well as in individual clones that were analyzed; in all cases the transcripts are appropriately initiated. This gene, containing a beta IVS 2 is inducible in contrast to the lack of induction seen with a gamma gene containing beta IVS 2. This inducibility can be explained by the presence of a beta 3' enhancer[10,11] in this construct (Table 1). Thus, the 3' beta enhancer contained in the gamma promoter beta fusion gene may serve to function in a manner homologous to gamma IVS 2 in induction of this gene. If this is the case, then the 3' beta enhancer is not specific for the beta gene per se but rather can serve to enhance the transcript accumulation of other globin genes or chimeric genes located in close proximity with this regulatory element.

To further define the role of 5' flanking regions and the potential effect of 3' beta enhancer sequences, hybrid genes containing globin and neo gene sequences were constructed (Figs. 2,3)[14]. The fusions of both gamma-neo and beta-neo were made so that the translational AUG used was contributed by the neo gene itself. In the first series of experiments the ability of both gamma-neo and beta-neo to generate G418 resistant colonies was compared to the efficiency when a promoterless neo gene (pSVOneo) and a neo gene driven by a SV40 promoter (pSV2neo) were used. Transfection frequency into K562 cells as measured by the number of G418 resistant clones was always higher with gamma-neo compared with beta-neo. In fact, the transfection efficiency with gamma-neo, in erythroid cells, was similar to that observed when the high efficiency SV40 promoter was driving the neo gene. When these same globin-neo contructs were transfected

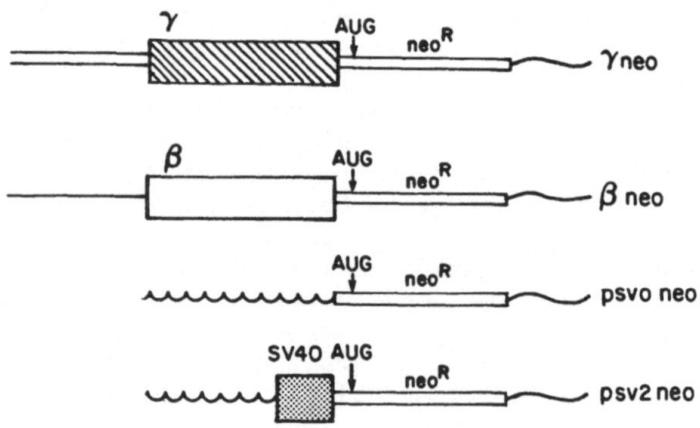

Fig. 2. Schematic representation of the promoters that were compared by the ability to generate G418 resistant clones following transfection into erythroid and nonerythroid cell lines.

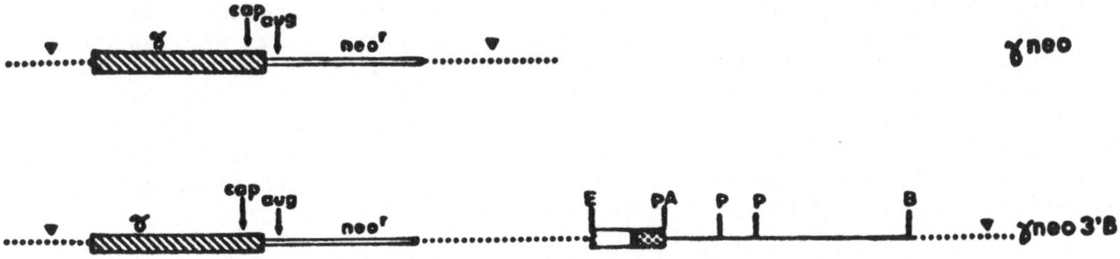

Fig. 3. Schematic representation of the gamma-neo fusion gene and the gamma-neo gene containing the beta 3' Eco RI-BGl II fragment ligated downstream. The gamma-neo gene is composed of the gamma promoter indicated by the hatched box fused to the neo gene indicated by the double line.

into hela cells, a non-erythroid cell line, no G418 resistant colonies were detected. However, pSV2neo was able to confer resistance to G418 in these cells with high efficiency. These data suggest that there are erythroid specific factors in K562 cells that preferentially interact with the gamma globin promoter and facilitate transcription of genes that contain this region 5' to coding sequence. There is a low level of clones selected with the beta-neo gene, higher than observed with the promoterless neo gene. The gamma 5' region is much better than the beta 5' region in generating clones suggesting that there is either an absence of positive factors or the presence of negative factors for beta gene expression in these cells.

To evaluate the expression of these hybrid genes, the transcripts produced were analyzed with Rnase protection using fusion probes complementary to the 5' end of the globin-neo transcripts. The gamma-neo gene was transcribed from the canonical cap in all lines examined. One line also utilized a downstream start site that has been reported by others[12]. There was, however, no induction when RNA from 12 of 13 lines grown in the presence of hemin was analyzed. The beta-neo fusion gene was also transcribed from the canonical cap start in most lines and in addition a downstream start site was also used. There was no induction of transcript accumulation observed with this gene.

To further define the relationship between a beta 3' enhancer and induction the beta Eco RI-Bgl II fragment containing this region (Figure 3) was ligated downstream of the gamma-neo fusion gene and G418 resistant clones were selected. When transcripts from this construction were analyzed both expression and induction were observed in most cases. To control for RNA concentration the transcription of the endogenous human gamma-actin gene was evaluated and expression of this gene was used as an internal control to currect for variations in mRNA loaded in each lane. Thus, the 3' beta enhancer appears to permit induction of the gamma-neo gene which by itself is not inducible. This further suggests that the 3' beta enhancer is not specific in affecting the beta gene alone, but can also increase gamma gene transcription as well. An enhancer 3' to the A-gamma globin gene has recently been described using a transient expression system[15]. The in vivo role of this gamma enhancer and the beta 3' enhancer on gamma gene expression remains to be determined.

In summary, the analysis of fusion gene transcription and induction in K562 cells indicates that gamma 5' flanking sequences are sufficient for tissue specific expression but not induction in K562 cells. Gamma IVS 2 seems to play a role in the ability of genes to be expressed and induced but the mechanism responsible for this effect is unclear. Beta 3' flanking sequences result in induced expression of a gamma-neo fusion gene suggesting that this enhancer element is not specific for beta gene induction. Thus, multiple genetic elements are involved in gamma gene expression in K562 cells. These experiments are continuing to localize the important cis-acting sequences to regions within the gamma globin gene, and to determine the details of their interaction with trans-acting factors in the control of fetal gene expression.

REFERENCES

1. Lozzio, C.B., and B.B. Lozzio. 1975. Human chronic myelogenous leukemia cell line with positive philadelphia chromosome. Blood 45: 321-324.
2. Rutherford, T., J.B. Clegg, D.R. Higgs, R.W. Jones, J. Thompson, and D.J. Witherall. 1981. Embryonic erythroid differentiation in the human leukemic cell line K562. Proc. Natl. Acad. Sci. 78(1): 348-352.
3. Miller, C., K. Young, D. Dumenil, B.P. Alter, J.M. Schofield, and A. Bank. 1984. Specific globin mRNAs in human erythroleukemia (K562) cells. Blood 63(1): 195-200.
4. Young, K., M. Donovan-Peluso, K. Bloom, M. Allan, J. Paul, and A. Bank. 1984. Stable transfer and expression of exogenous human globin genes in human erythroleukemia (K562) cells. Proc. Natl. Acad. Sci. 81: 5315-5319.
5. Young, K., M. Donovan-Peluso, R. Cubbon, and A. Bank. 1985. Trans-acting regulation of beta globin gene expression in erythroleukemi (K562) cells. Nucl. Acids Res. 13(14): 5203-5213.
6. Donovan-Peluso, M., S. Acuto, M. Swanson, C. Dobkin, and A. Bank. 1984. Erythroleukemia (K562) cells contain a functional beta globin gene. Mol. and Cell Biol. 4(11): 2553-2555.
7. Fordis, C.M., N.P. Anagnou, A. Dean, A.W. Nienhuis, and A.N. Schechter. 1984. A beta globin gene inactive in the K562 leukemic cell, functions normally in a heterologous expression system. Proc. Natl. Acad. Sci. 81: 4485-4489.
8. Southern, P.J., and P. Berg. 1982. Transformation of cells to antibiotic resistance with a bacterial gene under control of the SV40 promoter. J. Mol. Appl. Genet. 1, 327-341.
9. Kioussis, D., F. Wilson, K. Khazie, and F. Grosveld. 1985. Differential expression of human globin genes introduced into K562 cells. EMBO J. 4: 927-931.
10. Hesse, J.M., J.M. Nickol, M.R. Lieber, and G. Felsenfeld. 1986. Regulated gene expression in transfected primary chicken erythrocytes. Proc. Natl. Acad. Sci. 83: 4312-4316.
11. Trudel, M., L. Bruckner, and F. Costantini. 1987. A 3' enhancer con tributes to stage specific expression of human beta globin genes. Genes and Dev. 1: 954-961.
12. Khazaie, K., F. Gounari, M. Antoniou, E. de Boer, and F. Grosveld. 1986. β-globin gene promoter generates 5' truncated transcripts in the embryonic/fetal erythroid environment. Nucl. Acids Res. 14: 7199-7212.
13. Donovan-Peluso, M., S. Acuto, M. Swanson, C. Dobkin, and A. Bank. 1987. Expression of human gamma globin genes in human erythroleukemia (K562) cells. J. Biol. Chem. in press.
14. Acuto, S., M. Donovan-Peluso, N. Giambona, and A. Bank. 1987. The role of human globin gene promoters in the expression of hybrid genes in erythroid and non-erythroid cells.
15. Bodine, D.M., and T.J. Ley. 1987. An enhancer element lies 3' to the human A-gamma globin gene. The EMBO J. 6(10)" 2997-3004

HUMAN GENE EXPRESSION IN MURINE HEMOPOIETIC CELLS IN VIVO

Frederick A. Fletcher, Kateri A. Moore, Grant R. MacGregor,
John W. Belmont, and C. Thomas Caskey

Institute for Molecular Genetics and
Howard Hughes Medical Institute
Baylor College of Medicine
Houston, Texas 77030

INTRODUCTION

Somatic gene transfer offers the possibility of a new approach in the treatment of human genetic disease. Defects affecting the blood and blood forming tissues are candidates for therapies involving transfer of new genetic information into hemopoietic stem cells. One such defect, adenosine deaminase (ADA) deficiency, is being used as a model in which hemopoietic gene transfer techniques can be developed and evaluated. In this model, gene transfer is mediated by a retroviral vector. Retroviral vectors have been used extensively to deliver information to hemopoietic cells[1-14]. We have previously reported delivery and expression of human ADA (hADA) sequences in murine hemopoietic progenitors in vitro[15] and in vivo[16], but were not able to demonstrate long term stability of expression. We describe here the construction and testing of four new vectors, representing the first demonstration of efficient transfer and long term in vivo expression of hADA in murine hemopoietic cells.

RESULTS

Infection of Murine Hemopoietic Cells

Four new vectors were constructed and shown to transduce hADA into target cells in vitro[17]. These vectors (N2Fos2, XT1, deltaXT1, deltaNN2), illustrated in Figure 1, were subsequently used to infect primary mouse bone marrow. Cells were collected from femurs and tibias of 8-12 week C57Bl/6J donor mice. Infection was performed by co-cultivating bone marrow (10^6 cells/ml) with irradiated (1500R) virus-producing fibroblasts (10^6/ml bone marrow suspension) in supplemented IMDM (10% FCS, 10 mg/ml BSA, 10% WEHI 3B-conditioned IMDM, 10% 5637-conditioned IMDM, 100 µg/ml soy lipids, 300 µg/ml Fe-saturated transferrin, and 4 µg/ml polybrene). WEHI 3B-conditioned medium provides murine interleukin-3; 5637-conditioned medium provides human interleukins-1α and interleukin-6. Co-cultivation was continued for 24hr at 35°C and 5% CO_2. Infected marrow was either plated into semisolid medium (supplemented IMDM plus 1% methylcellulose) for enumeration and assay of CFU-C, or injected into lethally irradiated (1100R) syngeneic recipients. Some animals received non-reconstituting doses ($5x10^4$)

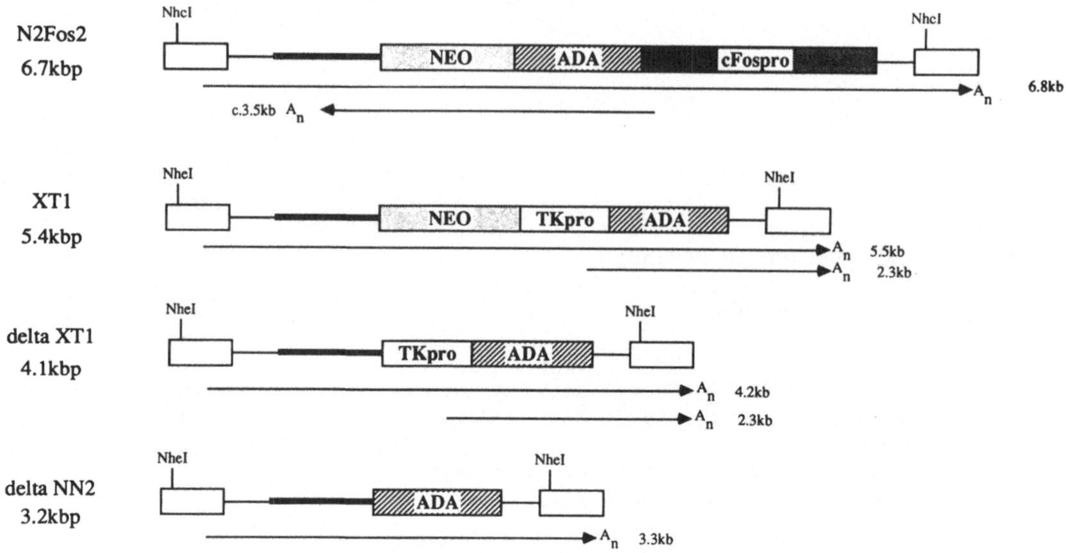

Fig. 1. Structure of hADA-transducing Vectors. The heavy line indicates the extended gag region (gag+[18]. Open boxes denote the Moloney murine leukemia virus LTRs. Arrows indicate possible RNA transcripts, along with their sizes. The plasmid pFC1, containing the cFos promoter and cap site, was kindly provided by R. Vincent[19]. The plasmid pXT1 was kindly provided by E. Wagner.

of infected cells, in order to study discrete CFU-S; other animals received reconstituting doses (10^6 cells) to allow studies of long-term expression of hADA.

Expression of Human ADA In Vivo

Results of the in vitro and CFU-S studies are indicated in Table 1.

Table 1.

Vector[a]	In Vitro		In Vivo[c]	
	% G418^r CFC[b]	IEF on CFC	Southern	IEF
N2Fos2	8-20%	neg	2/35 (6%)	0/2
XT1	1.6-8.5%	neg	11/18 (61%)	0/11
deltaXT1	NA	pos	5/26 (19%)	5/5
deltaNN2	NA	pos	31/45 (69%)	24/31

[a]See Figure 1 for description of vectors.
[b]Infected bone marrow cells were plated in supplemented IMDM with 1% methyl - cellulose as previously described[15]. CFC were counted at 10 days.
[c]CFU-S (day 14) or whole spleens from transplanted animals were analyzed by Southern blotting and IEF. Southern data are indicated as a ratio of positive samples over total samples analyzed. IEF data are indicated as a ratio of hADA positive samples over Southern positive samples. All IEF positive samples are also Southern positive (not shown). Survival splenectomy was performed using standard techniques[20], with Avertin anesthesia.

Only hemopoietic cells infected with delta XT1 or deltaNN2 (both lacking neo selectable marker) expressed hADA in vitro and in vivo. The hADA activity could be detected by IEF in pooled CFU-C in the absence of selection, indicating high infection efficiency. When individual CFU-S were analyzed by Southern blotting and IEF (data not shown) infection efficiencies between 20 and 85% were observed. Expression of hADA in the reconstituted animals was monitored by IEF analysis of periperal blood from serial bleedings. A representative series is shown in Fig. 2. Human ADA was initially detected in the blood of 100% of the mice reconstituted with deltaNN2- and deltaXT1-infected marrow, however, the level of activity decreased with time. No animals reconstituted with deltaXT1 continued to express hADA after 10 wk (14 animals total). Two animals reconstituted with deltaNN2-infected marrow have continued to express hADA, at reduced levels, for six months post-transplant (14 animals total from three separate experiments, data not shown).

DISCUSSION

Four new vectors (N2Fos2, XT1, deltaXT1, deltaNN2) fulfilled the criteria to enable their use in bone marrow experiments, which were the ability to be used to generate high titers of ecotropic virus and an ability to transduce hADA activity to target cells. Two of the four failed to express hADA in CFU-C or CFU-S despite efficient infection of those cells. In contrast, the vectors deltaXT1 and deltaNN2 both successfully infected and generated high levels of expression of hADA in CFU-C and CFU-S. At four weeks post-reconstitution, recipient animals had easily detectable levels of hADA in their blood. Over time, the levels decreased, suggestion that some expression originated from more mature progenitors infected with the virus. The average life-span of a murine RBC is 45 days[21,22] and expression started to fall around 60 days, suggesting some early expression may be derived from infected erythrocyte progenitors.

These experiments demonstrate long term expression of a human disease-related gene product in murine hemopoietic cells in vivo. A number of

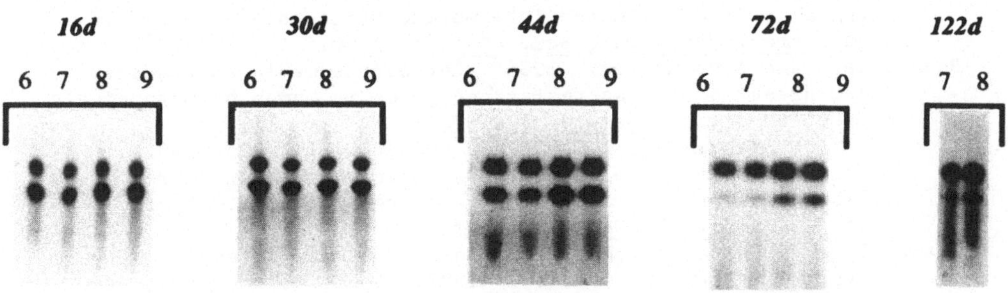

Fig. 2. Expression of hADA in Blood of Reconstituted Animals. 1×10^6 infected and viable bone marrow cells were injected retroorbitally into lethally irradiated (1100R) C57Bl/6J recipients. These animals were bled at intervals post-transplant and blood protein extracts analyzed by IEF. Time after transplantation is indicated above each assay. The four animals shown (donated 6,7,8,9) were all from a single deltaNN2 experiment. The upper band in each lane is the endogenous mouse ADA, the lower indicates hADA. Animals 6 and 9, while continuing to express hADA, were sacrificed for serial transplantation studies.

technical questions remain to be addressed before any attempt at human disease correction can be made. Long term stability and level of expression must be related to pluripotent stem cell infection. Also, cell purification protocols should be developed to improve infection efficiency of pluripotent stem cells. Future experiments will focus on optimization of stem cell infection by enrichment using monoclonal antibodies and cell sorting, and by hormonal stimulation. We are optimistic that improvements can be made that will allow use of retroviral mediated gene therapy in treatment of a variety of human genetic diseases.

REFERENCES

1. Joyner A., G. Keller, R.A. Phillips, and A. Bernstein. 1983. Retrovirus transfer of a bacterial gene into mouse haematopoietic progenitor cells. Nature 305: 556.
2. Miller, A.D., R.J. Eckner, D.J. Jolly, T. Friedmann, and I.M. Verma. 1984. Expression of a Retrovirus Encoding Human HPRT in Mice. Science 225: 630.
3. Williams, D.A., I.R. Lemischka, D.G. Nathan, and R.C. Mulligan. 1984. Introduction of new genetic material into pluripotent haematopoietic stem cells of the mouse. Nature 310: 476.
4. Dick, J.E., M.C. Magli, D. Huszar, R.A. Phillips, and A. Bernstein. 1985. Introduction of a selectable gene into primitive stem cells capable of long-term reconstitution of the hemopoietic system of W/Wv mice. Cell 42: 71.
5. Keller, G., C. Paige, E. Gilboa, and E.F. Wagner. 1985. Expression of a foreign gene in myeloid and lymphoid cells derived from multipotent heamatopoietic precursors. Nature 318: 149.
6. Eglitis, M.A., P. Kantoff, E. Gilboa, and W.F. Anderson. 1985. Gene Expression in Mice after High Efficiency Retroviral Mediated Gene Transfer. Science 230: 1395.
7. Lemischka, I.R., D.H. Raulet, and R.C. Mulligan. 1986. Developmental Potential and Dynamic Behavior of Hematopoietic Stem Cells. Cell 45: 917.
8. Williams, D.A., S.H. Orkin, and R.C. Mulligan. 1986. Retrovirus-mediated transfer of human adenosine deaminase gene sequences into cells in culture and into murine hematopoietic cells in vivo. Proc. Nat. Acad. Sci. USA 83: 2566.
9. Hock, R.A., and A.D. Miller. 1986. Retrovirus-mediated transfer and expression of drug resistance genes in human haematopoietic progenitor cells. Nature 320: 275.
10. Kwok, W.W., F. Schuening, R.B. Stead, and A.D. Miller. 1986. Retroviral transfer of genes into canine hemopoietic progenitor cells in culture: a model for human gene therapy. Proc. Natl. Acad. Sci. USA 55 83: 4552.
11. Chang, S.M., K. Wager Smith, T.Y. Tsao, J. Henkel Tigges., S. Vaishnav, and C.T. Caskey. 1987. Construction of a defective retrovirus containing the human hypoxanthine phosphoribosyltransferase cDNA and its expression in cultured cells and mouse bone marrow. Mol. Cell. Biol. 7: 854.
12. Hawley, R.G., L. Covarrubias, T. Hawley, and B. Mintz. 2987. Handicapped retroviral vectors efficiently transduce foreign genes into hematopoietic stem cells. Proc. Natl. Acad. Sci. USA. 84: 206.
13. Kantoff, P.W., A.P. Gillio, J.R. McLachlin, C. Bordignon, M.A. Eglitis, N.A. Kernan, R.C. Moen, D.B. Kohn, S.F. Yu, and E. Karson. 1987. Expression of human adenosine deaminase in nonhuman primates after retrovirus-mediated gene transfer. J. Exp. Med. 166: 219.
14. Magli, M-C., J.E. Dick, D. Huszar, A. Bernstein, and R.A. Phillips.

1987. Modulation of Gene Expression in Multiple hematopoietic cell lineages following retroviral gene transfer. Proc. Natl. Acad. Sci. USA 84: 789.
15. Belmont, J.W., J. Henkel Tigges, S.M. Chang, K. Wager Smith, R.E. Kellems, J.E. Dick, M.C. Magli, R.A. Phillips, A. Bernstein, and C.T. Caskey. 1986. Expression of human adenosine deaminase in murine haematopoietic progenitor cells following retroviral transfer. Nature 322: 385.
16. Belmont, J.W., J. Henkel Tigges, K. Wager Smith, SM-W. Chang, and C.T. Caskey. 1987. Transfer and Expression of Human Adenosine Deaminase Gene in Murine Bone Marrow Cells. In R.P. Gale, R. Champlin (eds): "Progress in Bone Marrow Transplantation," New York: Alan R. Liss, p 963.
17. Manuscript in preparation.
18. Bender, M.A., T.D. Palmer, R.E. Gelinas, and A.D. Miller. 1987. Evidence that the packaging signal of Moloney murine leukemia virus extends into the gag region. J. Virol. 61: 1639.
19. Deschamps, J., F. Meijlink, and I. Verma. 1985. Identification of a transcriptional enhancer element upstream from the proto oncogen fos. Science 230: 1174.
20. Hogan, B., F. Constantini, and E. Lacy. In Manipulating the Mouse Embryo. New York: Cold Spring Harbor Laboratory Press.
21. Bannerman, R.M. 1983. Hematology. In Foster, H.L., Small, J.D., Fox, J.G. (eds): "The Mouse in Biomedical Research:Volume III. New York Academic Press, p 293.
22. This work was supported by the Howard Hughes Medical Institute (HHMI) and by grants from the NIH (HD21452) and from the Cystic Fibrosis Foundation (R004 7-0351). GRM is the recipient of an Arthritis Foundation post-doctoral fellowship; JWB is an Assistant Investigator and CTC is an Investigator in HHMI. The authors wish to thank Jenny Henkel-Tigges, Dianne Houston-Hawkins, Michelle Rives, Deborah Villalon, and Karen Wager-Smith for technical assistance and Elsa Perez for help in preparation of the manuscript.

HOMING OF HEMOPOIETIC STEM CELLS TO HEMOPOIETIC STROMA

Cheryl L. Hardy and Mehdi Tavassoli

Department of Medicine
University of Mississippi Medical Center
Division of Hematology/Oncology
VA Medical Center
Jackson, Mississippi 39216

ABSTRACT

The existing knowledge of the molecular mechanism that underlies the successful engraftment of hemopoietic progenitor cells in their specific stromal microenvironment has been discussed. It appears that membrane lectins on the surface of the stem cell with specificity for galactose with and/or mannose-bearing glycoconjugates on the surface of the stromal cell are involved. This recognition and binding of hemopoietic stem cells which is called "homing" initiates the processes of differentiation, proliferation, and maturation of hemopoietic cells.

INTRODUCTION

Gene therapy involves the introduction of a desired gene into the desired cell, usually self-propagating hemopoietic stem cells. The desired gene can be modified so that the cell bearing it can have preferential proliferative capacity. It can be anticipated that the cell containing the new gene propagates preferentially within the body.

If hemopoietic progenitor cells bearing new genes which have replaced defective ones are to function normally, they must be recognized by and lodged correctly within the hemopoietic tissues. This lodgement, which is generally known as "homing", involves the migration, recognition and subsequent binding of hemopoietic progenitor cells to their specific microenvironmental niche. Homing is dependent upon molecular interactions at the cell surface. Thus, an understanding of the molecular nature of these specific cell surface interactions within hemopoietic tissues is critical to the success of gene therapy.

Stem Cell-Stromal Cell Recognition

Hemopoiesis is the result of interactions between hemopoietic progenitor cells and their supporting stroma. In hemopoietic tissues, there is a selective association of specific stromal cells with different lineages of hemopoietic cells. For instance, in the bone marrow, erythroid cells are generally clustered around an acid phosphatase-positive central macrophage[1] and these clusters are known as an erythroblastic island. Similarly, granulocytic cells are associated with an alkaline phosphatase-positive reticular cell[2,3], and megakaryocytes are in intimate contact with endothelial cells[4]. Similar specific associations have also been reported in the spleen of irradiated mice after infusion of marrow cells (CFU-S)[5].

The basic tenet of bone marrow transplantation is specific homing of hemopoietic precursor cells which are introduced into the circulation. This homing is necessary for sustained hemopoiesis[6]. The way in which the infused progenitor cells are recognized and bound by the stroma appears to be a two-step phenomenon. In the first step, progenitor cells are recognized by marrow endothelium and transported to the extravascular space where they come in contact with stromal cells. In the second step, each progenitor cell type recognizes and binds to specific supporting stromal cells through specific membrane recognition mechanisms. The binding initiates the processes of differentiation, proliferation, and maturation of hemopoietic cells. A specific membrane-surface recognition mechanism may be involved at each level, although the molecular basis of these events is unclear.

Work from our laboratory and others suggest that sugar-lectin interactions may be involved in the recognition event[7-11]. Glycoprotein components of the plasma membrane and other glycoconjugates can provide the sugar residues for these interactions. Mammalian or endogenous lectins have been described recently and may serve as the receptor for these glycoconjugates[12].

Involvement of Galactosyl-and Mannosyl-Specific Lectins In Homing Of Stem Cells To The Microenvironment

In our laboratory we have used synthetic neoglycoprotein molecules[7,13] to probe the stem cell-stromal cell interactions in murine long term bone marrow cultures (LTBMC). These probes are synthesized by covalent binding of various sugars to bovine serum albumin (BSA). Biologically relevant sugars (comprising the glycan moiety of glycoproteins as N-linked) are few in number and include mannose, galactose, fucose, N-acetyl-glucosamine and N-acetyl-galactosamine. In addition, sialic acids form the terminal sugar residue in most cases[14-16]. The method for the synthesis of these neoglycoproteins, developed in our laboratory, is based on several theoretical considerations. 1. The sugar must be linked to a larger molecule, such as a protein, because the small sugar molecules are diffusible across the cell membrane. Moreover, the protein can be easily radiolabeled, giving a very sensitive probe. 2. The linkage should be convalent to provide stability. 3. Sugars should retain the pyranose (ring) form necessary for lectin binding. 4. The protein itself should not have membrane receptors, to allow probing for sugar receptors. In this regard BSA serves well. 5. A high sugar to protein ratio should result. The method we have used fulfills all these criteria, and a sugar to protein ratio of about 50 results. In our method the p-aminophenyl derivative of the sugar in pyranose form is converted to an isothiocyanate derivative via thiophosgene. This derivative can then bind amino groups of the protein, particularly lysine residues.

We have found that specific binding between the stem cell and stromal in LTBMC cell involves a surface recognition mechanism with galactosyl and mannosyl specificities[7]. We have established murine LTBMC adherent layers and at 3 weeks recharged them with fresh stromal-free marrow cells in the presence of 1 mM neoglycoprotein probes. Control cultures were established without the probes. At weekly intervals the cultures were demidepopulated and refed with media containing the probes or control media.

CFU-S and CFU-C were determined every two weeks from adherent and supernate cells. Total cell production and cobblestone formation (representing differentiating and maturing colonies of progenitor cells) were monitored. We found that in the presence of galactosyl-BSA and/or mannosyl-BSA but not fucosyl-BSA or BSA alone the concentration of CFU-S and CFU-C in the supernate of LTBMC declined and gradually disappeared. Likewise, a similar decline in CFU-S and CFU-C in the adherent layer was observed, indicating that the decline was not a consequence of compartmentalization between the adherent layer and the supernate. The production of cobblestone areas in these cultures also was halted in the presence of galactosyl-BSA and/or mannosyl-BSA, but not fucosyl-BSA or BSA alone.

Additionally, these same probes will differentially agglutinate marrow CFU-C and CFU-S. ^{125}I labeling of galactosyl and mannosyl probes and of asialofetuin, a natural counterpart of the synthetic galactosyl-BSA, has identified binding sites for these sugars on the surface of two multipotential cell lines B6SUT and FDCP-1[17]. Thus, we hypothesize that in LTBMC the binding of stem cell to stromal cell involves a membrane lectin on the surface of the stem cell recognizing specific sugar configurations of membrane glycoconjugates on the surface of stromal cells.

In co-cultures of the multipotential cell line FDCP-1 with the stromal cell line D2X, we have found that asialofetuin will inhibit the binding that occurs between these two cells. Figure 1 shows this binding as seen by phase microscopy, and the resulting inhibition when co-cultures were established in the presence of 1 mM asialofetuin.

In vivo data have further substantiated this hypothesis in that preincubation of bone marrow cells with galactosyl or mannosyl probes or with galactosyl-terminating asialofetuin prevented engraftment of CFU-S and CFU-C within the bone marrow cells[8,9]. However, in the murine system in which the spleen is a hemopoietic organ, homing to the spleen was not inhibited by these probes, indicating that a surface recognition mechanism with different specificities may be operative within the spleen. Alternatively, architectural differences between these two hemopoietic tissues may account for the lack of inhibition of homing in the spleen.

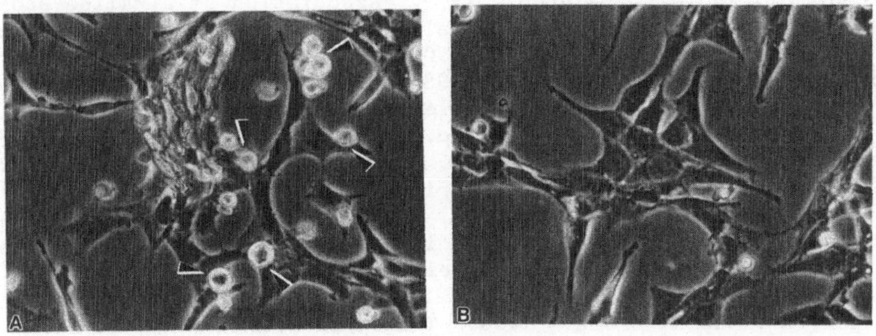

Fig. 1. Phase microscopy of co-cultures of stem cell and stromal cell lines. FDCP-1 was co-cultivated with D2X in the absence (A) and presence (B) of 1 mM asialofetuin. In panel A numerous round hemopoietic cells (arrows) are bound to elongated, spreading stromal cells. In panel B inhibition of binding of the stem cell lines to the stromal cell line occurred in the presence of asialofetuin, as evidenced by fewer round hemopoietic cells bound to stromal cells. (x200)

These architectural differences are particularly manifested in the patterns of blood circulation in the two organs: circulation of marrow is a closed one where the hemopoietic compartment (in which hemopoiesis takes place) is separated from the circulatory space by a non-fenestrated sinus endothelium. The entry of hemopoietic stem cells which have been introduced into the circulation then depends on their recognition by endothelium which transports them to the hemopoietic compartment[18]. By contrast, splenic circulation is largely an open one where penicillary arterioles open directly into the red pulp and the cells must find their way through a meshwork of stromal cells into the venous circulation[19]. This type of circulation precludes the need for cells to interact with sinus endothelium.

Lymphocyte Homing

In our sister field of lymphocyte biology identification of the receptor for lymphocyte homing to the surface of high-walled endothelium of postcapillary venules has also been accomplished[20]. This receptor is a glycoprotein which has been ubiquitinated by the cell. This ubiquitination is apparently necessary for homing[21,22]. This lymphocyte homing receptor is also a lectin with specificity for mannose-6-phosphate. Gallatin and colleagues have isolated a monoclonal antibody called MEL-14 directed against the lymphocyte lymph node homing receptor[23]. This probe has provided a tremendous momentum for the study of homing receptors in lymphocytes. Thus, other similarities may exist between lymphocyte homing and stem cell homing and require exploration.

ACKNOWLEDGEMENTS

This work was supported by NIH grant #DK-30142 to M. Tavassoli, and funds from the Veterans Administration and from the Ladies Auxiliary of the Veterans of Foreign Wars.

REFERENCES

1. Bessis, M. 1977. Erythrocytic series. In: Blood Smears Reinterpreted. M. Bessis, editor. Springer International, Berlin. pp. 25-53.
2. Westen, H., and D.F. Bainton. 1979. Association of alkaline-phosphatase-positive reticulum cells in bone marrow with granulocytic precursors. J. Exp. Med. 150: 919-937.
3. Ahmad, L.A., and D.F. Bainton. 1986. Presence of alkaline-phosphatase-positive reticulum cells in fetal liver and bone marrow of mice and their absence in yolk sac. Exp. Hematol. 14: 705-709.
4. Tavassoli, M., and M. Aoki. 1981. Migration of entire megakaryocytes through the marrow-blood barriers. Br. J. Hematol. 48: 25-29.
5. La Pushin, R.W., and J.J. Trentin. 1977. Identification of distinctive stromal elements in erythroid and neutrophil granuloid spleen colonies: light and electron microscopic study. Exp. Hematol. 5: 505-522.
6. Osogoe, B., and K. Omura. 1950. Transplantation of hematopoietic tissue with the circulating blood. Anat. Rec. 108: 663-681.
7. Aizawa, S., and M. Tavassoli. 1987. In vitro homing of hemopoietic stem cells is mediated by a recognition system with galactosyl and mannosyl specificities. Proc. Natl. Acad. Sci. USA 84: 4485--4489.
8. Aizawa, S., and M. Tavassoli. 1987. In vivo homing of transplanted marrow cells to marrow, but not spleen, is mediated by lectins with galactosyl and mannosyl specificities. Blood 70: 301a. (Abstract)

9. Aizawa, S., and M. Tavassoli. 1988. Molecular basis of the recognition of intravenously transplanted hemopoietic cells by bone marrow. Proc. Natl. Acad. Sci. USA (In press).
10. Samlowski, W.E., and R.A. Daynes. 1985. Bone marrow engraftment efficiency is enhanced by competitive inhibition of the hepatic asialoglycoprotein receptor. Proc. Natl. Acad. Sci. 82: 2508--2512.
11. Reisner, Y., L. Itzicovitch, A. Meshorer, and N. Sharon. 1978. Hemopoietic stem cell transplantation using mouse bone marrow and spleen cells fractionated by lectins. Proc. Natl. Acad. Sci. 75: 2933-2936.
12. Schaaf-La Fontaine, N., R.J. Hooghe, and F. Vander Plaetse. 1985. Modification of blood-borne arrest properties of lymphoma cells by inhibitors of protein glycosylation suggests the existence of endogenous lectins. Carbohydrates Res. 138: 315-323.
13. Kataoka, M., and M. Tavassoli. 1984. Synthetic neoglycoproteins: A class of reagents for detection of sugar recognizing substances. J. Histochem. Cytochem. 39: 1091-1098.
14. Sharon, N., and H. Lis. 1982. Glycoproteins: research booming on long-ignored ubiquitous compounds. Mol. Cell Biochem. 42: 167-187.
15. Sharon, N., and H. Lis. 1982. Glycoproteins. In: The proteins, 3rd edition, Vol. 5. H. Neurath and R.L. Hill, editors. Academic Press, New York. pp. 1-144.
16. Sharon, N. 1984. Glycoproteins. In: Trends in Biochemical Sciences. J. Hall, editor. Elsevier Science Publishing B.V., Amsterdam. pp. 198-202.
17. Matsuoka, T., and M. Tavassoli. 1988. Characterization of homing receptors in a cloned multipotential hemopoietic cell line. Clin. Res. 36: 16A. (Abstract)
18. Tavassoli, M. 1979. The marrow-blood barrier. Br. J. Haematol. 41: 297-302.
19. Weiss, L., and M. Tavassoli. 1970. Anatomical hazards to the passage of erythrocytes throughout the spleen. Sem. Hematol. 7: 372-380.
20. Gallatin, M., T.P. St. John, M. Siegelman, R. Reichert, E.C. Butcher, and I.L. Weissman. 1986. Lymphocyte homing receptors. Cell 44: 673-680.
21. St. John, T., W.M. Gallatin, M. Siegelman, H.T. Smith, V.A. Fried, and I.L. Weissman. 1986. Expression - linked cloning of a putative lymph node homing receptor cDNA: Ubiquitin is the reactive species. Science. 321: 845-849.
22. Siegelman, M., W.M. Bond, W.M. Gallatin, T. St.John, H.T. Smith, V.A. Fried, and I.L. Weissman. 1986. A putative lymphocyte homing receptor is a ubiquitinated branched-chain glycoprotein: additional cell surface proteins also appear ubiquitinated. Science 231: 823-829.
23. Gallatin, W.M., I.L. Weissman, and E.C. Butcher. 1983. A cell-surface molecule involved in organ-specific homing of lymphocytes. Nature 304: 30-34.

THE SIGNIFICANCE OF FREE RADICALS AND FREE RADICAL SCAVENGERS IN L1210 LEUKEMIA

A.C. Brown* and J.D. Lutton

Departments of Medicine and Anatomy*
New York Medical College
Valhalla, NY 10595

ABSTRACT

L1210 leukemia is a murine leukemia which is associated with anemia and marked neutrophilia. In order to determine the significance of free radicals (FR) in this disorder, we determined the presence and localization of free radical scavengers (FRS) and scavenger-like systems in L1210 leukemia cells obtained *in vivo* and from *in vitro* cultures. FR are metabolized or detoxified by certain FRS such as glutathione (GSH and GSSG), superoxide dismutase (SOD) and enzymes such as epoxide hydrolase (EH). In all cases specific fractions of L1210 cells, bone marrow and liver were examined for FR/FRS levels. Reduced (GSH) and oxidized (GSSG) glutathione were measured fluorometrically using o-opthalaldehyde (OPT). SOD was determined colorimetrically utilizing pyrogallol by substrate autolysis inhibition, and EH was determined by utilizing [^{3}H]styrene oxide as a substrate. Ratios of GSH/GSSG in fractions prepared from *in vivo* and *in vitro* L1210 cells showed a predominance of GSH-reductase with the highest activity in mitochondria (ratio = 15 vs. 10). Normal liver showed a similar pattern whereas, leukemic liver showed altered GSH/GSSG ratios in mitochrondria and microsomes. Leukemic bone marrow showed a predominance of GSH-reductase in all fractions. EH activity was highest in microsomal fractions obtained from L1210 cells grown *in vitro* and found to become increased in both the mitrochondrial (100%) and microsomal (200%) fractions when cells were exposed to retinoic acid (RA) in culture. SOD activity in the cytosolic (21.2 U SOD/mg) and mitochondrial (12 U SOD/mg) fractions whereas, leukemic liver showed a significant decrease in activity in all fractions compared to normals. SOD was determined in fractions taken from L1210 cells *in vivo* and *in vitro*. Results demonstrated detectable but reduced SOD activity in the L1210 cell fractions as contrasted with liver activity. Results from these studies indicate that certain FRS systems are functional in L1210 leukemic animals. Furthermore, variations in the ratios or levels may be of significance in the leukemic and hematological states.

INTRODUCTION

In recent years, free radicals (FR) have been implicated to play roles in a variety of the biological processes that occur during cellular transformations (for reviews, see 1-4). The generation of neoplastic cells by some chemical agents are thought to involve a series of stages which generate

FR, particularly those of molecular oxygen. Furthermore, normal cellular functions may become disturbed or altered when an abnormal balance of FR or free radical scavengers (FRS) are present in the cellular environment[5,6]. In this regard, it has been shown that nucleic acids, proteins, and some enzymes in the cell are attacked by FR[1,2]. Several kinds of phorbol esters such as 12-O-tetradecanoylphorbol12-acetate (TPA) have been shown to influence a series of chemical changes that could lead to DNA damage by FR[7-9]. A FR is a highly reactive chemical species that contains one or more unpaired electrons in its outer orbital. Examples include superoxide anion (O_2^-), hydroxyl radical (·OH), hydrogen peroxide (H_2O_2), singlet oxygen ($'O_2$) and organic hydroperoxides (eicosanoids).

Cells have multiple protective mechanisms against certain FR damage or toxic products in the form of FRS such as superoxide dismutase (SOD), glutathione, catalase, and FRS-like systems such as cytochrome P-450-epoxide hydrolase. The intracellular distribution of these systems may be important in the decomposition of toxic reactive intermediates. Anti-oxidants such as vitamins (A,D,E,C), mannitol and butylated biphenyls have been associated with FR activity in lipid peroxidation. Experimental results show that some FRS agents, including physiological ones, can influence and even enhance hematopoietic stem cell growth[10,11]. Furthermore, FR may act as endogenous inhibitors of hemopoietic bone marrow cells <u>in vivo</u> and <u>in vitro</u>, which may contribute to an impairment of hematopoiesis[12,13,26]. Thus, FR could trigger multiple events in cells and play roles in the growth and development of hematopoietic and leukemic cells.

In this study, we employed the L1210 murine lymphocytic leukemia model in order to evaluate the significance of FR and FRS in this type of disorder. This leukemic disease is a rapidly growing type and eventually involves progression into the bone marrow, lymph nodes and the spleen. As the disease progresses, marked hematological changes such as anemia and myelocytopoiesis are evident. Experiments have demonstrated that L1210 leukemia cells actually suppress the growth of normal clonogenic erythroid bone marrow cell (CFU-E)[14]. Numerous examples of leukemia cell-mediated inhibition of hematopoiesis have been demonstrated in cell lines and in animal models[15-17]. In addition, exposure of L1210 cells to scavenger-like differentiation inducing agents such as retinoic acid (RA) or 1,25-dihydroxycholecalciferol (vitamin D_3) have been shown to suppress some of the neoplastic characteristics[18-20]. It remains possible that FR and their scavengers may play significant roles in the hematological profile and virulence of this tumor.

The presence and distribution of FRS have been reported to become altered in some neoplastic cells; however, the significance of this is not yet clear[21-23]. Although the presence of FR in the metabolism of carcinogens such as benzene and benzo(a)pyrene is well established[24-26], the presence of an appropriate balance in FRS to counteract oncogenicity is not yet clearly defined. The ratio of certain FRS may be important in treating or altering the virulence of some tumors. For example, cisplatin is a drug now widely used in the chemotherapy of leukemia. The L1210/PAM cisplatin cell line of L1210 was found to reverse its resistance in cisplatin when depleted of glutathione by buthionine sulfoximide (BSO)[27]. As indicated previously, the resistance of cisplatin was suggested to be associated with the binding of BSO of glutathione[27-29].

In addition, the metabolism of various constituents within the cell may be carried out by a complex cytochrome P-450-epoxide hydrolase system. Epoxide hydrolase (EH) (EC 3.3.2.3) catalyzes the hydration of epoxides to diols[30], the latter of which, in some cases, have been shown to be toxic intermediates[31]. Cytochrome P-450 catalyzes the oxidative metabolism of a wide range of substrates such as drugs, carcinogens and a variety of other agents to different products. Although these systems have been investigated extensively in liver microsomes, their presence in leukemic states is nebulous.

Therefore, we examined the activity and presence of glutathione, EH, and SOD in L1210 and leukemic bone marrow cells to determine if a correlation exists in the levels of FR/FRS between the leukemia and hematological states.

MATERIALS AND METHODS

Animal And L1210 Tumors

Seven to 12 week-old SPF DBA/2 male mice, obtained from Charles River Breeding Colony (Wilmington, MA), are used for the experiments. L1210 leukemia is maintained in ascites from weekly intraperitoneal transfer of 10^5 cells/animal[32].

Cell Count And Viability

Cell count is routinely determined and viability is determined by the trypan blue exclusion technique. Bone marrow cells are obtained from mice by flushing the femur and tibia with medium[32].

L1210 Cell Cultures

In some cases, L1210 cells are transferred and cultured in tissue culture flasks for various time periods (1-7 days). Usually 75 cm^2 (200 ml) flasks are employed with culture medium innoculated with 10^5 cells/ml, and then incubated at 37°C with 5% CO_2 and 95% air.

Bone Marrow Cultures

Erythroid (BFU-E, CFU-E) and myeloid (CFU-GM) bone marrow cultures are done according to previously described procedures. Erythroid colonies are grown in both plasma clot and methylcellulose cultures in the presence of erythropoietin (0.4 U/ml, Toyobo). Granulocyte-macrophage colonies are grown in methylcellulose in the presence of colony-stimulating factor (GCT--CM, GIBCO)[32-34].

Cell Fractionation

Cells for enzyme determinations are harvested from culture flasks by standard procedure. In some cases, liver samples were also used. Briefly, cells are harvested, washed twice, counted, lysed by 3 freeze/thaw cycles, and homogenized in sucrose-tris buffer. A small volume of the homogenate is retained and labelled HOMOGENATE. Fractionation by differential centrifugation yields: A MITOCHONDRIAL PELLET (10,000 rpm, 10 min.); the MICROSOMES (supernatant spun at 105,000 x g, 90 min.) and the supernatant or CYTOSOL fraction[24].

Glutathione (GSH And GSSG)

Reduced (GSH) and oxidized (GSSG) forms are measured on sample fractions by the fluorometric method described by Hissin and Hilf[35]. O-ophthalaldehyde (OPT) and N-ethylmaleimide (NEM) are used as substrate and inhibitors, respectively. Flourescense is read in a Perkin-Elmer spectrophotometer at 420 nm after excitation at 350 nm. Results are expressed as nanograms (ng) (GSH or GSSG) per milligram (mg) protein <u>or</u> in some cases ng/10^9 cells.

Superoxide Dismutase (SOD)

SOD is determined by the method of Marklund and Marklund[36]. Briefly, pyrogallol substrate autoxidizes at a high pH with a color appearance which is linear for 3 to 4 min, when added to sample cuvettes and to controls without sample. SOD inhibits this autoxidation. One unit of SOD activity is the amount of enzyme that inhibits pyrogallol autoxidation by 50%.

Epoxide Hydrolase

Determination utilizes a radiometric assay which employs a radioligand [^{3}H]styrene oxide as the substrate as described by Oesch *et al.*[37]. In brief, the substrate is mixed with fractions and incubated at 37°C for 15 min; the reaction is terminated with ether and the labelled styrene glycol product extracted with ethyl acetate; aliquots are then counted in a scintillation counter.

RESULTS

Glutathione (GSH/GSSG) Levels In Normal vs. Leukemic Liver Fractions

In order to determine the subcellular distribution profile of glutathione activity in L1210 leukemia, we examined L1210 cells as well as L1210 leukemic and normal liver *in vivo* (direct) and *in vitro* (indirect). Determinations were made on subcellular fractions prepared from normal and L1210 leukemic liver. These experiments were done in order to determine if the livers of leukemia-bearing animals have a different balance in GSH/GSSG, which may, in turn, reflect FRS activity. The results are represented in Table 1 and demonstrate that mitochondria from controls have a high glutathione reductase activity (GSH/GSSG = 11.4). In contrast, leukemic livers have an equal balance between glutathione reductase and peroxidase (ratio = 1.0). Furthermore, microsomal activity appears to be different in leukemic livers as contrasted to normals (1.2 vs 0.83).

Glutathione (GSH/GSSG) Levels In L1210 Cells

In these experiments, GSH/GSSG levels were determined in subcellular fractions of L1210 cells (direct). In addition, determinations were made on fractions taken from L1210 cells grown *in vitro* (indirect), Table 2. Note that in homogenates, mitochondria, cytosol and microsomes, the ratios of GSH/GSSG favored a predominance of glutathione reductase (GSH/GSSG: 1.3, 15.0, 4.8, and 3.5). This was true for preparations taken from cells *in vivo* (direct) and *in vitro* (indirect). For example, the GSH/GSSG ratios after 3 days in culture were 10.0 in mitochondria and 6.5 in the cytosol compartments. These results also demonstrate that mitochondria had the

Table 1. GSH/GSSG glutathione in normal and L1210 leukemia liver subcellular fractions*

Fraction	Normal Liver (ng/mg Protein)		Ratio	Leukemic Liver (ng/mg Protein)		Ratio
	GSH	GSSG	GSH/GSSG	GSH	GSSG	GSH/GSSG
Homogenate	2,000	810	2.5	700	500	1.4
Mitochondria	2,000	175	11.4	100	100	1.0
Cytosol	2,000	550	3.6	2,000	600	3.3
Microsomes	250	300	0.83	350	300	1.2

* Liver tissue from 5 animals in each group were pooled and GSH/GSSG determinations made as described in Methods.

Table 2. Fractionation and subcellular localization of GSH-GSSG glutathione in L1210 leukemia cell <u>in vivo</u> (direct) and <u>in vitro</u> (indirect)*

Fraction	Direct In Vivo (ng/10^9 cells)		Ratio	Indirect Day 1(In Vitro) (ng/10^9 cells)		Ratio	Indirect Day 3(In Vitro) (ng/10^9 cells)		Ratio
	GSH	GSSG	GSH/GSSG	GSH	GSSG	GSH/GSSG	GSH	GSSG	GSH/GSSG
Homogenate	200	170	1.3	150	50	3.0	55	10	5.5
Mitochondria	150	10	15.0	48	ND*	---	100	10	10.0
Cytosol	720	150	4.8	530	60	8.8	130	20	6.5
Microsomes	350	100	3.5	210	ND*	---	125	ND*	---

*Approximately 25 animals or 10^9 cells were employed for each assay, n = 3.

**ND = none detected.

greatest glutathione reductase activity (GSH/GSSG ratio = 15 and 10).

Glutathione (GSH/GSSG) Levels In Leukemia Bone Marrow

As seen in Table 3, glutathione levels were highest in the mitochondria (ratio = 3.13). In addition, microsomes showed a predominance of glutathione reductase activity (ratio = 1.33). In each case, the GSH levels were always greater than the GSSG levels, indicating a greater activity of the steady state enzyme, glutathione reductase.

Epoxide Hydrolase (EH) Activity In Cell Fractions

When various subcellular fractions of 10^8 to 10^9 cells are fractioned as described in the Methods, the highest activity of EH was localized in the microsomal fraction (Tables 3,4). For example, note in Table 4 that the microsomal fraction yielded 74,404 cpm/mg protein (specific activity) as contrasted with significantly lower activity in mitochondria (30,081 cpm/mg protein).

EH Activity In Leukemic Bone Marrow

Table 3 shows the presence of EH levels in leukemic bone marrow cell fractions. Bone marrow fractions showed the highest EH activity in the microsomal fraction (46,355 cpm).

Table 3. Detection of glutathione and epoxide hydrolase activities in Leukemic L1210 murine bone marrow*

Fraction	Epoxide hydrolase	Glutathione (ng/10^9 cells)		Ratio
	(cpm)	GSH	GSSH	GSH/GSSG
Mitochondria	29,732	250	80	3.13
Cytosol	13,474	1,000	800	1.25
Microsomes	46,355	200	150	1.33

*Bone marrow was obtained from 25 animals, n = 3.

Table 4. Fractionation and localization of epoxide hydrolase activity in L1210 cells*

Fraction	Specific Activity (cpm/mg protein)
Homogenate	41,448
Mitochondria	30,081
Microsomes	74,404

*10^9 cells were utilized for each assay, n - 3.

The Effect Of Retinoic Acid (RA) On EH Activity In L1210 Cells

In these experiments, RA was employed in order to determine if RA could modify EH activity or the FR state in the leukemia cells. EH determinations were made on tumor cells (and fractions) immediately after removal from animals (direct), and on tumor cells cultured in vitro with RA (indirect). When RA (10^{-7} M) was added to L1210 cultures containing 2 x 10^5 cells/ml and then incubated at 37°C for 3 days, the EH activity was markedly increased in both mitochondria and microsomes. Note in Table 5 that mitochondria and microsomal activities in 3-day cultures were increased by 100-200% when RA was included in the culture. Thus, RA appears to significantly induce EH activity in L1210 cells.

In Vivo Induction Of L1210 Tumor-Bearing Mice With 3-MC

DBA/2 mice were injected with either 40 mg/kg or 80 mg/kg of 3-MC in corn oil, intraperitoneally, for 3 consecutive days. When tumor ascites cells were fractionated and assayed for EH activity, there was little or no induction using 40 mg/kg doses; however, with 80 mg/kg doses, there was a significant increase in the microsomal activity (Table 6). These results suggest that there is a potential for the induction of EH in the microsomal compartments of L1210 leukemia cells.

Superoxide Dismutase (SOD) Activity

Pyrogallol (1,2,3-benzenetriol, the substrate used in this assay, autoxi-

Table 5. Epoxide hydrolase activity in mitochondria from L1210 cells exposed to ritinoic acid** for 3 days

Fraction	Day 3 - In Vitro -RA	Day 3 - In Vitro +RA	Activity
Mitochondria	26,478	53,644	100%
Microsomes	19,224	62,062	200%

* 10^9 cells were utilized for each assay, n = 3.

**RA = retinoic acid (10^{-7} M).

Table 6. Determination of L1210 EH activity before and after induction with methylcholanthrene (3-MC)*

Fraction	Specific Activity (cpm/mg Protein)	
	Control	+ 3-MC
Homogenate	41,448	50,510
Mitochondria	30,081	44,558
Microsomes	74,404	99,740

*Mice (13/group) were injected with saline (control) or with 80 mg/kg of 3-MC, i.p., on 3 consecutive days prior to sacrifice.

dizes rapidly in aqueous environments with the formation of superoxide anions. Superoxide dismutase inhibits this autoxidation to 99% resulting in products with peak optical spectrum at 420 nm. Therefore, one unit of SOD activity is that amount of enzyme which inhibits pyrogallol autoxidation by 50%. Using this colorimetric method, the sample fractions required a 3-fold dilution for the detection of SOD activity. The results depicting SOD activity are represented in Tables 7-8.

SOD Activity In Normal And L1210 Leukemic Liver

Note in Table 7 that SOD activity was highest in the cytosolic fraction from DBA/2 mouse normal liver (21.2 Units SOD/mg protein) followed by the mitochondrial fraction (12 Units SOD/mg protein). In contrast, L1210 leukemic liver showed a significant decrease in activity in all subcellular compartments. For example, the cytosolic SOD decreased from 21.2 Units SOD/mg protein to 3.98 Units SOD/mg protein, whereas the mitochondrial SOD activity decreased from 12 Units SOD/mg protein to 1.13 Units SOD/mg protein.

SOD Activity In L1210 Cells In Vivo and In Vitro

Cells taken directly from the ascites of tumor-bearing mice and fractionated were examined. Table 8 shows the SOD activity in homogenates, mitochondria, cytosol and microsomes (3.11, 1.42, 1.05, 1.04). Cells after 3 days in culture differed in SOD activity (2.01, 1.43, 1.11, 1.20).

Table 7. SOD levels in normal and leukemic DBA mouse liver**

Fraction	Normal Liver		L1210 Leukemic Liver	
	Units SOD/ml	Units SOD/mg	Units SOD/ml	Units SOD/mg
Homogenate	270.5	11.7	95.0	5.0
Mitochondria	132.0	12.0	12.92	1.13
Cytosol	370.8	21.2	71.62	3.98
Microsomes	112.0	7.0	15.14	1.32

* Livers were obtained from 5 animals/group.

Table 8. SOD levels in L1210 cells in vivo and in vitro*

Fraction	In vivo (direct) L1210 cells		In vitro(indirect) L1210 cells	
	Units SOD/ml	Units SOD/mg	Units SOD/ml	Units SOD/mg
Homogenate	26.50	3.11	26.50	2.01
Mitochondria	16.56	1.42	16.06	1.43
Cytosol	18.93	1.05	13.94	1.11
Microsomes	18.92	1.04	17.66	1.20

*10^8 to 10^9 cells were utilized per assay, n = 2.

DISCUSSION

In this study, we have demonstrated the presence and distribution of FRS and scavenger-like systems in the L1210 leukemia model. The presence of FRS protective mechanisms may be essential for maintaining the proper physiological environment for cell functions. Variations in the levels or ratios of certain FRS could then contribute to changes in cellular functions such as may occur in certain disease states.

Since stem cells and stromal cell interactions are important for maintaining hematopoiesis[38-40], the abnormal hemopoiesis seen in the leukemic state may be related to the interactions of leukemic cells with bone marrow precursor and stromal cell populations[12,16]. Numerous studies with model systems have shown that reactive FR can chemically modify macromolecules in a way that metabolic disturbances and changes in cell structure become apparent. The hematological picture observed in L1210 leukemia is associated with marked anemia and changes in other cell functions. The interaction of leukemic cells with normal bone marrow cells are schematized in Figure 1. As seen in this figure, FR levels are modified by the cytochrome P-450 and epoxide hydrolase complex, glutathione system, catalase and superoxide dismutase. Also, the relationship of these FR/FRS to cells in the normal and abnormal environment is represented. As suggested by the diagram, leukemic cells in culture or in situ may in some way influence normal cells in the environment.

Cells in normal and abnormal environments generate FR and the ability to metabolize or detoxify FR in these cells are handled by scavengers. Therefore, the influence of leukemic cells on the growth and metabolism of other cells may come about, in part, by the generation of FR and availability of FRS. Thus, it remains possible that cell interactions may modulate hematopoiesis through these systems. This concept is strengthened by the fact that human serum contains circulating inhibitors of hematopoiesis that can be inactivated by FRS[14]. Helgestad et al.[12], demonstrated that a variety of FRS systems were able to enhance CFU-GM growth by inactivating certain FRS in human serum CSF. Our results suggest that agents that enhance FRS such as RA may have beneficial effects on cells. In this respect, RA was found to arrest L1210 cells in the G_1 phase[20], and enhance EH activity. Such conditions favor a loss in L1210 tumorogenicity[20] and would thus allow normal hemopoiesis to ensue.

The glutathione FRS system has been shown to have high activity in liver and erythrocytes[16]. Peroxidase is necessary for the oxidation of GSH to GSSG; reductase is required for the reduction of GSSG to GSH[20].

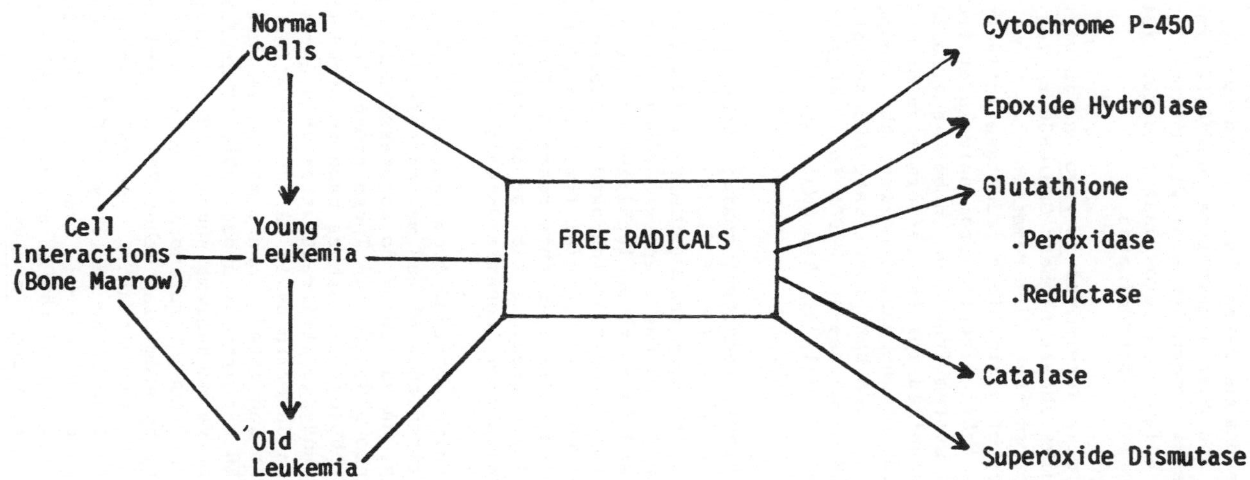

Fig. 1. RELATIONSHIPS OF FREE RADICALS AND FREE RADICAL SCAVENGERS IN NORMAL AND LEUKEMIC STATES

Glutathione is a tripeptide (α-glutamylcysteinylglycine) found in most mammalian cells. It plays many roles, along with various other anabolic and catabolic enzymes, in cellular expression[41]. In particular, levels of glutathione and the proper balance of GSH/GSSG, along with the activation of certain protein kinases, may act as key regulators of cell cycle events[42]. Thus, mitotic activity may be directly influenced by this system. Initial studies with liver cells demonstrated that glutathione FRS could be determined by our assay procedure, and that glutathione (GSH/GSSG) ratios in normal liver mitochondria showed a predominance of glutathione reductase, whereas leukemic liver showed a balance in peroxidase-reductase ratios. In microsomes, the peroxidase activity predominated.

Glutathione levels were determined in vivo and in vitro in L1210 leukemia cells. These studies showed highest GSH/GSSG ratios in the mitochondrial and cytosolic compartments, which remained high after 3 days in culture, indicating a predominance of glutathione reductase activity. Thus, similar GSH/GSSG ratios in cells taken directly from animals or from tissue culture show that this scavenger system functions in both environments. Our results suggest that reductase required for the re-formation of GSG was prevalent. Furthermore, these results can be explained from the viewpoint that high levels of hydrogen peroxide can both inactivate SOD and stimulate GSSG formation, which activates GSH reductase and NADPH in an attempt to maintain a balance in the glutathione cycle[42]. The exact mechanism for this pattern remains unclear.

Since epoxides are highly reactive, toxic, metabolic intermediates that are metabolized by the hydration of diols[43], certain epoxides can have damaging effects on cell structure and function via FR chain reactions resulting in lipid peroxidation[44]. EH has been extensively studied in rat and rabbit liver microsomes and found to be localized in the endoplasmic reticulum in association with the mixed-function oxidase system[45]. Similar patterns of EH activity were detected in subcellular fractions of L1210 cells taken in situ or from in vitro cultures. In this respect, EH activity was always highest in the microsomal fraction. A similar EH pattern was also obtained from bone marrow cells taken from leukemic animals.

Of interest were results obtained from RA exposed cells. In previous studies, we have demonstrated that RA has marked effects on L1210 leukemia[20,32]. In this regard, RA was found to suppress tumor colony growth and to shift the cells into a G_1 phase. These properties were associated with a change in cell morphology and others have reported that certain leukemic cells may actually undergo differentiation when exposed to RA. Therefore, we are interested in determining if RA had any effect on EH or other FRS in L1210 leukemia. As indicated, when tumor cells were exposed to RA for 3 days, RA enhanced EH activity by about 200%. Therefore, it appears that RA may induce protective mechanisms such as EH and other FRS, and thus favor a better cellular environment for normal hemopoiesis. We are currently exploring this possibility by examining other FRS systems and various scavenger agents such as 1,25-D_3 and vitamin E. Animals receiving injections of 3-MC, an agent known to induce the activity of EH[43], demonstrated a 75% increase in EH activity in the microsomal fraction. This is further proof of the presence and inducibility of the EH enzyme in culture and an indication of the potential of this enzyme in vivo.

A rich source of the synthesis of all enzymes is, of course, the liver. Subcellular fractions prepared from normal and L1210 leukemic liver showed that SOD activity decreased by more than half in the leukemic liver. In addition, or preliminary data is in agreement with that of other investigators show showed a decrease in the mitochondrial SOD activity in certain leukemic cells[46]. We also demonstrated a decrease in SOD activity in all other subcellular liver fractions in the leukemic state. Therefore, SOD activity is measurable and appears to show altered activity in our leukemic model. Measurements of L1210 subcellular fractions showed highest SOD activity in both the homogenate and mitochondria. The activity was reduced in the homogenate and remained the same in the mitochondria after 3 days in culture. These results may suggest that a normalization of SOD activity within the mitochondria may occur after in vitro culture. The slight decrease in SOD activity in other cellular compartments may not be significant and conclusions cannot be fully evaluated at this time. By contrast, the GSH/GSSG ratios were highest in the mitochondrial and cytosolic compartments in L1210 ascites and remained high in 3-day cultures.

Leukemia bone marrow cell GSH/GSSG ratios showed a predominance of the reductase activity in all fractions tested and may suggest a pattern consistent with this type of leukemia. In this regard, it has been shown that L1210 cells co-cultured with normal bone marrow cells suppress erythropoiesis in a way not due to crowding out, nutrient exhaustion, rather to cell products[14]. This could be due, in part, to FR generation or changes in FRS ratios contributed by the leukemia cells. Additional experiments are under way in this laboratory to establish the FRS mechanism in the leukemia bone marrow cells.

In summary, certain FRS and scavenger-like systems are demonstratable in subcellular fractions from L1210 leukemia. The glutathione ratios seen in the mitochondria of L1210 ascites cells after 3 days in culture are similar to normals. In contrast, leukemic liver and bone marrow showed very little glutathione presence. The cytosolic and microsomal glutathione levels were low in normal and leukemic liver as well as L1210 cells and leukemic bone marrow. With such high glutathione ratios in L1210 ascites cells, it is forseeable that they may be influencing both the liver and bone marrow adversely in situ. Microsomes from both in vivo and in vitro subcellular fractions showed the highest EH activity in all microsomes examined in this study. SOD activity was highest in normal liver cytosol and significantly lower in leukemic liver and L1210 leukemia cells.

In conclusion, we suggest that L1210 cells may be contributing to the FR/FRS imbalance seen in leukemic bone marrow and this imbalance may have adverse effects on the normal hemopoietic microenvironment.

REFERENCES

1. Halliwell, B. and J.M.C. Gutteridge. 1984. Oxygen toxicity, oxygen radicals, transition metal and disease. Biochem. J. 219: 1-14.
2. Slater, T.F. 1984. Free-radical mechanisms in tissue injury. Biochem. J. 222:1-15.
3. Ames, B.N. 1983. Dietary carcinogens and anticarcinogens: Oxygen radicals and degenerative diseases. Science 221: 1256-1264.
4. Fridovich, I. 1986. Biological effects of superoxide radical. Arch. Biochem. Biophys. 247: 1-11.
5. Kensler, T.W. and B.G. Taffe. 1986. Free radicals in tumor promotion. Adv. in Free Radical Biology and Medicine. 2: 347-387.
6. Slaga, T.J., A.J.P. Klein-Szanto, L.L. Triplett, L.P. Yotti, and J.E. Trosko. 1981. Skin tumor promoting activity of benzoyl peroxide, a widely used free radical generating compound. Science 213: 1023-1025.
7. Slaga, T.J., V. Solanski, and M. Logani. 1983. Studies on the mechanism of action of antitumor promoting agents: Suggested evidence for the involvement of free radicals in promotion. In: Radioprotectors and anticarcinogens. (Eds., O.F. Nygaard and M.G. Simic), pp. 471-485. Academic Press, New York.
8. Klein-Szanto, A.J.P. and T.J. Slaga. 1982. Effects of peroxides on

rodent skin: Epidermal hyperplasia and tumor promotion. J. Invest. Dermatol. 79: 30-34.
9. Kurokawa, Y., N. Takamura, Y. Matsushima, I. Takayoshi, and Y. Hayashi. 1984. Studies on the promoting and complete carcinogenic activities of some oxidizing chemicals in skin carcinogenesis. Cancer Lett. 24: 299-304.
10. Storm-Mathisen, I., J. Helgestad, and S.O. Lie. 1986. Inhibitory activity in mouse lung-conditioned medium studied in the agar assay for bone marrow colony-forming cells: Removal by vitamin C. Scand. J. Clin. Lab. Invest. 45: 67-76.
11. Helgestad, J., I. Storm-Mathisen, and S.O. Lie. 1986. Vitamin C and thiols reagents promote the in vitro growth of murine granulocyte/macrophage progenitor cells by neutralizing endogenous inhibitor(s). Blut 52: 1-8.
12. Helgestad, J., I. Storm-Mathisen and Lie, S.O. 1986. Endogenous inhibitors of human granulocyte/macrophage colony forming cells in vitro are inactivated by free radical scavengers. Scand. J. Haematol. 37: 395-403.
13. Snyder, R., E.W. Lee, J.J. Kocsis, and C.M. Witmer. 1977. Bone marrow depressant and leukomogenic actions of benzene. Minireview. Life Sciences 21: 1709-1722.
14. Austin, T., A. Ma, M.C. Datta, J.A. Ortega, N.A. Shore, and P. Dukes. 1979. Suppression of mouse erythroid colony formation by L1210 leukemia cells. Exp. Hematol. 7: 63-69.
15. Chiyoda, S., H. Mizoguchi, S. Asanu, F. Takaku, and Y. Miura. 1976. Influence of leukemic cells on the colony formation of human bone marrow cells in vitro. II. Suppressive effects of leukemic cells extracts. Br. J. Cancer 33: 379-384.
16. Miller, A.M., P.L. Page, B.L. Hartwell, and S.H. Robinson. 1977. Inhibition of growth of normal murine granulocytes by cocultured acute leukemic cells. Blood 50: 799-809.
17. Quesenberry, P.J., J.M. Rappaport, A. Fountebuoni, R. Sullivan, L. Zuckerman, and M. Ryan. 1978. Inhibition of normal hematopoiesis by leukemic cells. N. Eng. J. Med. 299: 71-75.
18. Sidell, N. 1982. Retinoic acid-induced growth inhibition and morphological differentiation of human neuroblastoma cells in vitro. J. Natl. Cancer Inst. 68: 589-593.
19. Flynn, J.P., J.W. Miller, J.D. Weisdorf, C.D. Arthur, R. Brunning, and F.R. Branda. 1983. Retinoic acid treatment of acute promyelocytic leukemia, in vitro and in vivo observation. Blood 62: 1211-1217.
20. Moqattash, S., J.D. Lutton, J.W. Chiao, and R.D. Levere. 1985. Abolition of L1210 clonogeneticy and G1 arrest by retinoic acid and 1,25-dihydroxyvitamin D3. Cancer Lett. 27: 125-134.
21. Solanski, V., R.S. Rana, and T.J. Slaga. 1986. Diminution of mouse epidermal superoxide dismutase and catalase activities by tumor promoters. Carcinogenesis 2: 1141-1146.
22. Taffe, B.G. and T.W. Kensler. 1986. Modification of cellular oxidant defense mechanisms in mouse skin by multiple applications of TPA. Proc. Am. Assoc. Cancer Res. 27: 148.
23. Perchellet, J.P., E.M. Perchellet, D.K. Orten, and B.A. Schneider. 1986. Decreased ratio of reduced/oxidized glutathione in mouse epidermal cells treated with tumor promoters. Carcinogenesis 7: 503-506.
24. Abraham, N.G., J.D. Lutton, and R.D. Levere. 1985. Benzene modulation of bone marrow hematopoietic and drug metabolizing system. Biochem. Arch. 1: 85-96.
25. Harigaya, K., M.E. Miller, E.P. Cronkite, and R.T. Drew. 1981. The detection of in vivo hematoxicity of benzene by in vitro liquid bone marrow cultures. Toxicol. Appl. Pharmacol. 60: 346-353.
26. Cronkite, E.P. 1985. Regulation and structure of hemopoiesis: Its

application in toxicology. In: Toxicology of The Blood and Bone Marrow. (Ed., R.D. Irons) pp. 17-38.
27. Hromas, R.A., B. Barlogie, R.E. Meyn, P.A. Andrews, and C.P. Burns. 1985. Diverse mechanisms and methods of over-coming cis-platinum resistance in L1210 leukemia cells. Proc. Am. Assoc. Cancer Res. 26: 261.
28. Hromas, R.A., P.A. Andrews, M.P. Murphy, and C.P. Burns. 1987. Glutathione depletion reverses cis-platin resistance in murine L1210 leukemia cells. Cancer Lett. 34: 9-13.
29. Suzukake, K., B.J. Petro, and D.T. Vistica. 1982. Reduction in glutathione content of L-PAM resistant L1210 cells confers drug sensitivity. Biochem. Pharm. 31: 121-124.
30. Oesch, F. 1983. Drug detoxification: Epoxide hydrolase. In: Developmental Pharmacology, Alan Liss, Inc., New York, pp. 81-105.
31. Oesch, F. 1979. In: Progress in Drug Metabolism, Vol. 3, (Eds. J.W. Bridges and L.F. Chasseaud), pp. 253-301.
32. Moqattash, S.T. 1985. "Modulatory effects of murine L1210 leukemia on hematopoiesis in vitro and in vivo." Ph.D. dissertation, New York Medical College.
33. Abraham, N.G., J.D. Lutton, and R.D. Levere. 1982. The role of haem biosynthetic and degradative enzymes in erythroid colony development: The effect of haemin. Br. J. Haematol. 50: 17-28.
34. Lutton, J.D., N.G. Abraham, M. Friedland, and R.D. Levere. 1984. The toxic effects of heavy metals on rat bone marrow in vitro erythropoiesis: Protective role of hemin and zinc. Envir. Res. 35: 97-103.
35. Hissin, P.J. and R.A. Hilf. 1976. A fluorometric method for the determination of oxidized and reduced glutathione in tissue. Anal. Biochem. 74: 214-226.
36. Marklund, S. and G. Marklund. 1974. Involvement of the superoxide radical in the autoxidation of pyrogallol and a convenient assay for superoxide dismutase. Eur. J. Biochem. 47: 469-474.
37. Oesch, F., D.M. Jerina, and J. Daly. 1971. A radiometric assay for hepatic epoxide hydrolase activity with [7-^{3}H]styrene oxide. Biochim. Biophys. Acta 227: 685-691.
38. Cronkite, E.A., M.E. Miller, A. Garnett, and K. Harigaya. 1982. Regulation of hematopoiesis: Inhibition and stimulators produced by a murine bone marrow stromal cell H-1). In: Haemopoietic Stem Cells, (S.V. Skillman), (Ed., E.P. Cronkite, C.N. Muller Borat), Munksgaard Pub. Co., Copenhagen, pp. 266-282.
39. Zuckerman, K.S., G.C. Bagby, E. McCall, B. Sparks, J. Wells, V. Phatei, and D. Goodrum. 1985. A monokine stimulates production of human erythroid-burst-promoting activity by endothelial cells in vitro. J. Clin. Invest. 75: 722-725.
40. Broudy, V.C., K.S. Zuckerman, S. Jetmalani, J.H. Fitchen, and G.C.Bagby. 1986. Monocytes stimulate fibroblastoid bone marrow stromal cells to produce multilineage hematopoietic growth factors. Blood 68: 530-534.
41. Meister, A., and M.E. Anderson. 1983. Glutathione. Ann Rev. Biochem. 52: 711-760.
42. Luthman, M., S. Eriksson, A. Holmgren, and L. Thelander. 1979. Glutathione-dependent hydrogen donor system for calf thymus ribonucleoside-diphosphate reductase. Proc. Nat. Acad. Sc. 76: 2158-2162.
43. Lu, A.Y.H. and G.T. Miwa. 1980. Molecular properties and biological functions of microsomal epoxide hydrolase. Ann. Rev. Pharmacol. 20: 513-531.
44. Burk, R.F. 1983. Glutathione-dependent protection by rat liver microsomal protein against lipid peroxidation. Biochim. Biophys. Acta 757: 21-28.
45. Walker, C.H., C.W. Timms, C.R. Wolf, and F. Oesch. 1986. The

hydration of sterically hindered epoxides by epoxide hydrolase of the rat and rabbit. Biochem. Pharmacol. 35: 499-503.
46. Peskin, A.V., Y.M. Koen, and I.B. Zborsky. 1977. Superoxide dismutase and glutathione peroxidase activities in tumors. FEBS Lett. 78: 41-45.

EXOGENOUS AND ENDOGENOUS REGULATIONS OF HUMAN MEGAKARYOCYTOPOIESIS

Alan M. Gewirtz

Section of Hematology/Oncology, Departments of Medicine, Pathology, and Thrombosis Research Center
Temple University School of Medicine
Philadelphia, PA 19140

INTRODUCTION

Megakaryocytopoiesis may be viewed as a developmental continuum which begins when an undifferentiated hematopoietic stem cell commits to maturation within this lineage (for recent review see ref.#1) The process of commitment results in the generation of megakaryocyte progenitor cells (CFU-Meg). When appropriately stimulated, CFU-Meg are capable of intense proliferative activity which determines the numbers of megakaryocytes which will ultimately populate the bone marrow. As proliferative activity declines, CFU-Meg progeny mature into precursor cells which terminally differentiate into polyploid megakaryocytes capable of releasing several thousand platelets each into the circulation.

Regulation of the above described events has become the object of intense scrutiny. The process of stem cell commitment remains poorly understood, but once it occurs further development is undoubtedly highly regulated. In vitro culture system designed to study megakaryocyte development have demonstrated the existence of several candidate, though not necessarily lineage specific, regulatory molecules. Some of these regulatory molecules promote megakaryocyte development. These include the CFU-Meg progenitor cell stimulators megakaryocyte colony stimulatory factor (Meg-CSF)[2], granulocyte-macrophage colony stimulatory factor (GM-CSF)[3], and interleukin-3 (IL-3)[4]. They also comprise the cell maturation promoters thrombopoietin (TPO)[5,6], megakaryocyte stimulatory factor (MSF)[7,8], and erythropoietin[9]. Other regulatory factors inhibit megakaryocyte development. For example, immunocytes, and their products, have been documented to cause clinically significant suppression of megakaryocyte production[11,12]. Several groups have also provided evidence that megakaryocyte colony growth is inferior in serum when compared to growth in platelet poor plasma[13,14], and that specific platelet constituents can reproducibly inhibit megakaryocyte development in vitro[15,16].

In toto, the observations described above suggest that megakaryocytopoiesis is governed by externally derived (exogenous) and autocrine (endogenous) regulatory signals. I present herein recent studies from my laboratory which investigate the importance of monocytes, T lymphocytes, natural killer cells, and megakaryocyte constituents in the regulation of human megakaryocytopoiesis in vitro.

MATERIALS AND METHODS

Isolation of Mature Human Marrow Megakaryocytes

Mature (morphologically recognizable) megakaryocytes were isolated from the marrow of normal bone marrow donors by the process of counterflow centrifugal elutriation (CCE) as previously described[17,18]. Lysates prepared from such cells were then utilized for co-culture experiments with autologous progenitor cells.

Megakaryocyte Progenitor Cell Assay

Megakaryocyte colonies were grown in plasma clot cultures as previously described[19,20]. The progenitor cell population cultured consisted of either unseparated MNC, or MNC depleted of adherent monocyte-macrophages (MO), T lymphocytes, and B lymphocytes (NALD MNC) prepared as described[20]. To provide essential growth factors, all cultures were supplemented with normal human AB serum (25% v/v; derived from the platelet poor plasma of a single donor), and if indicated, with aplastic anemia serum (5% v/v).

The co-culture studies reported were performed in a totally autologous system as previously described[21]. A variable number of effector cells were added directly to a fixed number of target MNC (5×10^5/ml).

Megakaryocyte colonies were scored as previously described using a rabbit anti-human platelet glycoprotein antiserum as a megakaryocyte probe in an indirect immunofluorescence assay[21]. Unless otherwise stated, all data are reported as the mean ± S.E.M. of colonies enumerated.

Isolation, And Enrichment of Monocyte-Macrophages (MO), Natural Killer Cells, And T Lymphocytes

Monocyte-macrophages (MO) were depleted from unseparated MNC by adherence to serum coated plastic petri dishes as described[20]. MO were recovered for autologous co-culture studies by adding 0.2% EDTA in phosphate buffered saline (PBS) to the dishes and then incubating them for 30 minutes at 4° C. MO were >90% non-specific esterase positive, and ∿92% Leu M3 positive.

NK cells were enriched on discontinuous Percoll (polyvinylpyrrolidone-coated colloidal silica particles; Pharmacia) gradients as previously described[22], and further depleted of residual T lymphocytes by high affinity E rosette formation using the method of Timonen et al[23].

NK cells of highest purity (Leu 11b[+]:85±2%) were obtained by immune rosetting with chronic chloride treated sheep erythocytes conjugated to goat anti-mouse 1gG or 1gM monoclonal antibody (MoAb-erythrocyte complexes)[24]. The complexes were then reacted with enriched MNC which had been incubated with NK specific MoAb Leu 11b. Leu 11b positive cells formed rosettes with the MoAb-erythrocyte complexes and could therefore be isolated, and enriched by concentration on a Ficoll-Hypaque gradient and subsequent lysis of the sheep RBC.

T lymphocytes (T cells) were obtained by two different methods. One method was utilized to obtain T cells employed as simultaneous controls for NK co-culture experiments. T cell rich, NK depleted, cells from high density (>1.075 gm/ml) Percoli fractions (HDPF) were pooled, washed 3X, and co-cultured with target cells identical to those to which NK cells had been added. The other method employed for obtaining T cells was a standard rosetting technique with neuraminidase treated sheep red blood cells (RBC)[25]. Rosette forming cells were >98% T cells as assessed by immunofluorescence staining with OKT 11 (Ortho Pharmaceuticals, Raritan, N.J.). T-MNC contained >2% residual T cells as determined by OKT 11 staining.

Helper/inducer (Leu 3a+), and cytotoxic/supressor (Leu2a+) T cells were isolated by standard solid phase immunoadsorption ("panning") as described previously[21]. Viability of recovered cells and purity, as assessed by indirect immunofluorescence with the appropriate probe antibody, was routinely found to exceed 95%.

Characterization of NK, T Lymphocyte, And Monocyte Enriched Effector Cell Fractions

Effector cells were characterized by their reactivity with monoclonal antibodies and histochemical stains as previously described[21]. In brief, cells were typed by morphologic and histochemical analysis after reaction with Wright-Giemsa or alpha-naphthyl acetate (non-specific) esterase stains, or after reaction with a panel of monoclonal antibodies consisting of OKT3 (pan T cell), OKT11 (E rosette receptor), OKT10 (lymphocyte activation antigen), OKT9 (transferrin receptor), OKM1 (monocyte-macrophage lineage), Leu M3 (monocyte-macrophage lineage), HLA-DR (1a antigen), Leu 7 (NK çell), Leu 11a (NK cell), and Leu 12 (B lymphocyte). "OK" monoclonal antibodies (MoAb) were obtained from Ortho Pharmaceuticals, Raritan, NJ; "Leu," and HLA-DR MoAb from Becton-Dickinson, Mountainview, CA.

Preparation of Conditioned Media

Cell free conditioned media were prepared by incubating NK cells (2×10^6/ml) in supplemented alpha medium containing 20% fetal calf serum for 18 hrs at 37°C. Cells were then removed by centrifugation (550 g x ten minutes) and the cell free conditioned medium stored at -20%°C until assayed for colony stimulating activity. Supplemented alpha medium with 20% fetal calf serum without cells was processed concurrently to serve as a control.

Preparation of Human Platelet Factor 4 (PF4)

Human PF4 was purified essentially as published[26]. In brief, outdated human platelets were thrombin stimulated. The resulting supernatant was applied to a heparin-agarose column. The column was washed, and the PF4 specifically eluted with 1.5M NaCl. Some PF4 preparations were purified from the supernatants of stored, outdated platelets by combination of sephadex G-75 gel filtration, and heparin-agarose column chromatography. Eluted fractions were pooled, and then stored lyophilized. Such preparations gave a single band on SDS-Polyacrylamide gells, and were judged to <95% pure.

Statistical Analysis

Statistical significance of differences between groups was tested using a two-tailed Student's T Test for unpaired observations.

RESULTS

Exogenous regulators- Role of Immunocompetent cells.

Effect of NK Enriched Cell Fractions on CFU-Meg Cloning Efficiency

Patients with a proliferation of T γlymphocytes and increased NK activity are often neutropenic, and anemic but are not commonly thrombocytopenic[27,28]. Given this clinical observation, and the fact that the impact of NK cells on human megakaryocyte colony formation had not been explored, we investigated the ability of enriched autologous NK cells to modulate the growth of megakaryocyte progenitor cells (CFU-Meg) in a plasma clot culture system.

A total of twelve separate co-culture experiments with autologous target and NK cells were performed. The aggregate results of these studies are shown in Figure 1. In these experiments, target cells alone gave rise to 43±7 megakaryocyte colonies. When plated at the maximum number (5×10^5/ml) added to the target cell population (1:1 ratio), the NK enriched cells gave rise to 8±2 megakaryocyte colonies. At [effector-(NK)]:[target] cell ratios of 1:100, 1:20, 1:10, and 1:5, stimulation of colony growth was occasionally observed in the individual experiments but such stimulation was typically low grade, and not usually of statistical significance ($p<.05$). In contrast, at [effector]:[target] cell ratios of 1:2 and 1:1 stimulation of megakaryocyte colony formation in comparison to baseline growth was observed in eleven of the twelve separate experiments performed, and was found to be statistically significant ($p>.05$) in eight of these eleven. In aggregate, at the 1:2 ratio a mean (±SEM) of 126±21 (<.0009) megakaryocyte colonies were enumerated, while at the 1:1 ratio a mean of 95 ± 21 (p=,005) megakaryocyte colonies were observed in the culture plates.

<u>Co-culture Experiments With Highly Purified NK Cells and Progenitor Enriched Mononuclear Target Cells</u>

The results described above suggested that CFU-Meg cloning efficiency might be increased when progenitor cells were cultured in the presence of a sufficient number of NK cells. To more rigorously test this hypothesis two types of experiments were performed.

In the first series of experiments, the NK:MNC co-culture studies were repeated on three separate occasions using highly purified NK effector cells and an MNC target cell population which had been extensively depleted of T cells, B cells and adherent monocytes (Table I). The NK cells employed for these experiments were prepared by immune rosetting as described, and were ∿85% pure. In each of the three experiments a significant augmentation ($p<.05$, $P<.001$, $p<0.1$ respectively) in megakaryocyte colony formation was noted at NK/target cell ratios of 1:2. One experiment was also performed at an NK/target cell ratio of 1:10 (Table I). In contrast to results obtained with less pure effector and target cells, a three fold increase ($p=.012$) in colony formation, in comparison to baseline, was observed at this ratio.

In the second set of experiments, we determined the ability of the NK enriched cells to stimulate megakaryocyte colony formation after being incubated with Leu 11b monoclonal antibody on two different occasions. An NK enriched cell population was divided into two aliquots. One aliquot was co-cultured directly with adherent cell, T cell depleted target cells as described above. The other aliquot was incubated with Leu 11b (0.1μg antibody/10^6 cells) for 30 minutes at room temperature. The cells were then immediately admixed with the target cell population and plated as usual. Control cells gave rise to 6±1 colonies. At an NK/Target cell ratio of 1:2, 13±2 colonies were enumerated ($p=.003$). The number of colonies formed after co-culturing Leu 11b treated NK cells with the identical target cell population was 7±1 ($p=.258$). Therefore, leu 11b treatment blocked the ability of NK cells to stimulate megakaryocytopoiesis in vitro.

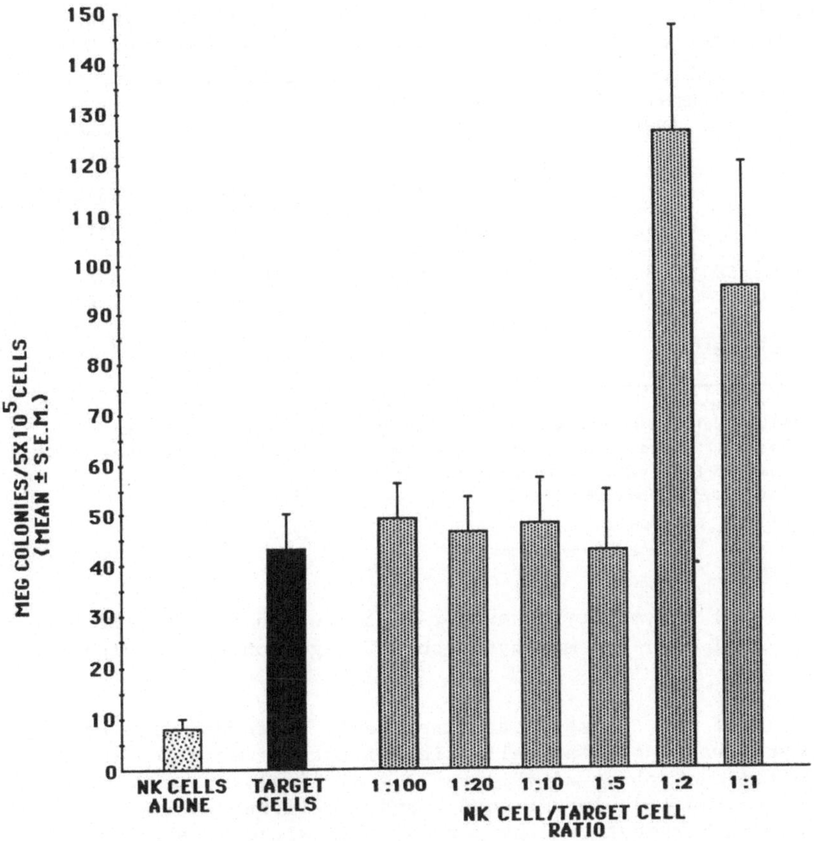

Fig. 1. Effect of co-culturing human marrow mononuclear cells autologous Natural Killer (NK) cells on megakaryocyte colony formation. NK enriched, light density peripheral blood cells were isolated by Percoll density gradient sedimentation and co-cultured with a fixed number of autologous, unseparated bone marrow mononuclear cells obtained by Ficoll-Hypaque density gradient sedimentation. Effector and target cells were mixed together at the ratios indicated and then plated directly into plasma clot cultures. Resulting megakaryocyte colonies were enumerated twelve days later by indirect immunofluorescence microscopy using a polyclonal rabbit anti-human platelet glycoprotein antiserum as probe for cells of the megakaryocyte lineage. Aggregate results of twelve separate experiments are graphically displayed. At NK/Target cell ratios of 2:1 and 1:1 statistically significant augmentation (;<.0009, and p<,005 respectively) in megakaryocyte colony formation was observed. Cells of $5x10^5$/ml, i.e. the maximum number co-cultured with the target cells (1:1). [From reference #21-with permission].

Effect of T lymphocytes and Monocytes on CFU-Meg Cloning Efficiency

In order to provide a frame of reference for the NK cell studies, and to further document that NK stimulation of CFU-Meg was not an artifact

Table I. Effect of Purified NK (Effector) cells on megakaryocyte Colony Formation by Enriched Marrow Mononuclear (Target) Cells

Exp No.	Effector (E) Cells *	Target (T) Cells *	T/E 10:1 ‡	T/E 2:1
1	15 ± 1 §	41 ± 6	---	264 ± 40
2	---	30 ± 7	96 ± 29	106 ± 8
3	10 ± 3	31 ± 12	---	171 ± 5
mean ± SE ‖	12 ± 2	33 ± 4	96 ± 29	182 ± 28
(n) ¶	(4)	(8)	(4)	(10)
df **	---	---	10	16
p value ††	---	---	.012	<.0009

* NK cells prepared by immune rosetting with purity of 85 ± 2%- number plated 5 x 10^5/ml
‡ adherent cell depleted, T cell depleted MNC- number plated held constant at 5 x 10^5/ml
§ mean ± SEM of megakaryocyte colonies enumerated per culture performed in duplicate to quadruplicate
‖ mean ± SEM of all cultures performed in group
¶ total number of observations in group
** degrees of freedom
†† compared to growth in target cell group

due to residual T lymphocytes and/or monocytes in the NK cell preparations, the effect of T cell and monocytes on CFU-Meg cloning efficiency was assessed.

In none of the ten separate experiments performed with non-activated T lymphocytes was a statistically significant augmentation in CFU-Meg cloning efficiency observed (Figure 2). Similarly, when we examined the effect of Leu3(+)helper/inducer, and Leu2(+)suppressor/cytotoxic T cells on megakaryocyte colony formation in vitro no statistically significant effect on megakaryocyte colony formation could be demonstrated with either sub-set when cultured with target cells at target:effector ratios of 2:1, or 1:1 (Table II).

MO isolated by adherence to serum coated plastic dishes also failed to increase the number of megakaryocyte colonies enumerated when co-cultured with autologous adherent cell depleted MNC in four separate experiments. In fact, the addition of MO to the cultures resulted in a slight, though statistically insignificant, inhibition of megakaryocyte colony formation when compared to colonies formed by target cells (adherent cell depleted MNC) alone.

<u>Mechanism of NK Mediated Stimulation of Megakaryocyte Colony Formation</u>

NK mediated stimulation of CFU-Meg cloning efficiency could have been due to a cell contact phenomenon and/or the elaboration of a short range megakaryocyte colony stimulating activity. To explore these possibilities the following experiments were performed.

We first examined the effect of direct effector:target cell contact on CFUOMeg cloning efficiency. This was done by centrifuging effector and target cells together (1:2 ratio), and then pre-incubating the cell pellet for three hours 37°C, 5% CO_2) prior to re-suspending the cells and plating as usual. This manipulation resulted in the loss of significant ($p>.05$) augmentation of megakaryocyte colony formation by unstimulated NK cells. Further, in the case of interferon treated NK cells not only was stimulation abrogated but a clear trend towards CFU-Meg growth inhibition was discerned. In the control T cell co-culture studies pre-incubation with T cells obtained by either method neither caused, nor suggested, any discernible effect on CFU-Meg development. Accordingly, it was unlikely that CFU-Meg stimulation was the result of direct cell contact.

We then assayed NK cell conditioned media (NKCM), prepared as described in the methods section, for the presence of CFU-Meg stimulatory activity.

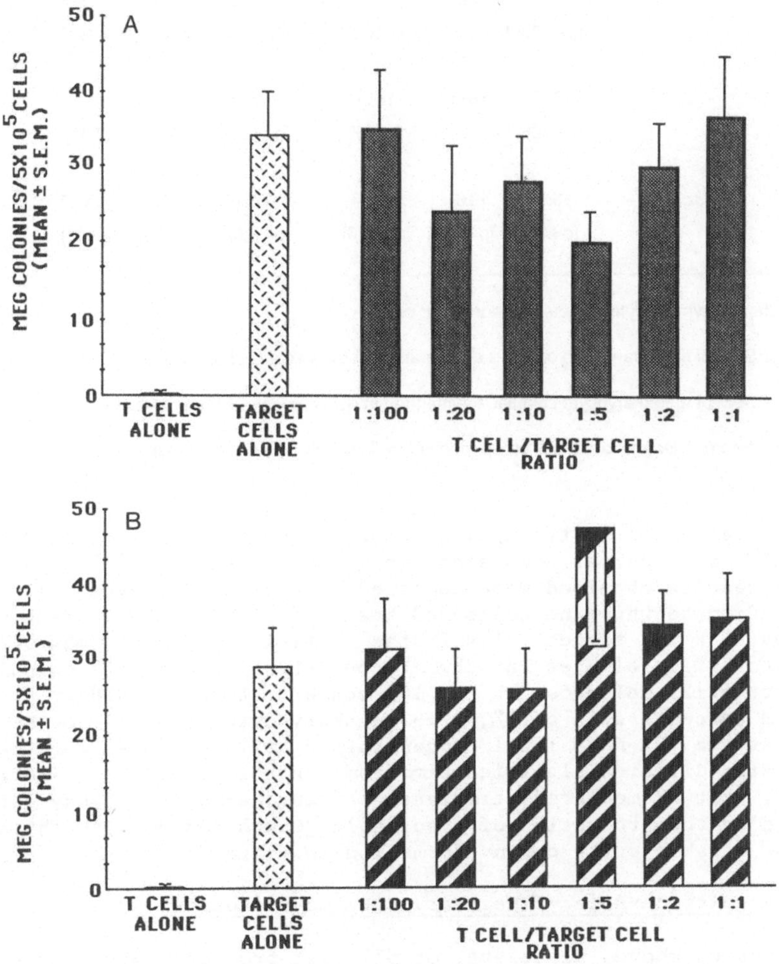

Fig. 2. Effect of co-culturing human marrow mononuclear cells with autologous T lymphocytes on megakaryocyte colony formation. T cells were obtained by rosetting with neuramindase treated sheep red blood cells[Panel A], or by concentration on Percoll density gradients [Panel B], as described in methods section. T cells were then co-cultured with a fixed number of autologous light density marrow mononuclear cells, and resulting megakaryocyte colonies enumerated as outlined in the legend for figure 1. Panel A represents the aggregate results from eight separate experiments, while Panel B depicts the result of ten separate experiments. No statistically significant change ($p>.05$) in megakaryocyte colony formation was observed with T cells derived from either source at any of the ratios tested. [From reference #21-with permission].

Table II. Effect of Lymphocyte Subsets on CFU-Meg Cloning Efficiency

EXP. No.	TARGET CELLS ALONE*	TARGET/ Leu 3(+) T CELLS 10/1	2/1	1/1	TARGET/ Leu 2(+) T CELLS 10/1	2/1	1/1
1	19±1†	12±7 §	22±4	27±7	20±3	31±3	37±1
		[p=.288]¶	[p=.402]	[p=.162]	[p=.744]	[p=.002]	[p=<.001]
2	12±2	16±4	16±1	14±1	11±2	14±3	16±3
		[p=.496]	[p=.209]	[p=.467]	[p=.509]	[p=.617]	[p=.287]
3	18±2	15±4	19±2	21±2	22±4	24±4	26±5
		[p=.361]	[p=.903]	[p=.436]	[p=.353]	[p=.222]	[p=.106]

* non-adherent, lymphocyte depleted mononuclear cells

† mean ± SEM of megakaryocyte colonies enumerated in sextuplicate culture dishes

§ mean ± SEM of megakaryocyte colonies enumerated in quadruplicate culture dishes

¶ p value in comparison to target cell group- derived by Two Tailed Student's T test

Conditioned media were tested against bone marrow mononuclear cells obtained from four different normal donors at final concentrations of 5% and 10% (v/v). The results obtained were compared to growth in tissue culture (control) medium to which no cells had been added (Figure 3). Mean colony formation per 5 x 10^5 target cells in the control experiments was 34±7 (range 2±2 to 57±12 colonies) at the 5% concentration and 36±7 colonies (range 2±1 to 62±21 colonies) at the 10% concentration (p=.828). At a final NKCM concentration of 5%, 62±9 megakaryocyte colonies formed (p=.025). At the 10% NKCM final concentration 84±15 colonies were detected. This value also differed significantly from control (p=0.009). Therefore, it is likely that NK mediated stimulation of megakaryocyte colony formation was mediated by the production of a soluable growth factor with the ability to stimulate megakaryocyte colony formation in vitro.

Endogenous Regulation-Role of Megakaryocyte Constituents

As discussed above, platelets, or platelet products, may play a role in regulating human megakaryocytopoiesis. However, nether the constitutents, nor the mechanism responsible for inducing inhibition of megakaryocyte development have been fully defined. In addition, it is also unknown if megakaryocytes themselves contribute to this putative autoregulatory process. We have attempted to address these important issues by co-culturing viable intact megakaryocytes, megakaryocyte lysates, or megakaryocyte conditioned medium with autologous light density marrow mononuclear cells (MNC) in plasma clot cultures supplemented with human serum prepared from platelet poor plasma. The effect of these preparations, and similarly prepared leucocyte control material, on subsequent megakaryocyte colony formation was then determined.

When intact megakaryocytes were co-cultured with a fixed number of target MNC (2x10^5/ml) at MNC:megakaryocyte ratios >100:1, megakaryocyte colony formation was unchanged when compared to colonies formed by MNC alone (p>.05). At MNC:megakaryocyte ratios <50:1, megakaryocyte colony formation was inhibited a mean (±SE) of 62±5% in five of seven separate experiments. To determine if megakaryocytes exerted a direct inhibitory

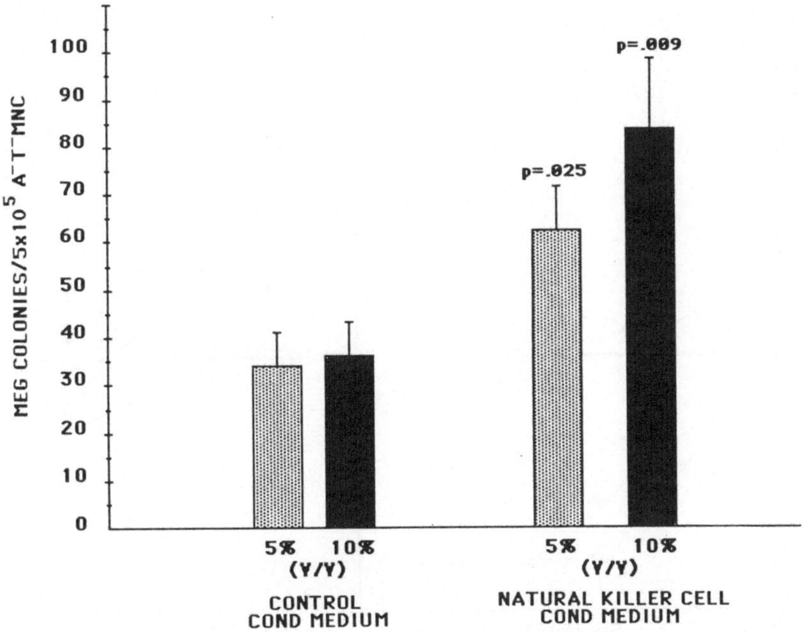

Fig. 3. Effect of NK cell conditioned medium, and control tissue culture medium on CFU-Meg cloning efficiency. Media were tested against four different marrow donor cells at final concentration (v/v) of 5% and 10%. Aggregate results of the four separate experiments are displayed. p values shown above the bars were generated with a two tailed Student's T test after comparing results obtained with NK vs. control medium at the concentrations indicated. [From reference #21-with permission].

effect on progenitor cell growth, MNC ($2x10^5$/ml) depleted (<2%) of adherent cells and T lymphocytes were cultured with megakaryocyte conditioned medium and megakaryocyte lysates ($\sim 1x10^5$ megakaryocytes/ml) at a final concentration of 10% (v/v). Colony formation in this experiment was as follows: baseline = 35±9; in 10% megakaryocyte CM= 17±8; in 10% megakaryocyte LYS= 7±4 ($p<.02$). Similar results were obtained in two further experiments (Figure 4). At equivalent concentrations, neither control conditioned medium, nor control cells lysate had any effect on megakaryocyte colony formation [Baseline colonies=35±9; in 10% control conditioned medium = 34±13; in 10% control cell lysate = 38±17]. Megakaryocyte colony inhibition was lineage specific since none of the megakaryocyte preparations tested inhibited either CFU-E or BFU-E derived colony formation. These results definitely suggested that megakaryocyte progenitor cell growth in vitro is susceptible to negative feedback inhibition by mature megakaryocytes. However the nature of the substance or substances responsible for exerting this inhibition were still undefined. We therefore investigated whether the platelet specific alpha granule protein PF4 could be at least partially responsible for this effect.

Platelet factor 4 (PF4) is an alpha granule protein which can modulate T lymphocyte function. Since we have previously demonstrated that, under the appropriate circumstances T cells may help regulate megakaryocytopoiesis[20], we hypothesized that T cell-PF4 interactions might play a role in autoregulating marrow megakaryocyte (MEG) production. To test this

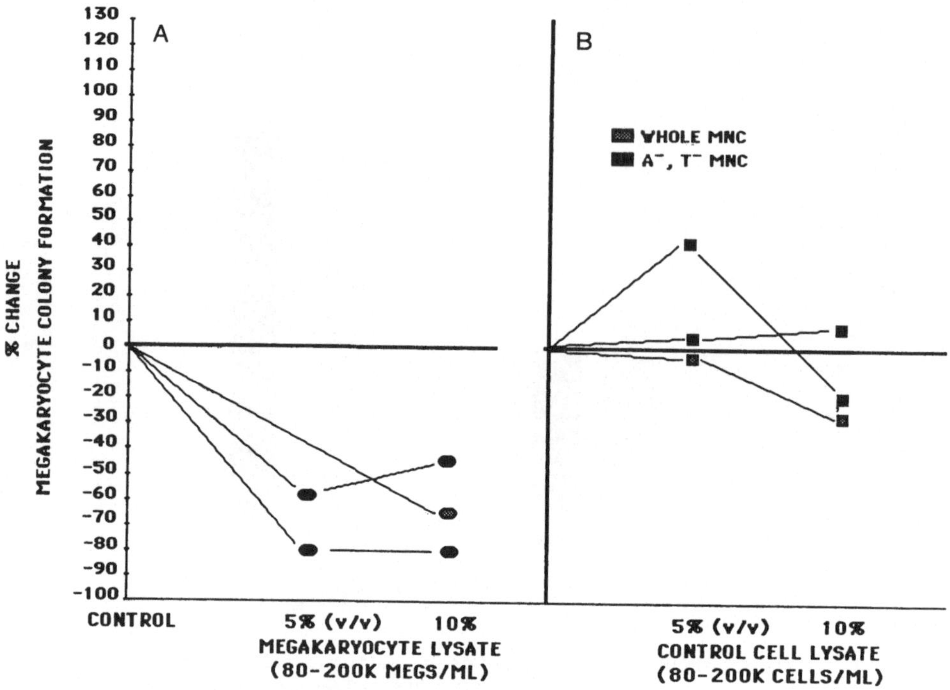

Fig. 4. Effect of megakaryocyte [Panel A], and control cell [Panel B] lysates on the formation of megakaryocyte colonies in plasma clot culture. Cultures were not supplemented with exogenous growth factors and contained only those nutrients present in normal human AB serum and the tissue culture medium. Results are expressed as % change in colony growth in comparison to control cultures containing no lysates (arbitrary zero point). Each line represents the result of a single experiment performed in duplicate or quadruplicate. Megakaryocyte lysate inhibited colony formation by ∿65% while control cell lysate had no significant effect on megakaryocyte colony formation in any of the three experiments performed.

idea, we studied MEG colony formation in plasma clot cultures containing human serum derived solely from platelet poor normal AB plasma, enriched hematopoietic progenitor cells (HPC), autologous T cells, and exogenous PF4. Highly purified PF4 (single band on SDS gel) was prepared as detailed in the methods section. HPC were prepared by depleting normal light density marrow mononuclear cells of adherent monocytes, and T cells which were further fractionated into helper (Leu 3^+) and suppressor (Leu 2^+) subtypes by solid phase immunoabsorption ("panning"). MEG colonies were enumerated by indirect immunofluorescence with an anti-human platelet glycoprotein antiserum. HPC ($5x10^5$/ml) were co-cultured with Leu 3^+, or Leu 2^+ T cells at target:T cell ratios of 2:1(n=3; n=4 respectively) and 1:1(n=4; n=4) in the presence of 2.5 μg/ml PF4. Under these growth conditions, MEG colony formation was unchanged (p>0.5) when compared to colonies formed by HPC alone. When the above experiments sere repeated (n=2-3/condition) at a higher PF4 concentration [25 μg/ml], MEG colony formation was markedly (>60%) inhibited.

To determine if PF4 directly inhibited MEG or erythroid progenitor cell growth (CFU-Meg;CFU-E) in vitro, HPC were cloned in PF4 [25 μg/ml] without added T cells. Mean ±SEM of MEG colonies formed (**without** vs with PF4) was as follows: Exp#1-5±1 vs 0±0 [p=.002]; Exp#2-6±0 vs 2±0 [p=.006]; Exp#3-116±16 vs 54±6 [p=.064]; Exp#4-33±6 vs 6±2 [p=.044). CFU-E growth was unaffected: Exp#1-1003±91 vs 1232±128 [p=.316]; Exp#2-1082±6 vs 1496±162 [p=.125]. These results suggested that PF4 may be a non-T cell dependent, lineage specific inhibitor of CFU-MEG, and that PF4 may play a role in autoregulating human megakaryocytopoiesis.

DISCUSSION

Megakaryocyte development is currently thought to be dependent on two humoral activities which are defined by their ability to promote CFU-Meg proliferation and precursor cell maturation. These activities have been designated megakaryocyte colony stimulatory factor (Meg-CSF)[2] and thrombopoietin (TPO) respectively[5,6]. TPO is thought to be elaborated after thrombocytopenia is detected by an unknown medhanism. It then acts on existing marrow megakaryocytes to stimulate maturation and platelet formation leading to an increase in the number of circulating platelets. If this response is sufficient to restore homeostasis, TPO levels fall. If the thrombocytopenic stress is severe enough however, and of sufficient duration, the number of mature megakaryocytes in the marrow is effectively diminished as those that matured released their platelets. An increase in Meg-CSF levels is therefore triggered. Increased Meg-CSF in turn stimulate proliferation of CFU-Meg thereby providing an increase in megakaryocyte precursor cells which can then complete their maturation routine and repopulate the marrow. Once the marrow is repopulated by mature megakaryocytes the stimulus for Meg-CSF production ends and homeostasis is restored.

This conceptualization has a number of problems however, and importantly fails to explain the development of megakaryocyte hyperplasia which develops in such disparate disorders as idiopathic autoimmune thrombocytopenic purpura (ATP), and reactive thrombocytosis (RT). In both ATP and RT increased numbers of marrow megakaryocytes are observed. Based on the preceding discussion one would expect for example that Meg-CSF levels would either be normal or low in such patients, a finding that has in fact been reported[19]. If Meg-CSF levels are not elevated in these patients some undefined factor(s) must be acting to maintain the megakaryocyte hyperplasia. Clearly then, the regulatory schema developed above is in need of further development.

One possible explanation for the above discrepancies is that the influence of other exogenous regulators, including those derived from ancillary bone marrow cells, on megakaryocytopoiesis has not been considered in previously proposed regulatory schema. Our study on the regulatory role of immunocompetant cells are therefore of interest. We have found that when monocytes, T lymphocytes, and NK cells are directly, and simultaneously compared for their ability to stimulate megakaryocytopoiesis in our culture system, it is only NK cells which are consistently active in this regard. The mechanism of NK cell mediated stimulation of hematopoietic progenitor cell proliferation is not completely defined but our data suggest that the CFU-Meg stimulatory effect may be mediated via the production of a soluble growth factor(s). How control over the elaboration of such putative growth factors is effected is unclear but may involve the Fc receptor since blockade with Leu 11b resulted in a loss of stimulatory activity (Figure 3).

Our studies suggest a number of interesting possibilities regarding the cellular regulation of human megakaryocytopoiesis, and perhaps hematopoiesis in general. The T cell and MO data presented above in conjunction with the data derived from the study of other hematopoietic lineage[29], and the clinical observation that megakaryocytopoiesis [hematopoiesis] is unimpeded in almost all immune deficiency syndromes[30] suggest that neither T cells nor MO play a significant role in regulating basal CFU-Meg [hematopoietic progenitor cell] proliferation. After receiving a specific "programming/activation" stimulus however, either T lymphocytes or monocytes might augment, or inhibit the growth of CFU-Meg as we have recently shown[20]. It has not yet been demonstrated that NK cells play a physiologically relevant role in the basal regulation of megakaryocytopoiesis, but our in vitro data suggest that such a role is possible. It is attractive to postulate the existence of an ancillary cell population with such capabilities since these cells could then perform a useful function under baseline conditions without the need for specific programming. Furthermore, if the NK cell compartment became expanded or specifically activated, an exaggeration of their stimulatory capabilities might be manifested as a bone marrow megakaryocyte hyperplasia. Accordingly, it will be interesting to determine if these cells can be implicated in the generation and maintenance of the megakaryocyte hyperplasia and reactive thrombocytosis that often accompanies neoplastic conditions, and chronic inflammatory states[19].

The regulatory schema described above also fails to consider the possibility that megakaryocytopoiesis is in some way autoregulated (Figure 5). Ebbe and Phalen first suggested this possibility based on the observation that hydroxyurea induced megakaryocytopenia, in the absence of thrombocytopenia, resulted in the formation of macrocytosis in the remaining marrow megakaryocytes[31]. This response, induced by an unknown mechanism, was postulated to be a compensatory mechanism for maintaining platelet

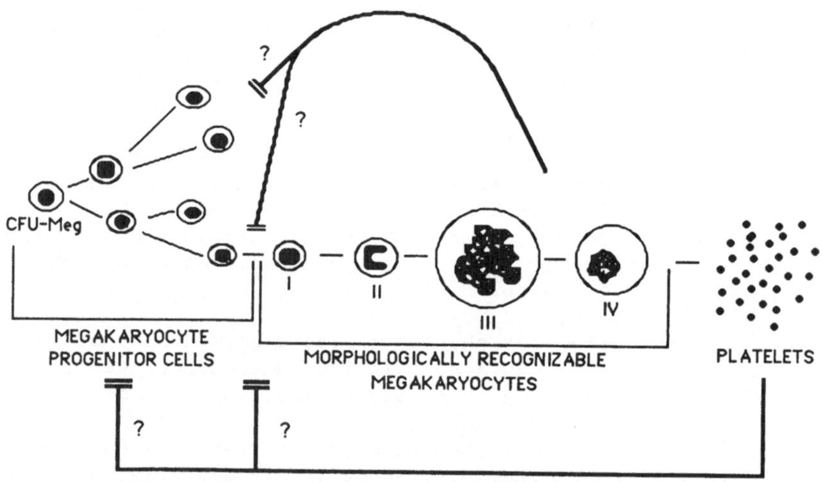

Fig. 5. Cartoon depicting proposed autoregulatory circuit for governing production of marrow megakaryocytes. In this schema, platelets and mature megakaryocytes are postulated to elaborate an inhibitory factor(s) which impedes megakaryocytopoiesis at one, or both, of two critical developmental levels. The first level is that occupied by proliferating progenitor cells (CFU-Meg). The second level is occupied by very early precursor cells which are just beginning to enter their terminal maturation pathway. Inhibition of either CFU-Meg proliferation, or megakaryocyte precursor cell maturation would decrease the numbers of mature megakaryocytes in the marrow, resulting in an effective negative feedback loop.

count. We have generated both clinical[32], and in vitro data (see above) that suggest that the megakaryocyte progenitor cell compartment may also be subject to autoregulation.

Our preliminary studies with PF4 suggest that we may have found one potential mediator of this phenomenon. We initially became interested in PF4 when Katz et al reported that platelet alpha granule releaseates had immunomodulatory activity[33,34]. This group found that such releaseates could augment the immune response of SJL mice to TNP-Ficoll and sheep red blood cells, and reverse the immunosuppression caused by transplantation of a syngenic lymphoma or allogeneic T lymphocytes. The responsible factor was subsequently identified as PF4. Since several groups have reported that $CD4^+$ "helper" T lymphocytes can augment megakaryocytopoiesis[35,36] we expect that the addition of PF4 to our cultures might result in an augmentation in numbers of colonies formed. Alternatively, since our own studies with T lymphocytes and T cell subsets suggest that non-activated cells have little impact on megakaryopoiesis[21] we believed that no change in colony formation might also be observed. Therefore, we were suprised when the addition of PF4 to our cultures caused suppression of colony formation. Further experiments suggested that the suppressive effect observed did not require the cooperation of T lymphocytes, and was likely due to a direct effect on cells of the megakaryocyte lineage. Since PF4 is synthesized by megakaryocytes, and binds to platelets[megakaryocytes] in a reversible and saturable manner ($Kd \sim 2.7 \times 10^{-8}M$)[37], it is a bona fide autocrine regulator[38]. These results contrast in an important way with the more general inhibition of hematopoiesis reported to be caused by TGF-β[15].

The regulation of human megakaryocytopoiesis is clearly more complex than has heretofore been postulated. In addition to the more classically accepted humoral regulators of this process, work from a number of latoratories[20,21,35,36] suggest that a variety of exogenous and endogenously derived regulators exert a significant effect on this process, at least in vitro. Whether, the phenomena described will ultimately be shown to have physiologic significance remains to be determined, but the rapid progress being made in this area suggest that an answer to this, and other questions pertaining to regulation of this compartment will soon be at hand.

ACKNOWLEDGEMENTS

Supported in part by DHHS grant RO1-CA-36896, from the National Cancer Institute, and a Grant-In-Aid from the Delaware Chapter of the American Heart Association. Dr. Gewirtz is the recipient of a Research Career Development Award (RO4-CA-01324). I gratefully acknowledge the contribution of my collaborators to these studies: Drs. Kenneth F. Mangan, Bruno Callabretta, Boguslaw Rucinski, and Stefan Niewiarowski, Depts. of Medicine, Pathology, Physiology and Thrombosis Research Center, Temple University School of Medicine, Philadelphia, PA. I also thank Michelle Keefer Sacchetti for excellent technical assistance.

REFERENCES

1. Gewirtz, A.M. 1986. Human megakaryocytopoiesis. Seminars in Hematology 23: 27-42.
2. Hoffman, R., H.H. Yank, E. Bruno, and J. Straneva. 1985. Purification and partial characterization of a megakaryocyte colony stimulation factor from human plasma. J. Clin. Invest. 75: 1174-1182.
3. Kaushansky, K., P. O'Hara, K. Berner, G.M. Segal, F.S. Hagen,

and J. W. Adamson. 1986. Genomic cloning, characterization, and multilineage growth-promoting activity of human granulocyte-macrophage colony-stimulating activity. Proc. Natl. Acad. Sci. USA 83: 3101-3105.
4. Leary, A.G., Y.C. Yang, S.C. Clark, J.C. Gasson, D.W., Golde, and M. Ogawa. 1987. Recombinant Gibbon Interleukin-3 supports formation of human multilineage colonies and blast cell colonies in culture: Comparison with recombinant human granulocyte-macrophage colony stimulating factor. Blood 70: 1343-1348.
5. McDonald, T.P., M. Cottrell, and R.D. Clift. 1985. Studies on the purification of thrombopoietin from kidney cell culture medium. J. Lab. Clin. Med. 106: 162-174.
6. Hill, R., and J. Levin. 1986. Partial Purification of Thrombopoietin using lectin chromatography. Exp. Hematol. 14: 752-759.
7. Greenberg, S.M., D.J. Kuter, and R.D. Rosenberg. 1987. In vitro stimulation of megakaryocyte maturation by megakaryocyte stimulatory factor. J. Biol. Chem. 262: 3269-3277.
8. Tayrien, G. and R.D. Rosenberg. 1987. Purification and properties of a megakaryocyte stimulatory factor present both in the serum--free conditioned medium of human embryonic kidney cells and in thrombocytopenic plasma. J. Biol. Chem. 262: 3262-3268.
9. Ishibashi, T., S.A. Kozial, and S.A. Burstein. 1987. Human recombinant erythropoietin promotes differentiation of murine megakaryocytes in vitro. J. Clin. Invest. 79: 286-289.
10. Nagasawa, T., T. Sakuri, H. Kashiwagi, and T. Abe. 1986. Cell-mediated amegakaryocytic thrombocytopenia associated with systemic lupus erythematosus. Blood 67: 479-483.
11. Gewirtz, A.M., M.C. Sacchitti, R. Bien, and W.E. Barry. 1986. Cell mediated supression of megakaryocytopoiesis in Acquired Amegakaryocytic Thrombocytopenic Purpura. Blood 68: 619-626.
12. Vainchenker, W., J. Chapman, J.F. Deschamps, G. Vinci, J. Bouguet, M. Titeux, and J. Breton-Gorius. 1982. Normal human serum contains a factor(s) capable of inhibiting megakaryocyte colony formation. Exp. Hematol. 10: 650-660.
13. Messner, H.A., N. Jamal, and C. Izaquirre. 1982. The growth of large megakaryocyte colonies from human bone marrow. J. Cell Physiol. [Suppl.] 1: 45-51.
14. Kimura, H. S. A. Burstein, D. Thorning, J.S. Powell, L.A. Harker, P.J. Failkow, and J. W. Adamson. 1984. Human megakaryocyte progenitors (CFU-M) assayed in methylcellulose: Physical characteristics and requirements for growth. J. Cell Physiol. 118-87-94.
15. Ishibashi, T., S.L. Miller, and S.A. Burstein. 1987. Type β transforming growth factor is a potent inhibitor of murine megakaryocytopoiesis in vitro. Blood 69- 1737-1741.
16. Dessypris, E.N., J.H. Gleaton, S.T. Sawyer, and O.L. Armstrong. 1987. Suppression of maturation of megakaryocyte colony forming unit in vitro by a platelet-released glycoprotein. J. Cell. Physiol. 130: 361-368.
17. Gewirtz, A., M. Keefer, K. Doshi, A. Annamali, H.C. Chiu, and R.W. Colman. 1986. Biology of Human Megakaryocyte Factor V. Blood 67: 1639-1648.
18. Gewirtz, A. 1987. Recent methodologic advances in the study of human megakaryocyte development and function. In Modern methods of pharmacology: Methods for studying platelet and megakaryocytes (R.W. Colman and B.J. Smith, eds.) pp 1-18, Alan R. Liss, Inc., New York.
19. Gewirtz, A.M., E. Bruno, J. Elwell, and R. Hoffman. 1983. In vitro studies of megakaryocytopoiesis in thrombocytotic disorders of man. Blood 61: 384-389.
20. Gewirtz, A.M., M.K. Sacchetti, R. Bien, and W. Barry. 1986. Cell mediated supression of megakaryocytopoiesis in acquired

amegakaryocytic thrombocytopenic purpura. Blood 68: 619-626.
21. Gewirtz, A.M., W.Y. Xu, and K.F. Mangan. 1987. Role of natural killer cells, in comparison with T lymphocytes and monocytes, in the regulation of normal human megakaryocytopoiesis in vitro. J. Immunol. 139: 2915-2925.
22. Mangan, K.F., M. Hartnett, S.A. Matis, A. Winkelstein, and T. Abo. 1984. Natural killer cells suppress human erythroid stem cell proliferation in vitro. Blood 63: 260.
23. Timonen, T., C. W. Reynolds, J.R. Ortaldo, and R.B. Herberman. 1982. Isolation of human and rat natural killer cells. J. Immunol. Methods. 51: 269.
24. Indiveri, F., B.S. Wilson, M.A. Pellegrino, and S. Ferrone. 1979. Detection of human histocompatibility (HLA) antigens with an indirect rosette microassay. J. Immunol. Methods. 29: 101.
25. Weiner, W.S., C. Bianco, V. Nussenzweig. 1973. Enhanced binding of neuraminidase treated sheep erythrocytes to human T lymphocytes. Blood 42: 939.
26. Rucinski, B., S. Niewwiarowski, P. James, D.A. Walz, and A. Budzynski. 1979. Antiheparin proteins secreted by human platelets. Purification, characterization, and radioimmunoassay. Blood 53: 47-62.
27. Neland, A.C., D. Catovsky, D. Linch, J.C. Cawley, P. Beverley, J.F. San Miguel, E.C. Gordon-Smith, T.E. Blecher, S. Shahriari, and S. Varadi. 1984. Chronic T cell lymphocytosis: a review of 21 cases. Br. J. Haematol. 58: 433.
28. Reynolds, C.W., and K.A. Foon. 1984. T-lymphoproliferative disease and related disorders in humans and experimental animals: A review of the clinical, cellular, and functional characteristics. Blood 64: 1146.
29. Levitt, L., T.J. Kipps, E.G. Engleman, and P.L. Greenberg. 1985. Human bone marrow and peripheral blood T lymphocyte depletion: Efficacy and effects of both T lymphocytes and monocytes on growth of hematopoietic progenitors. Blood 65: 663-679.
30. Lipton, J.M., and D. Nathan. 1985. Interaction between lymphocytes and macrophages in hematopoiesis. In Hematopoietic Stem Cells. D.W. Golde, and F. Takaku, eds. Marcel Dekker, New York, pp 145-202.
31. Ebbe, S., and E. Phalen. 1979. Does autoregulation of megakaryocytopoiesis occur? Blood Cells 5: 123-138.
32. Gewirtz, A.M., and R. Hoffman. 1986. Transitory hypomegakaryocytic thrombocytopenic purpura. Etiologic association with ethanol abuse and implication regarding regulation of human megakaryocytopoiesis. Br. J. Haematol. 62: 333-344.
33. Katz, I.R., M.K. Hoffmann, M.B. Zucker, M.K. Bell, and G.J. Thorbecke. 1985. A platelet derived immunoregulatory serum factor with T cell affinity. J. Immunol. 134: 3199-3203.
34. Katz, I.R., G.J. Thorbecke, M.K. Bell, Z. Yin, D. Clarke, and M.B. Zucker. 1986. Protease-induced immunoregulatory activity of platelet factor 4. Proc. Natl. Acad. Sci. USA. 83: 3491-3495.
35. Geissler, D., G. Konwalinka, C. Peschel, K. Grunewald, R. Odavic, and H. Braunsteiner. 1985. A regulatory role of activated T lymphocytes on human megakaryocytopoiesis. Br. J. Haematol. 60: 233-238.
36. Geissler, D., L. Lu, E. Bruno, H. Yank, H. Broxmeyer, and R. Hoffman. 1986. The influence of T lymphocyte sub-sets and humoral factors on colony formation by human bone marrow and blood megakaryocyte progenitor cells in vitro. J. Immunol. 137: 2508-2513.
37. Capitanio, A.M., S. Niewiarowski, B. Rucinski, G.P. Tuszynki, C.S. Cierniewski, D. Hershock, and E. Kornecki. 1985. Interaction

of platelet factor 4 with human platelets. Biochem Biophys Acta. 839: 161-173.
38. Sporn, M.B. and A.B. Roberts. 1985. Autocrine growth factors and cancer. Nature 313: 745-747.

DETECTION OF A HUMAN HEMATOPOIETIC PROGENITOR CELL CAPABLE OF FORMING BLAST CELL CONTAINING COLONIES *IN VITRO*

John Brandt, Li Lu, Edwin B. Walker, and Ronald Hoffman

Division of Hematology/Oncology and
Indiana Elks Cancer Research Center
Indiana University School of Medicine
541 Clinical Drive, Rm. 379
Indianapolis, IN 46223

ABSTRACT

A progenitor cell CFU-Bl (blast cell colony forming unit) present in human bone marrow and capable of producing blast cell containing colonies *in vitro* was detected using a serum containing semisolid culture system. The CFU-Bl has the capacity not only to undergo self-renewal, but also commitment to a number of hematopoietic lineages. This progenitor cell therefore has characteristics which suggest that it is identical to or closely related to the human pluripotent hematopoietic stem cell. Pretreatment of marrow cells with 5 fluorouracil facilitated detection of CFU-Bl derived colonies. The formation of CFU-Bl derived colonies was dependent upon the addition of media conditioned by the human bladder carcinoma cell line 5637. The ability of 5637 CM (conditioned media) to support blast cell colony formation was in part but not totally ablated by pretreatment of the CM with an IL-1α (interleukin-1) neutralizing antibody. This data suggests that IL-1α plays a role in the regulation of primitive events occuring during human hematopoiesis. IL-1α might be exerting these effects by either acting directly on the CFU-Bl, causing marrow accessory cells to elaborate other cytokines or by synergizing with cytokines already present in 5637 CM.

INTRODUCTION

Primitive hematopoietic progenitor cells capable *in vitro* of producing colonies composed of blast cells (CFU-Bl) have been detected in murine bone marrow[1-6]. These cells appear to possess many of the features of stem cells in that they are capable of self-renewal and also commitment to a number of hematopoietic lineage[1-6].

Human marrow is composed of heterogeneous populations of hematopoietic precursor and progenitor cells. A variety of *in vitro* clonal assay systems have been utilized to detect hematopoietic progenitor cells in the presence of various hematopoietic growth factors[7-9]. Each of these progenitor cells have been shown to produce colonies which are composed of differentiated cellular elements with limited self renewal capacity[10-12]. Since the frequency of various hematopoietic progenitor cells appears to be directly related to the degree of their cellular differentiation, one would hypothesize that the frequency in normal bone marrow of CFU-Bl is extremely low. In order to facilitate the enrichment of CFU-Bl so that these cells might be more carefully characterized, a number of physical and immunological characteristics of these cells could be utilized.[13-28]. Another means of obtaining enriched populations of CFU-Bl might be to pharmacologically purge the marrow of more differentiated elements by _in vitro_ exposure to chemotherapeutic agents[29-30]. Van Zant has previously shown that _in vitro_ exposure of murine marrow cells to 5-fluorouracil (5-FU) results in the direct, rapid kill of more differentiated progenitor cells and the selective sparing of primitive pluripotent hematopoietic stem cells[30]. 5-FU does not require _in vivo_ metabolism for this effect to occur. Utilizing these two approaches we have developed an assay for the human CFU-Bl.

MATERIALS AND METHODS

Cell Separation Techniques

Bone marrow aspirates were obtained from the posterior iliac crest of hematologically normal volunteers after informed consent was obtained according to guidelines established by the Human Investigation Committee of the Indiana University School of Medicine. The samples were immediately diluted 1:1 with Iscove's Modified Dulbecco's Medium (IMDM, Grand Island Biological Company, Grand Island, NY) containing 20 units/ml sodium heparin. This mixture was passed through a 150 uM screen and layered over an equal volume of Ficoll-Paque (specific gravity 1.077 g/cm^3, Pharmacia Fine Chemicals, Piscataway, NJ). Density centrifugation was performed using a Beckman model TJ-6R centrifuge (Beckman Instruments, Inc., Palo Alto, CA) at 500 x g for 25 min. at 4°C and the interface layer of low density mononuclear cells (LD) was collected, washed, and resuspended in PBS-EDTA (Phosphate-buffered saline pH 7.4 containing 5% fetal bovine serum, v/v 0.01% ETDA, w/v, and 1.0 g/L D-glucose). The cells were then injected into a Beckman Elutriator System with a standard separation chamber (Beckman Instruments, Inc., Palo Alto, CA) which had previously been sterilized with a 70% ethanol solution, washed with sterile distilled water and primed with PBS-EDTA. Rotor speed and temperature were maintained at 1950 rpm and 10°C throughout the elutriation. After loading, 200 ml of effluent was collected at a flow rate of 10 ml/min and 100 ml were collected at flow rates of 11, 12, 14, 16, 18, 20, 22, 24, 28, and 32 ml/min, after which the rotor was stopped and the final fraction flushed from the separation chamber[31].

Treatment With A Chemotherapeutic Agent

5-fluorouracil (5-FU) (Sigma Chemical Co., St. Louis, MO) was dissolved in IMDM. Cell numbers were enumerated using a Coulter model ZBI (Coulter Electronics, Hialeah, FL). The cells in individual fractions were incubated for 24-72 hours at a concentration of 1 x 10^6 cells/ml at 37°C in 100% humidified air with 5% CO_2 in IMDM containing 10% FBS (fetal bovine serum) and varying concentrations of the above drugs. After appropriate incubation periods, the cells were washed x 3 with IMDM and assayed for their ability to form hematopoietic colonies.

Preparation of 5637 CM

Human bladder carcinoma cell ine 5637 was grown to semi-confluence in Corning cell culture flasks (Corning Glass Works, Corning, NY) in RPMI 1640 with 1% FBS at 37°C with 5% CO_2 in 100% humidified air. The supernatant was discarded and new medium incubated over the cells for seven days. Supernatant was then collected, spun at 500 x g to remove cells, and stored at -20°C.

CFU-Bl Assay

1 x 10^5 drug-treated or control cells were suspended in 35 mm plastic tissue culture dishes containing a 1 ml mixture of IMDM, 1.1% methylcellulose, 30% FBS, 5 x 10^{-5}M 2-mercaptoethanol, 1 unit of Epo, and 10% 5637 CM. The culture dishes were incubated at 37°C in a 100% humidified atmosphere of 5% CO_2 in air and at 21 days scanned using an inverted light microscope in order to detect blast cell colonies.

Colonies composed of undifferentiated cellular elements were plucked from the methylcellulose under direct microscopic visualization using sterile 100 ul glass micropipets (Drummond Scientific Co., Broomall, PA) and suspended 200 ul IMDM in sterile 1.5 ml microcentrifuge tubes (Tekmar Co., Cincinnati, OH). 100 ul of this volume was suspended in a 350 ul mixture of IMDM with 1.1% methylcellulose, 30% FBS, 5 x 10^{-5}M 2-mercaptoethanol, 0.35 unit Epo, and 10% 5637 CM or PHA-LCM in 24 well tissue culture plates (Corning Glass Works, Corning, NY). The remaining 100 ul was transferred to slides on a Cytospin 2 (Shandon Southern Instruments, Inc., Sewickly, PA) at 750 rpm for 5 min for Wright Giemsa staining and microscopic examination. Culture plates were incubated at 37°C in 100% humidified air with 5% CO_2 for 14 days, after which colonies were plucked from the methylcellulose, resuspended in 100 ul PBS, and transferred to slides for staining and examination.

Thymocyte Comitogenic Assay

The details of this standard assay have been recently described at length[32]. Briefly, 1 x 10^6 thymocytes from four to six week old C3H/Hej female mice are placed in 200 ul (final volume) of supplemented RPMI 1640 medium (2mM L-glutamine, penicillin/stretomycin, 10^{-6}M 2-mercaptoethanol and 0.1 mg/ml sodium pyruvate, 5% fetal calf serum (Hyclone; Logan, UT) plus 5637 conditioned medium of different concentrations, and cultured in 96 well plates at 37°C, 5% CO_2 for 48 hrs. in the presence of 0.5 ug/ml Con A. After 48 hrs., one uCi of H-thymidine is added to each culture well and the cells incubated for an additional 20 hrs. Cells from each culture well are collected on a Brandel automated cell harvester onto glass fiber filters and counted by means of a liquid scintillation counter. Recombinant IL-1α (Genzyme, Boston MA) (500 units/ml) as added to the assay system as the positive control. Neutralization studies were carried out by titrating a purified rabbit antihuman IL-1α IgG preparation (Genzyme, Boston MA) or an affinity purified rabbit immunoglobulin for their ability to neutralize a known amount of purified IL-1α or the IL-1 activity present in 5637 CM.

Statistical Analysis

All results are presented as the mean ± 1 S.D. Levels of significance for comparison between samples were determined using the Student's t test distribution.

RESULTS

In order to obtain populations of marrow cells that might be enriched for primitive hematopoietic progenitors, fractions of LD marrow cells were obtained by counterflow centrifugal elutriation. The largest numbers of CFU-GEMM derived colonies were localized to fractions that were elutriated at a velocity of 12-14 ml/min. Pooling of these two fractions resulted in a 5 fold enrichment for CFU-GEMM derived colonies enumerated on day 14 and a 10 fold enrichment of CFU-GEMM derived colonies enumerated after 21 days of incubation when compared to colony formation by LD cells. In preliminary studies, various doses and times of exposure to 5-FU were evaluated in order to obtain maximal kill of more differentiated progenitor cells. It was hoped that this might facilitate the detection of more quiescent primitive hematopoietic stem cells. Based on these experiments, however, we arrived at the conclusion that a prolonged exposure (24-36 hours) to low levels of 5-FU would be optimal for our purposes. Such treatment with 5-FU resulted in almost total kill of all hematopoietic progenitor cells assayed from low density cells. Yet, when FR 12-14 cells were exposed for 24 hours to 25 ug of 5-FU, small numbers of CFU-GEMM derived colonies that were enumerated after 21 days of incubation survived. We assumed that these surviving CFU-GEMM represented a more primitive population of CFU-GEMM than normally detected on day 14.

The conditions described above were then employed to develop an assay for the human CFU-Bl. LD marrows were fractionated by centrifugal elutriation and FR 12-14 cells were exposed to 25 ug/ml of 5-FU for 24 to 36 hours. After extensive washing, these cells were then plated in methylcellulose cultures in the presence of 10% 5637 CM. Between day 21 to day 25, colonies of round, agranular, nonpigmented cells were observed. These colonies were composed of cells that did not have any evidence of myeloid or erythroid maturation. The cells in those colonies were predominantly non-granular blast cells with basophilic cytoplasm and multiple nucleoli. Between 28% and 90% of the cells comprising these colonies were actually blasts that did not demonstrate either specific or nonspecific esterase activity. None of these CFU-Bl derived colonies were composed exclusively of blasts. There were also varying numbers of monocytes, promyelocytes and segmented neutrophils that frequently represented minority cell populations in these colonies. CFU-Bl derived colony formation was dependent on the presence of 5637 CM. No CFU-Bl derived colonies were detected when 5-FU treated FR 12-14 cells were plated in the absence of CM. Since IL-1α is known to be present in 5637 CM we attempted to define whether this cytokine was the constituent in 5637 CM which promoted CFU-Bl derived colony formation. For this purpose we utilized a rabbit anti-human IL-1α antibody to neutralize IL-1α activity which was quantitated using a thymocyte comitogenic assay. In Figure 1, one can see that a known amount of IL-1α could be neutralized by a 1:100 dilution of the rabbit anti-human IL-1α IgG but not by an incidential rabbit IgG. Using this same assay system a significant amount of IL-1α activity was detected in 5637 CM which was neutralizable by the anti IL-1α IgG. In Table 1 one can observe that the ability of 5637 CM to promote blast cell colony formation was diminished in part but not totally by pretreatment with a 1:100 dilution of the anti IL-1α antibody.

DISCUSSION

An _in vitro_ assay system with which to examine the biological characteristics of human pluripotent hematopoietic stem cells that have a capacity for self renewal and to differentiate along multiple cellular lines has been a long term goal of a number of groups working in the field of hematopoiesis. Recently, a great deal of progress has been achieved toward accomplishing this objective. Initially, Nakahata and Ogawa described such an assay system using human umbilical cord blood as a source of cells[3]. More recently, Rowley and coworkers[4], Leary and Ogawa[5], and Gordon and coworkers[6] have each described assay systems with which to detect these primitive hematopoietic progenitor cells in human marrow.

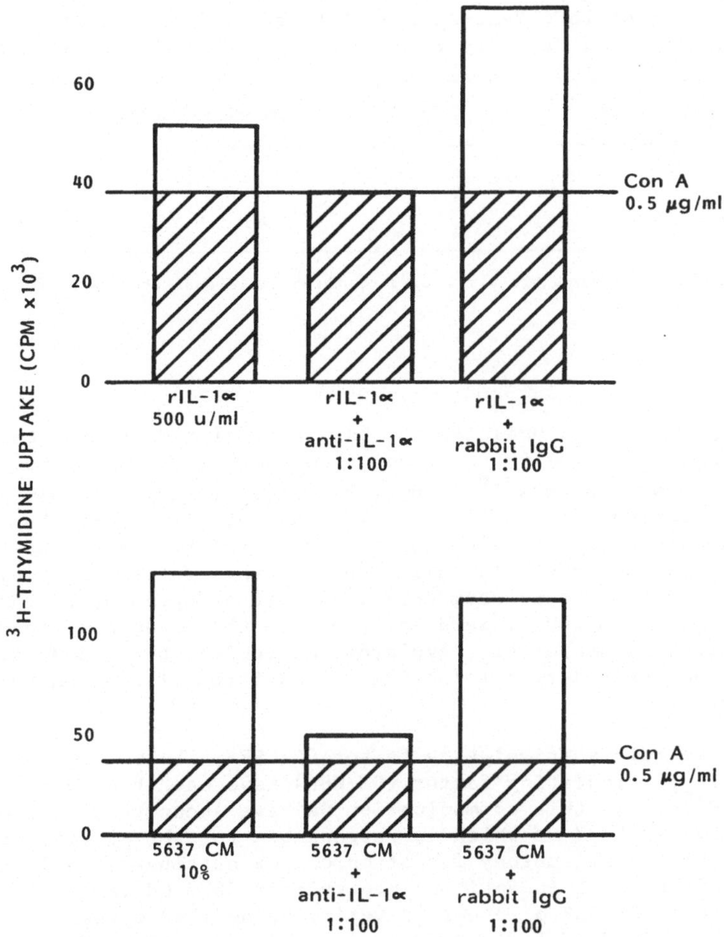

Fig. 1. Abrogation of IL-1 activity by anti-IL-1α on C_3H/Hej thymocyte proliferation. Effect of interleukin-1 and 5637 CM on thymocyte proliferation in the presence of Con A. The CPM values represent the mean of 3H-thymidine incorporation performed in triplicate on two occasions following a twenty hour pulse with the ^{3}H-thymidine added to 1.0 uCi per culture well. Total culture time was 72 hours. Thymocytes from C_3H/Hej mice were used throughout these experiments.

A number of groups have demonstrated that a population of primitive hematopoietic stem cells are capable of surviving either in vitro or in vivo exposure to chemotherapeutic agents such as 5_FU, 4-HC and hydroxyurea[29,30,33,34]. This chemoselection of primitive hematopoietic cells was successfully exploited in the present studies in order to eliminate overgrowth of colonies derived from more differentiated progenitor cells. The inability to detect CFU-Bl derived colonies from non drug treated LD cells or non drug treated FR 12-14 should not be interpreted as a reflection of the absence of these cells from these fractions, but rather to the difficulty of isolating CFU-Bl derived colonies amongst other hematopoietic colonies present in the culture dish.

Table I. Effect of IL-1 Neutralizing Antibody on the Ability of 5637 Conditioned Media to Promote Blast Cell Colony Formation

Treatment of 5637 CM	CFU-Bl/10^5 Cells
None	$2.5 \pm 0.9^{+}$
Anti-IL-1 Antibody	1.0 ± 0.5

$^{+}$ Each point represents the $\bar{x} \pm$ S.E.M. of three individual studies performed in quadruplicate.

$p < 0.05$.

A large amount of information has been accumulated to show that the process of blood formation is sustained throughout life by a number of hematopoietic growth factors[7-9]. These hormones are collectively known as colony-stimulating factors (CSFs). The availability of the human CFU-Bl assay has allowed us to determine the dependence of the CFU-Bl on the addition of exogenous growth factors. Our findings suggest that even this primitive hematopoietic stem cell is in vitro dependent on the addition of exogenous CSFs. No CFU-Bl were assayable in the absence of conditioned media or purified hematopoietic regulatory molecules, but CFU-Bl derived colony formation was observed following the addition of 5637 CM to the culture mixture.

Granulocyte colony stimulating factor (G-CSF), IL-1α, and granulocyte-macrophage colony stimulating factor (GM-CSF) have each been purified from 5637 CM[35-39]. In this communication we have shown that part of the activity present in 5637 CM which is responsible for the ability of this CM to support blast cell colony formation can be attributed to IL-1α. The residual CSA present in antibody neutralized 5637 CM is likely due to the presence of GM-CSF or other as yet to be defined cytokines. The IL-1α activity in 5637 might be promoting blast cell colony formation by synergizing with G-CSF or GM-CSF present in 5637 CM[40], by promoting marrow accessory cell populations such as fibroblasts, endothelial cells or macrophages to elaborate G- and GM-CSF[41-42], or by acting directly on the CFU-Bl. Further studies designed to define the mechanism by which IL-1α might affect CFU-Bl are currently being performed in our laboratory.

REFERENCES

1. Nakahata T and M. Ogawa. 1982. Identification in culture of a new class of hemopoietic colony forming units with extreme capability to self-renew and generate multipotential colonies. Proc. Natl. Acad. Sci. USA 79: 3843-3847.
2. Keller, G.M. and R.A. Phillips. 1982. Detection in vitro of a unique, multipotent hemopoietic progenitor. J. Cell. Physiol. Supp. 1: 31-36.
3. Nakahata, T. and M. Ogawa. 1982. Hemopoietic colony-forming cells in umbilical cord blood with extensive capability to generate mono and multipotential hemopoietic progenitors. J. Clin. Invest. 70: 1324-1328.
4. Rowley, S.D., S.J. Sharkis, C. Hattenburg, and L.L. Sensenbrenner. 1987. Culture from human bone marrow of blast progenitor cells with an extensive proliferative capacity. Blood 69: 804-808.
5. Leary, A.G. and M. Ogawa. 1987. Blast cell colony assay for umbilical cord blood and adult bone marrow progenitors. Blood 69: 953-956.

6. Gordon, M.Y., C.R. Dowding, G.P. Riley and M.F. Greaves. 1987. Characterization of stroma-dependent blast colony-forming cells in human marrow. J. Cell Physiol. 130: 150-156.
7. Sieff, C.A. 1987. Hematopoietic growth factors. J. Clin. Invest. 79: 1549-1557.
8. Metcalf, D. 1985. The granulocyte-macrophage colony stimulating factors. Science 229: 16-22.
9. Clark, S.C. and R. Kamen. 1987. The human hematopoietic colony-stimulating factors. Science 236: 1229-1237.
10. Humphries, R.K., A.C. Eaves and C.J. Eaves. 1981. Self-renewal of hemopoietic stem cells during mixed colony formation in vitro. Proc. Natl. Acad. Sci. USA 78: 3629-3633.
11. Johnson, G.R. and D. Metcalf. 1977. Pure and mixed erythroid colony formation in vitro stimulated by spleen conditioned medium with no detectable erythropoietin. Proc. Nat. Acad. Sci. USA 74: 3879-3882.
12. Ash, R.C., R.A. Detrick and E.D. Zanjani. 1981. Studies of human pluripotential hemopoietic stem cells. Blood 58: 309-316.
13. Morstyn, G., N.A. Nicola and D. Metcalf. 1980. Purification of hematopoietic progenitor cells from human marrow using a fucose-binding lectin and cell sorting. Blood 56: 798-806.
14. Civin, C.I., L.C. Strauss, C. Brovall, M.J. Fackler, J.F. Schwartz and J.H. Shaper. 1984. Antigenic analysis of hematopoiesis III. A hematopoietic progenitor cell surface antigen defined by a monoclonal antibody raised against KG-Ia cells. J. Immunol. 133: 157-165.
15. Civin, C.I., M.L. Banguenigo, L.S. Strauss and M.R. Loken. 1987. Antigenic analysis of Hematopoiesis. VI. Flow cytometric characterization of My10-positive progenitor cells in normal human bone marrow. Exp. Hematol. 15: 10-17.
16. Bodger, M.P., C.A. Izaguirre, H.A. Blacklock, and A.V. Hoffbrand. 1983. Surface antigenic determinants on human pluripotent and unipotent hematopoietic progenitor cells. Blood 61: 1006-1010.
17. Katz, F.E., R. Tindle, D.R. Sutherland and M.F. Greaves. 1985. Identification of a membrane glycoprotein associated with haemopoietic progenitor cells. Leukemia Res. 9: 191-198.
18. Andrew, R.G., J.W. Singer and I.D. Bernstein. 1986. Monoclonal antibody 12-8 recognizes a 115 KD molecule present on both unipotent and multipotent hematopoietic colony forming cells and their precursors. Blood 67: 842-845.
19. Griffin, J.D., K.D. Sabbath, F. Herrmann, P. Larcom, K. Nichols, M. Kornachi, H. Levine and S.A. Cannistra. Differential expression of HLA-DR antigens in subsets of human CFU-GM. Blood 66: 788-795.
20. Nicola, N.A., D. Metcalf, H.V. Melchner and A.W. Burgess. 1984. Isolation of murine fetal hemopoietic progenitor cells and selective fractionation of various erythroid precursors. Blood 68: 376-384.
21. Williams, D.E., J.E. Straneva, R.N. Shen and H.E. Broxmeyer. 1987. Purification of murine bone marrow derived granulocyte-macrophage colony forming cells. Exp. Hematol. 15: 243-250.
22. Visser, J.W.M., J.G.J. Bauman, A.H. Mulder, J.E. Eliason, A.M. DeLeluw. 1984. Isolation of murine pluripotent hematopoietic stem cells. J. Exp. Med. 59: 1576-1590.
23. Emerson, S.G., C.A. Sieff, E.A. Wang, G.G. Wong, S.C. Clark and D.G. Nathan. 1985. Purification of fetal hematopoietic progenitors and demonstration of recombinant multipotential colony stimulating activity. J. Clin. Invest. 76: 1286-1290.
24. Griffin, J.D., R.P. Beveridge and S.F. Schlossman. 1982. Isolation of myeloid progenitor cells from peripheral blood of chronic myelogenous leukemia patients. Blood 60: 30-37.

25. Ferrero, D., H.E. Broxmeyer, G.L. Paglidari, S. Ventura, S. Lange, S. Pessano and G. Rovera. 1983. Antigenically distinct subpopulations of myeloid progenitor cells (CGU-GM) in human periopheral blood and marrow. Proc. Natl. Acad. Sci. USA 80: 4114-4118.
26. Winchester, R.J., G.D. Ross, C.I. Jarowski, C.Y. Wang, J. Halper, and H.E. Broxmeyer. 1977. Expression of Ia-like antigen molecules on human granulocytes during early phases of differentiation. Proc. Natl. Acad. Sci. USA 74: 4012-4019.
27. Janossy G., G.E. Francis, D. Capellaro, A.H. Goldstone, and M.F. Greaves. 1978. Cell sorter analysis of leukemia-associated antigens on human myeloid precursors. Nature (Lond) 276: 176--178.
28. Lu, L., D. Walker, H.E. Broxmeyer, R. Hoffman, W. Hu, and E. Walker. 1987. Characterization of adult human hematopoietic progenitors highly enriched by two-color sorting with My10 and major histocompatibility class II monoclonal antibodies. J. Immunol. In Press.
29. Hodgson, G.S., and T.R. Bradley. 1979. Properties of hematopoietic stem cells surviving 5-fluorouracil treatment: evidence for a pre-CFU-S cell? Nature (Lond) 281: 381-382.
30. VanZant, G. 1984. Studies of hematopoietic stem cells spared by 5-fluorouracil. J. Exp. Med. 159: 679-690.
31. Berkow, R.L., J.E. Straneva, E. Bruno, G.S. Beyer, J.S. Burgess, and R. Hoffman. 1984. The isolation of human megakaryocytes by density centrifugation and counterflow centrifugal elutriation. J. Lab. Clin. Med. 163: 811-818.
32. Walker, E.D., Leemhuis, T. 1987. The same cloned murine B lymphoma cell lines can be selectively induced to release either interleukin-1 or interleukin-2-like factor activity. Lymphokine Res. 6: 71-81.
33. Korbling, M., A.D. Hess, P.J. Tutschka, H. Kaizer, O.M. Colvin, and G.W. Santos. 1982. 4-hydroperoxycyclophosphamide: a mode for eliminating residual tumor cells and T lymphocytes from the bone marrow graft. Br. J. Haematol. 52: 89-96.
34. Rowley, S.D., O.M. Colvin, and R.K. Stuart. 1985. Human multilineage progenitor cell sensitivity to 4-hydroperoxycyclophosphamide. Exp. Hematol. 13: 295-298.
35. Welte, K., E. Platzer, L. Lu, J. Gabrilove, E. Levi, R. Mertelsmann, and M.A.S. Moore. 1985. Purification and biochemical characterization of human pluripotent hematopoietic colony stimulating factor. Proc. Natl. Acad. Sci. USA 82: 1526-1530.
36. Platzer, E., K. Welte, J. Gabrilove, L. Lu, P.M. Harris, R. Mertelsmann, and M.A.S. Moore. 1985. Biological activity of a human pluripotent hematopoietic colony stimulating factor on normal and leukemia cells. J. Exp. Med. 162: 1788-1801.
37. Gabrilove, J., K. Welke, P. Harris, E. Platzer, L. Lu, E. Levi, R. Mertelsmann, and M.A.S. Moore. 1986. Pluripoietin: A second human hematopoietic colony-stimulating factor produced by the human bladder carcinoma cell line 5637. Proc. Natl. Acad. Sci. USA 83: 2478-2482.
38. Jubinsky, P.T. and E.R. Stanley. 1985. Purification of hemopoietin-1 a multilineage hematopoietic growth factor. Proc. Natl. Acad. Sci. USA 82: 2764-2768.
39. Mochizuki, D.Y., J.R. Eisenman, P.J. Conlon, A.P. Larsen, and R.J. Tushinski. 1987. Interleukin 1 regulates hematopoietic activity, a role previously ascribed to hemopoietin 1. Proc. Natl. Acad. Sci. USA 84: 5267-5271.
40. Moore, M.A.S., and D.J. Warren. 1987. Synergy of interleukin 1 and granulocyte colony stimulating factor: In vitro stimulation of stem cell recovery and hematopoietic regeneration following 5-fluoruracil treatment of mice. Proc. Natl. Acad. Sci. USA 84: 7134-7138.

41. Bagby G.C., C.A. Dinarello, P. Wallace, C. Wagner, S. Hefeneider, and E. McCall. 1986. Interleukin 1 stimulates granulocyte-macrophage colony-stimulating activity release by vascular endothelial cells. J. Clin. Invest. 78: 1316-1323.
42. Fibbe, W.E., J. VanDamme, A. Billiau, P.J. Voogt, N. Duinkerken, P.M.C. Kluck and J.F. Falkenburg. 1986. Interleukin-1 (22-K factor) induces release of granulocyte-macrophage colony stimulating activity from human mononuclear phagocytes. Blood 68: 1316-1321.

STUDIES OF MURINE MEGAKARYOCYTE

COLONY SIZE AND PLOIDIZATION

Gerald M. Segal

Division of Hematology and Medical Oncology
Department of Medicine
Oregon Health Science University
Portland, Oregon 97201

INTRODUCTION

Megakaryocytes are the product of a series of proliferative and differentiative steps involving a complex cellular hierarchy. Pluripotent stem cells undergo a progressive narrowing in their differentiative potential eventually giving rise to unipotent progenitor cells committed to the megakaryocytic lineage. These cells, which are operationally termed colony-forming units - megakaryocyte (CFU-Meg), undergo varying numbers of mitotic divisions, giving rise to a class of cells often referred to as promegakaryoblasts. These diploid cells cease cell division but, instead, undergo a series of synchronous nuclear endoreduplications or endomitoses, giving rise to the unique polyploid cells megakaryocytes. Megakaryocyte colony formation represents an *in vitro* model of these processes in which colony size reflects the mitotic activity of megakaryocytic progenitor cells while the ploidy of colony cells is determined by the endomitotic activity of promegakaryoblasts and their descendants. Therefore, it might be expected that the analysis of megakaryocytic colony size and the ploidy of colony cells could provide insight into the cellular responses which underlie megakaryocytopoietic regulation *in vivo*.

In performing these studies, we sought to address two major issues. First, can subsets of megakaryocyte colony-forming cells with distinct *in vitro* growth characteristics be recognized? This has been an area of some controversy[1-4]. Second, we were interested in examining the effects of growth factors on the activities of the mitotic and endomitotic pathways which are reflected, as we have seen, by megakaryocyte colony size and the ploidy values of colony cells.

EXPERIMENTAL DESIGN

In planning our studies, we adapted an experimental design described first by Paulus *et al*[3] and later modified by Chatelain and Burstein[4]. In brief, marrow cells from 6-8 week-old male C57BL/6 mice were cultured in agar[5] at a concentration of 3×10^4 cells per ml. This relatively low cell concentration was chosen to minimize the overlapping of colonies, a problem which would interfere with our ability to analyze all the megakaryocyte colonies which had developed in culture. After five days incubation, the entire agar cultures were carefully removed, applied to glass slides, washed and then fixed with 2% paraformaldehyde. After further washing, the slides were stained for two hours with the Karnofsky-Roots reagent for demonstration of acetylcholinesterase (AchE) activity, a relatively specific marker of the megakaryocytic lineage in the mouse[6]. In preliminary experiments, we found that a two-hour staining time was sufficient to identify all the megakaryocyte colonies. Longer staining times resulted in heavier staining of the colony cells but did not reveal additional colonies. The slides were then stained with propidium iodide (25 mcg/ml) and bovine pancreatic ribonuclease-A (2 mg/ml) in PBS for 30 minutes. Excess stain was then blotted off, glass cover slips were applied, and the edges sealed.

Each slide was scanned at 100x magnification under transmission light in order to identify megakaryocyte colonies, which we defined as consisting of two or more AchE-positive cells. The ploidy of colony cells was then measured by quantitative cytofluorometry[4]. Under phase contrast at 400x magnification, individual colony cells were identified and positioned within the measuring aperture of a Leitz Orthoplan microscope equipped with an MPV compact photometer (Leitz, Wetzlar, West Germany). Cytofluorometry was performed with a filter combination optimal for propidium iodide excitation. Using an automatic shutter sequencing device, the fluorescent emission of the cell during a 250 millisecond exposure was measured by the photometer and the digitalized readout recorded. A background reading for each cell was obtained by positioning the measuring aperture over an area directly adjacent to the cell. For each slide, the fluorescent emissions of 20 granulocytes were determined and the mean value taken as the diploid or 2N value. The fluorescent emission minus background of each cell is directly proportional to its DNA content. The ploidy of each megakaryocyte was simply calculated from the following formula:

$$\text{Ploidy} = \frac{\text{Fluorescent emission of megakaryocyte}}{\text{Mean fluorescent emission of granulocyte}} \times 2.$$

RESULTS AND DISCUSSION

In our first set of experiments, we examined megakaryocyte colony formation stimulated by pokeweed mitogen-stimulated spleen cell-conditioned medium (PWM-SCM)[5]. In the presence of an optimally stimulatory concentration of PWM-SCM, 3×10^4 marrow cells gave rise to an average of 11 megakaryocyte colonies and 40 granulocyte-macrophage colonies. We performed ploidy determinations on a total of 1098 megakaryocytes which comprised 146 megakaryocyte colonies. Clear peaks were observed at ploidy values corresponding to 2N, 4N, 8N, 16N, 32N, 64N, and 128N, accounting for 5%, 11%, 21%, 33%, 23%, 6%, and 1% of colony cells, respectively. The geometric mean ploidy of all the megakaryocytes analyzed was 14N. The mean colony size was 7.5 cells with a range of 2-53 cells.

We next analyzed the data to try to discern relationships which described the distribution of colony sizes and the ploidy of colony megakaryocytes. The distribution of colony size was best described by an inverse exponential relationship (Figure 1). A similar inverse exponential relationship was originally reported by Paulus <u>et al</u>[3]. This result implies that, regardless of how many divisions a diploid precursor cell has undergone, the probability that its most recent mitotic division will be its last is constant. In other words, this relationship is consistent with a model in which the process of switching from mitosis to endomitosis is random.

The relationship between colony size and ploidization is shown in Figure 2. There was a highly significant inverse relationship ($p < 10^{-5}$) between the logarithms of the size of a colony and its geometric mean ploidy. Similar observations linking colony size and ploidy have been reported previously by Chatelain and Burstein[4]. Our data provides no evidence for the existence of different classes of megakaryocyte colony-forming cells, at least as defined by the growth characteristics of day

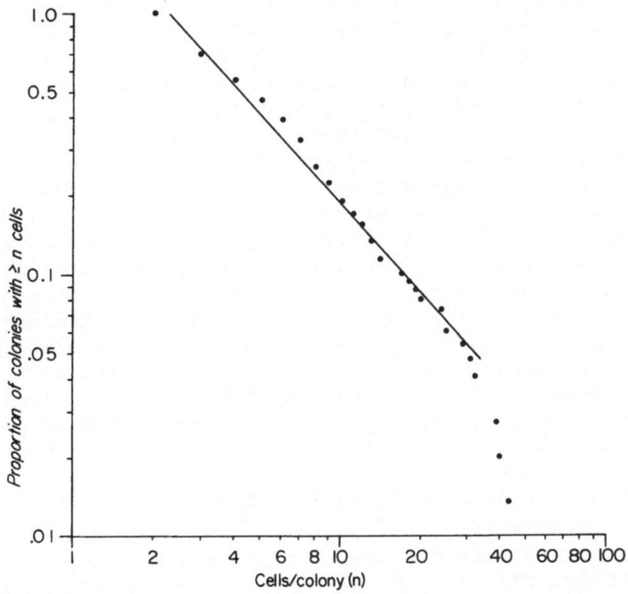

Fig. 1. The distribution of megakaryocyte colony size.

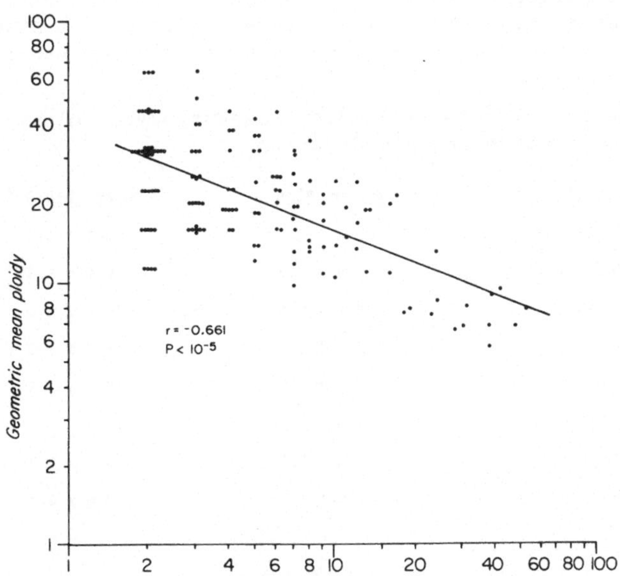

Fig. 2. The relationship between megakaryocyte colony size and ploidy. Each data point represents an individual colony.

5 murine megakaryocyte colonies. The size and ploidy characteristics of these colonies appear to represent a continuum.

PWM-SCM is a crude supernatant containing a variety of factors which might modulate megakaryocyte colony growth in a highly complex fashion. We, therefore, next examined the effects of a defined growth factor interleukin-3 (IL-3), in concentrations between 0.5 u/ml and 80 u/ml, on the size and ploidization of megakaryocyte colonies. The IL-3 used in these studies was provided by Dr. James Ihle and had been purified to homogeneity from media conditioned by WEHI-3B cells[7]. At concentrations between 0.5 u/ml and 10 u/ml, IL-3 stimulated a dose-dependent increase in megakaryocyte colony formation and in granulocyte-macrophage colony formation (not shown). Likewise, IL-3, in concentrations up to 10 u/ml, stimulated a dose-dependent increase in megakaryocyte colony size (Table I). IL-3 concentrations above 10 u/ml did not stimulate further increases in megakaryocyte and granulocyt-macrophage colony numbers or megakaryocyte colony size. Megakaryocyte colony size at each concentration of IL-3 was described by an inverse exponential relationship similar to that observed with PWM-SCM (not shown). There were no significant differences in the geometric mean ploidy of colony megakaryocytes (Table I) or in ploidy distributions (not shown) at any of the IL-3 concentrations tested. The modal ploidy value was 16N at all concentrations of IL-3 except at 5 u/ml and 80 u/ml, where 8N cells were slightly more frequent that 16N cells. Finally, at each concentration of IL-3, as with PWM-SCM, we observed significant inverse correlations between colony size and geometric mean ploidy (not shown).

That we did not observe any significant concentration-dependent effect of IL-3 on megakaryocyte ploidization seems in conflict with a report by Ishibashi and Burstein that IL-3 promotes increases in the size and, presumably, ploidy of single mouse megakaryocytes cultured under serum-free conditions[8]. There are several features of our assay system which might obscure any effects of IL-3 on megakaryocyte ploidization. Two obvious possibilities are serum effects and accessory cell interactions. Our cultures were performed in horse serum and, clearly, certain serum factors can influence megakaryocytic differentiation. For example, transforming growth factor-beta, a platelet alpha-granule-derived protein

Table I. Effects of IL-3 on Megakaryocyte Colony Size and Ploidization

IL-3 (u/ml)	Colony Size[1]	Colony Ploidy[2]
0.5 (21)[3]	3.2 ± 0.4	12.3 ± 2.1
1 (43)	4.4 ± 0.9	10.2 ± 2.3
2.5 (48)	4.9 ± 0.6	11.2 ± 2.0
5 (72)	5.1 ± 0.7	9.9 ± 2.0
10 (34)	7.6 ± 1.5	10.0 ± 2.1
20 (30)	7.6 ± 1.2	10.3 ± 2.3
40 (51)	7.2 ± 1.2	9.4 ± 2.2
80 (42)	7.5 ± 1.3	9.6 ± 2.3

[1]Cells/colony (mean ± SEM)

[2]Geometric mean ± SD

[3]Number of colonies analyzed

present in high concentration in serum, has been reported to inhibit IL-3-induced increases in megakaryocyte size[9]. Erythropoietin, on the other hand, can promote megakaryocytic differentiation[10]. Other serum factors capable of influencing megakaryocyte development <u>in vitro</u> might exist. Serum factors with sufficiently potent stimulatory or inhibitory effects on megakaryocytic differentiation might overwhelm or neutralize an IL-3-mediated effect. The same might be true with respect to accessory cell interactions. Although our cultures were performed at a relatively low cell concentration, accessory cell interactions with megakaryocytes and their precursors might well influence proliferation and differentiation events. For example, Williams <u>et al</u> have shown that a population of adherent marrow cells can elaborate an activity which promotes terminal megakaryocytic differentiation[11]. Such activity, termed "megakaryocyte potentiator," might overwhelm and obscure an IL-3-associated differentiation signal.

To try to overcome the potential problem of serum factors, a serum-free megakaryocyte colony growth assay, now under development in our laboratory, will be required. In order to address the accessory cell problem, we are using marrow cell targets depleted of various classes of accessory cells. As a first approach, we examined megakaryocyte colony growth from mouse marrow cells depleted of adherent cells[11] by a 90 minute incubation in plastic tissue culture dishes. The nonadherent cells were removed, washed, and then resuspended to their original volume. The nonadherent cells were then cultured at a concentration equivalent to that of the original whole marrow cells from which the nonadherent cells were obtained. When cultured in the presence of PWM-SCM, megakaryocyte and granulocyte-macrophage colony growth (mean ± SEM/3 x 10^4 "whole" marrow cells) from nonadherent marrow cells (14 ± 3 and 44 ± 1, respectively) was equivalent to that observed with the original whole marrow cells (12 ± 2 and 45 ± 4, respectively). These data demonstrate that there was no significant loss of colony-forming cells with the 90 minute adherence step. As shown in Figure 3, megakaryocyte and granulocyte-macrophage colony growth in the presence of 5, 10, 20, or 40 u/ml IL-3 were similar in cultures of

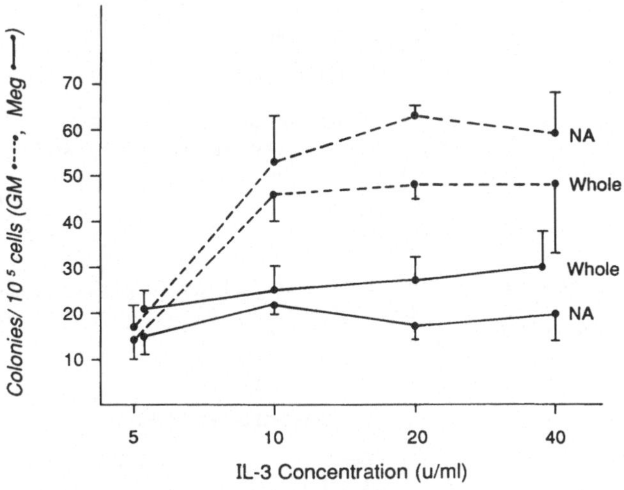

Fig. 3. Megakaryocyte (solid lines) and granulocyte-macrophage (broken lines) colony growth (mean ±SEM) from whole and nonadherent (NA) marrow cells cultured in the presence of various concentrations of IL-3.

nonadherent cells were, on the average, somewhat smaller than the colonies which developed from whole marrow cells, these differences did not reach statistical significance (Figure 4). Likewise, the geometric mean ploidy of colony megakaryocytes was similar in the whole marrow and nonadherent marrow cell cultures (Table II).

We conclude that, under conditions of maximal stimulation with PWM-SCM and at all concentrations of IL-3 tested, megakaryocyte colony size and mean ploidy are inversely related. We were unable to distinguish distinct subsets of megakaryocyte progenitors on the basis of the size and ploidy distributions of day 5 megakaryocyte colonies. The size and ploidy characteristics of day 5 megakaryocyte colonies each appear to be represented by a continuum. Increasing concentrations of IL-3 stimulated a dose-dependent increase in the mean size and numbers of megakaryocyte colonies but did not significantly alter their ploidy distribution. Adherent cell depletion does not significantly modify these relationships.

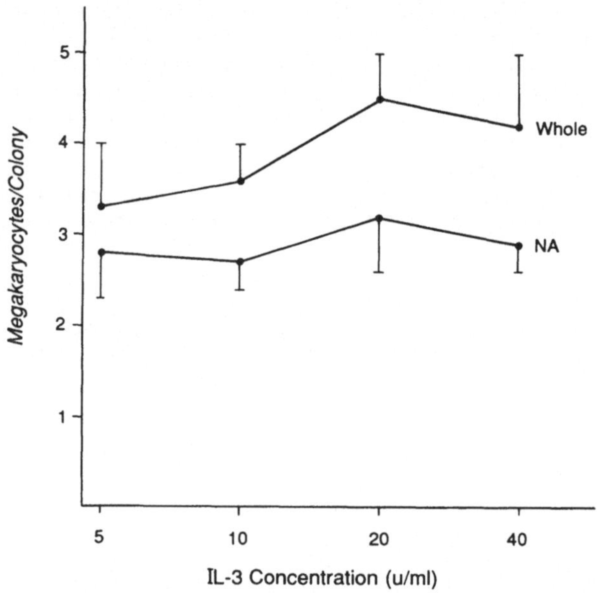

Fig. 4. Megakaryocyte colony size (mean ± SEM) in IL-3-stimulated cultures of whole and nonadherent (NA) marrow cells

Table II. Geometric Mean Ploidy of Colony Megakaryocytes in IL-3-Stimulated Cultures of Nonadherent and Whole Marrow Cells

	PLOIDY (Geometric Mean ± SD)	
IL-3 (u/ml)	Whole	Nonadherent
5	12.6 ± 2.1	14.5 ± 1.7
10	14.4 ± 2.1	12.9 ± 1.7
20	10.9 ± 2.0	12.7 ± 1.8
40	12.2 ± 2.1	13.4 ± 1.7

These studies have revealed relationships which describe the link megakaryocyte colony size and ploidization. Future studies will focus on how candidate regulatory factors and cells can modulate the mitotic and endomitotic activity of megakaryocyte precursors and, thereby, alter these relationships.

ACKNOWLEDGEMENTS

The author thanks Ms J. Jensen for help with preparation of the manuscript. This work was supported by research grants HL-31641 and AM-19410 from the NIH, DHHS and the Medical Research Foundation of Oregon.

REFERENCES

1. Levin, J., F.C. Levin, D.G. Penington, and D. Metcalf. 1981. Measurement of ploidy distribution in megakaryocyte colonies obtained from culture: With studies of the effects of thrombocytopenia. Blood 57: 287-297.
2. Levin, J. 1983. Murine megakaryocytopoiesis in vitro: An analysis of culture systems used for the study of megakaryocyte colony-forming cells and of the characteristics of megakaryocyte colonies. Blood 61: 617-623.
3. Paulus, J.M., M. Prenant, J.F. Deschanps, and M. Henry-Amer. 1982. Polyploid megakaryocytes developed randomly from a multicompartmental system of committed progenitors. Proc. Natl. Acad. Sci. USA. 79: 4410-4414.
4. Chatelain, C., and S.A. Burstein. 1984. Fluorescence cytophotometric analysis of megakaryocytic ploidy in culture: Studies of normal and thrombocytopenic mice. Blood 64: 1193-1199.
5. Burstein, S.A., J.W. Adamson, D. Thorning, and L.A. Harker. 1979. Characteristics of murine megakaryocytic colonies in vitro. Blood 54: 169-179.
6. Jackson, C.W. 1973. Cholinesterase as a possible marker for early cells of the megakaryocytic series. Blood 42: 413-421.
7. Ihle, J.N., J. Keller, and L. Henderson. 1982. Procedures for the purification of interleukin-3 to homogeneity. J. Immunol. 1290: 2431-2435.
8. Ishibashi, T., and S.A. Burstein. 1986. Interleukin-3 promotes the differentiation of isolated single megakaryocytes. Blood 67: 1512-1514.
9. Ishibashi, T., S.L. Miller, and S.A. Burstein. 1987. Type beta transforming growth factor is a potent inhibitor of murine megakaryocytopoiesis in vitro. Blood 69: 1737-1741.
10. Ishibashi, T., J.A. Koziol, and S.A. Burstein. 1987. Human recombinant erythropoietin promotes differentiation of murine megakaryocytes in vitro. J. Clin. Invest. 79: 286-289.
11. Williams, N., S.H. Oon, H. Jackson, and R. Lim. 1986. Studies on megakaryocyte potentiator: Its production and some biochemical characteristics. In Megakaryocyte Development and Function. R.F. Levine, N. Williams, J. Levin, and B.L. Evatt, editors. Alan R. Liss, New York. Pages 91-103.

FACTORS REGULATING MEGAKARYOCYTOPOIESIS AND PLATELET FORMATION

Tsutomu Abe[1], J. Peter Fuhrer[1], Marvin D. Bregman[1], Atsushi Kuramoto[2], and Martin J. Murphy, Jr.[1]

[1]Hipple Cancer Research Center, Daytorn, Ohio
[2]Research Institute for Nuclear Medicine and Biology Hiroshima University, Hiroshima, Japan

During the past two decades evidence has accumulated which indicates that platelet production from megakaryocytes or megakaryocyte precursor cells is regulated by humoral factors[1-4]. Many factors which are suspected to be involved in megakaryocytopoiesis and/or platelet formation have been reported (Tables I and II). Some of these factors have been purified to homogeneity or molecularly cloned, and their biochemical natures and biological activities on hematopoietic cells, including cells of the megakaryocyte lineage, have been investigated. Among these factors are megakaryocyte colony-stimulating factor (Meg-CSF)[5-7], thrombocytopoiesis-stimulating factor or thrombopoietin (Tpo)[8-10], interleukin 3 (IL-3)[11,12], multi-colony--stimulating factor (multi-CSF)[13-17], granulocyte/macrophage colony-stimulating factor (GM-CSF)[12,18,19] and erythropoietin (Epo)[5,20,21].

The limited availability of the native, molecularly homogeneous factors necessary to execute experiments both in vitro and in vivo, and lack of an adequate culture method to induce megakaryocytes to produce platelets in vitro, makes it difficult to elucidate the mechanism(s) of megakaryocyte colony proliferation, maturation, and platelet production. However, recent advances in molecular biology are enabling us to understand some of these mechanisms in more detail.

Thrombocytopoiesis-stimulating factor (TSF) or thrombopoietin (Tpo) is believed to affect platelet production in vivo. This effect can be detected by counting the increased number of circulating platelets or bone marrow megakaryocytes, or by measuring the increased level of selenomethionine-^{75}Se or $Na_2{}^{35}SO_4$ in the platelets of recipient animals[4,22,23]. TSF or Tpo is believed to influence the proliferation[24] and maturation[6,10] of megakaryocytes in vitro and, when added to cultures containing megakaryocyte colony-stimulating factor, to increase the number of CFU-meg. On the other hand, following the establishment of in vitro megakaryocyte colony culture methods[5,25,26], a number of factors which are thought to be involved in megakaryocytopoiesis, or megakaryocyte maturation and platelet production in vitro have been reported.

At least two factors and two separate steps are thought to be involved in late-stage megakaryocytopoiesis or thrombocytopoiesis in experimental animal and human in vitro cultures[6,27]. One factor, termed megakaryocyte colony-stimulating factor (Meg-CSF), may preferentially promote the prolifera-

Table I. Thrombopoietin or megakaryocyte colony-stimulating factors

Name or Function	Source (origin)	M.W. (KD)	Stimulating Activity in vitro	Stimulating Activity in vivo	Reference
1. Megakaryocyte colony-stimulating factor (Meg-CSF)	WEHI-3 CM (murine)	35	+		Williams (1982)[6]
2. Megakaryocyte potentiator	WEHI-3 CM (murine)	100	+		Williams (1982)[6]
3. Meg-CSF Thrombopoietin (Tpo)	Aplastic Anemia, ITP urine (human)	45, 155, 76	+	+	Kawakita (1982)[53]
4. Thrombopoietin	Thrombocytopenic serum (rabbit)		+	+	Levin (1982)[42]
5. Thrombocytopoiesis-stimulating factor (TSF) or Tpo	Embryonic kidney cell line (human)	32	+	+	McDonald (1985)[52]
6. Meg-CSF	Hypomegakaryocytic plasma (human)	46	+		Hoffman (1985)[54]
7. Megakaryocyte potentiator (Meg-POT)	P388D1 cell line (murine)	21	+		Huat (1986)[55]
8. Meg-CSF	T lymphocyte cell line Mo (human)	40-50	+		Bagnara (1987)[56]

tion of megakaryocyte progenitor cells (CFU-meg) to form colonies in culture. The other growth factor, called "megakaryocyte potentiator," when added to cultures containing either Meg-CSF or a source of Meg-CSF, produces a synergistic increase in both the size and ploidy of megakaryocyte colonies[6,10,28]. Although some functional similarities have been reported between thrombocytopoiesis-stimulating factor or thrombopoietin and megakaryocyte potentiator, it is not known whether or not these two factors are identical molecules. Further investigation is needed to answer this question.

Among multi-colony-stimulating factors, interleukin 3 (IL-3) was initially defined as a factor in conditioned medium from Concanavalin A-stimulated lymphocytes (ConACM) that induced the enzyme 20-α-hydroxysteroid dehydrogenase (20αSDH) in cultures of nude mouse splenic lymphocytes. It was then purified to homogeneity and shown to have various of stimulating activities, including P cell-stimulating activity, WEHI-3 growth factor activity, histamine-producing cell activity and hematopoietic stimulating activity[11,29,30]. IL-3 also has been demonstrated as having a stimulatory effect on megakaryocyte colony formation in vitro[12]. Moreover, murine recombinant IL-3 has shown a broad range of hematopoietic colony-stimulating activities on granulocyte, macrophage, granulocyte/macrophage, megakaryocyte, and mixed megakaryocyte colonies[12]. IL-3 has been used to promote the differentiation of single isolated megakaryocytes in vitro[31], and also stimulates hematopoietic progenitor cells in vivo[32-34]. A number of authors report that multi-colony-stimulating factors which differ from murine multi-CSF, granulocyte-macrophage colony-stimulating factor (GM-CSF) and interleukin 4 (IL-4), have hematopoietic stimulating activities, including all of the megakaryocyte lineage[34-40,47] both in vitro and in vivo (Table II).

The role of erythropoietin (Epo) in megakaryocytopoiesis or thrombocytopoiesis has long been controversial. Some investigators have reported a role for Epo in megakaryocytopoiesis or thrombocytopoiesis[5,20,41] while

Table II. Pluripotent colony-stimulating factors

Name or Function	Source (origin)	M.W. (KD)	In vitro activity	In vivo activity	Target cells, Function	Reference
1. Pluripotent stem cell activator	T lymphocyte cell Mo (human)	40	+		CFU-mix	Fauser (1983)[35]
2. CFU-GEMM-stimulating factor	PHA-LCM (human)	35-40	+		CFU-mix, BFU-e CFU-c	Ruppert (1983)[13]
3. P-cell-stimulating factor (PSF)	WEHI-3 CM (murine)	30	+		P-cell, CFU-s	Schrader (1983)[48] Clark-Lewis (1984)[49]
4. Colony-stimulating factor (CSF)	IL-2 dependent T lymphocyte (human)		+		CFU-mix, CFU-gm	Greenberger (1984)[50]
5. Interleukin 3	WEHI-3 (murine)	28	+		Colony-stimulating factor, WEHI-3 growth factor, P-cell, mast cell, eosinophil, CFU-gm, -mix, -meg, histamine-producing cell	Ihle (1983)[11] Quesenberry (1985)[12]
6. Multilineage hemopoietic growth-stimulating factor	Melanoma cell line HS0294 (human)	21	+		Murine: CFU-mix, BFU-e, CFU-gm, CFU-meg, Human: BFU-e	Kraft (1986)[17]
7. Recombinant multi-CSF (r-IL-3)	WEHI-3 (murine)		+	+	Eosinophil, NK cell, CFU-mix, -gm, -meg, BFU-e,, CFCs, mast cell progenitor	Metcalf (1986)[51] (1987)[34] Kindler (1986)[33]
8. Recombinant multi-CSF	T lymphocyte (murine)		+		CFU-mix,-meg, BFU-e, mast cell progenitor	Greenberger (1985)[16]
9. Multi-CSF	Urine (human)	32	+	+	CFU-mix,-gm,-e, -meg, BFU-e	Abe (1987)[39,40]

others have refuted any Epo involvement[42,43]. In some studies which described Epo involvement in megakaryocytopoiesis and thrombocytopoiesis, effects of contaminated Meg-CSF or Tpo were possible[20,41]. However, recent studies using highly purified Epo in serum or serum-free medium, or using recombinant Epo have shown that megakaryocyte colony stimulation[42,43], megakaryocyte maturation in vitro[44] or platelet production in vivo[43,45] occurred only with high concentrations of Epo.

Although several factors affecting megakaryocytopoiesis and thrombocytopoiesis have been proposed as mentioned above, a number of unresolved questions remain. One question is whether most of these factors, which are produced under special conditions, such as lectin activation of lymphocyte cell lines or are obtained from murine or human tumor cell lines, actually function in normal or steady-state megakaryocytopoiesis or thrombocytopoiesis in mice or human in vivo. Another question is whether each of these factors has a specific receptor on its target cell or do they share some common receptor on megakaryocyte progenitor cells or on megakaryocytes. Do these factors stimulate target cells directly or work via other cell mechanisms? And, perhaps most important, do these factors which have a stimulatory effect in vitro also work in vivo?

In order to answer these questions, we have been studying the urine of patients with aplastic anemia (AA), known to be good source of factors which stimulate megakaryocytopoiesis or thrombocytopoiesis not only in vitro but in vivo[21]. We have purified several factors from urine and have reported our findings[39,40].

One factor has shown wide-range colony-stimulating activity on murine hematopoietic progenitor cells in vitro and in vivo, including CFU-mix, BFU-e, CFU-gm, CFU-e and CFU-meg, and caused a marked increase in platelets, white blood cells and hematocrit in circulating murine blood. This factor exhibits an apparent molecular weight of 32 KD in SDS-polyacrylamide gel electrophoresis, is relatively stable under heat, acid, base and neuraminidase treatment, and is thought to be a different molecule from murine IL-3, or human multi-colony-stimulating factor reported before. We believe it to be a naturally occurring human multicolony-stimulating factor. Another factor purified from AA urine has a lower molecular weight and has shown a stimulatory activity on cells committed to megakaryocytes in vitro and in vivo (data not shown)

Many problems regarding thrombocytopoiesis and megakaryocytopoiesis remain unresolved. It is hoped that the use of more sophisticated culture systems which will allow platelet production in vitro, and the preparation of more homogeneous factors will enable these complex mechanisms to be elucidated.

ACKNOWLEDGEMENTS

This research was supported in part by a grant from the American Honda Foundation. The authors wish to thank M. A. Odell for help in preparing the manuscript.

REFERENCES

1. Abildgaard, C.F., and J.V. Simone. 1967. Thrombopoiesis. Semin Hematol. 1967. 4: 424-452.
2. Ebbe S. 1968. Megakaryocytosis and platelet turn over. Ser Hematol. 1: 65-98.
3. Cooper, G.W. 1970. The regulation of thrombopoiesis. In Regulation

of Hematopoiesis. A.S. Gordon, ed., Appleton-Century-Crofts, New York. Vol 2: 1611-1629.

4. Levin, J., and B.L. Evatt. 1979. Humoral control of thrombopoiesis. Blood Cells 5: 105-121.
5. McLeod, D.L., M.M. Shreeve, and A.A. Axelrad. 1976. Induction of megakaryocyte colonies with platelet formation. Nature 261: 492-494.
6. Williams, N., R.R. Eger, H.M. Jackson, and D.J. Nelson. 1982. Two factor requirement for murine megakaryocyte colony formation. J. Cell Physiol. 110: 101-104.
7. Hoffman, R., H.H. Yank, E. Bruno, and J.E. Straneva. 1985. Purification and partial characterization of a megakaryocyte colony-stimulating factor from human plasma. J. Clin. Invest. 75: 1174-1182.
8. Evatt, B.L., and J. Levin. 1969. Measurement of thrombopoiesis in rabbits using ^{75}selenomethionine. J. Clin. Invest. 48: 1615-1626.
9. McDonald, T.P., R. Clift, R.D. Lane, C. Nolan, I.I.E. Tribby, and G.H. Barlow. 1975. Thrombopoietin production by human embryonic kidney cells in culture. J. Lab Clin. Med. 85: 59-66.
10. Williams, N., T.P. McDonald, and E.M. Rabellino. 1979. Maturation and regulation of megakaryocytopoiesis. Blood Cells 5: 43-55.
11. Ihle, J.N., J. Keller, T. Oroszlan, L.E. Henderson, T.D. Copeland, F. Fitch, M.B. Prystowsky, E. Goldwasser, J.W. Schrader, E. Palazynski, M. Dy, and B. Lebel. 1983. Biological properties of homogeneous interleukin 3. J. Immunol. 131: 282-289.
12. Quesenberry, P.J., J.N. Ihle, and E. McGrath. 1985. The effect of interleukin 3 and GM-CSA-2 on megakaryocyte and myeloid clonal colony formation. Blood 65: 214-217.
13. Ruppert, S., G.W. Lohr, and A.A. Fauser. 1983. Characterization of stimulatory activity for human pluripotent stem cell (CFU_{GEMM}). Exp. Hematol. 11: 154-161.
14. Burgess, A.W., D. Metcalf, J. Koza, R.J. Simpson, G. Variro, J.A. Hamilton, and E.C. Nice. 1985. Purification of two forms of colony stimulating factor from mouse L-cell-conditioned medium. J. Biol. Chem. 260: 6004-6011.
15. Culter, R.L., D. Metcalf, N.A. Nicola, and G.M. Johnson. 1985. Purification of a multipotential colony-stimulating factor from pokeweed mitogen-stimulated mouse spleen cell conditioned medium. J. Biol. Chem. 260: 6579-6587.
16. Greenberger, J.S., R.K. Humphries, H. Messner, D.M. Reid, and M.A. Sakakeeny. 1985. Molecular cloned and expressed murine T-cell gene product is biologically similar to interleukin-3. Exp. Hematol. 13: 249-260.
17. Kraft, A.S., and K.S. Zucherman. 1986. Production of multilineage hemopoietic growth-stimulating activities by a human melanoma cell line. Exp. Hematol. 14: 867-872.
18. Yokota, T., F. Lee, D. Rennick, C. Hall, N. Arai, T. Mosmann, I.G. Nabel, H. Cantor, and K. Arai. 1984. Isolation and characterization of a full-length cDNA for mast cell growth factor from a mouse T-cell clone: Expressed in monkey cells. Proc. Natl. Acad. Sci. USA. 81: 1070-1074.
19. Gough, N.M., G. Gough, D. Metcalf, A. Kelso, D. Grail, N.A. Nicola, A. W. Burgess, and A.R. Dunn. 1984. Molecular cloning of cDNA encoding a murine haemopoietic growth regulator, granulocyte-macrophage colony stimulating factor. Nature 309: 763-767.
20. Vainchenker, W., J. Bouguett, J. Guichard, and J. Breton-Gorius. 1979. Megakaryocyte colony formation from human bone marrow precursors. Blood 54: 940-945.
21. Enomoto, K., M. Kawakita, S. Kishimoto, N. Katayama, and T. Miyake. 1980. Thrombopoiesis and megakaryocyte colony stimulating factor in the urine of patients with aplastic anemia. Br. J. Haematol. 45: 551-556.

22. Ebbe, S., F. Stohlman, Jr., J. Donovan, and J. Overcash. 1968. Megakaryocyte maturation rate in thrombocytopenic rats. 32: 787-795.
23. Odell, T.T., Jr., C.W. Jackson, T.J. Friday, and D.E. Charsha. 1969. Effects of thrombocytopenia on megakaryocytosis. Br. J. Haematol. 17: 91-101.
24. Freedman, M.H., T.P. McDonald, and E.F. Saunders. 1981. Differentiation of murine marrow megakaryocyte progenitors (CFU-m): humoral control in vitro. Cell Tissue Kinet 14: 53-58.
25. Metcalf, D., H.R. MacDonald, N. Odartchenko, and B. Sordat. 1975. Growth of mouse megakaryocyte in colonies in vitro. Proc. Natl. Acad. Sci. USA. 72: 1744-1748.
26. Nakeff, A., and S. Daniels-McQueen. 1976. Colony-Forming Unit Megakaryocyte (CFU-M). Proc. Soc. Exp. Biol. Med. 151: 587-590.
27. Straneva, J.E., H.H. Yank, S.L. Hui, E. Bruno, and R. Hoffman. 1987. Effects of megakaryocyte colony-stimulating factor on terminal cytoplasmic maturation of human megakaryocyte. Exp. Hematol. 15: 657-663.
28. Williams, N., H. Jackson, P. Ralph, and I. Nakoinez. 1981. Cell interactions influencing murine marrow megakaryocytes: Nature of the progenitor cell in bone marrow. Blood 57: 157-163.
29. Ihle, J.N., L. Pepersack, and L. Rebar. 1981. Regulation of T cell differentiation: in vitro induction of 20α-hydroxysteroid dehydrogenase in splenic lymphocytes from athymic mice by a unique lymphokine. J. Immunol. 126: 2183-2189.
30. Dy, M., B. Lebel, P. Kamoun, and J. Hamburger. 1984. Histamine production during the anti-allograft response. Demonstration of a new lymphokine enhancing histamine synthesis. J. Exp. Med. 153: 293-309.
31. Ishibashi, T., and S.A. Burstein. 1986. Interleukin 3 promotes the differentiation of isolated single megakaryocyte. Blood. 67: 1512-1514.
32. Broxmeyer, H.E., D.E. Williams, S. Cooper, R. Shadduck, S. Gillis, A. Waheed, D. L. Urdal, and D.C. Bicknell. 1987. Comparative effects in vivo of recombinant murine interleukin 3, natural murine colony-stimulating factor-1, and recombinant murine granulocyte-macrophage colony-stimulating factor on myelopoiesis in mice. J. Clin. Invest. 79: 721-730.
33. Kindler, V., B. Thorens, S.D. Kossodo, B. Allet, J.F. Eliason, D. Thatcher, N. Farber, P. Vassalli. Stimulation of hematopoiesis in vivo by recombinant bacterial murine interleukin 3. Proc. Natl. Acad. Sci. USA. 1986. 83: 1001-1005.
34. Metcalf, D., Begley, C.G., N.A. Nicola, and G.R. Johnson. 1987. Quantitative responsiveness of murine hemopoietic populations in vitro and in vivo to recombinant multi-CSF (IL-3). Exp. Hematol. 15: 288-295.
35. Fauser, A.A., Messner, H.A., Lusis, A.J., and Golde, D.W. 1981. Stimulatory activity for human pluripotent hemopoietic progenitors produced by a human T-lymphocyte cell line. Stem Cells 1: 73-80.
36. Welte, K., E. Platzer, L. Lu, J.L. Gabrilove, E. Levi, R., Mertelsmann, and M.A. Moore. 1985. Purification and biochemical characterization of human pluripotent hematopoietic colony-stimulating factor. Proc. Natl. Acad. Sci. USA. 82: 1526-1530.
37. Quesenberry, P., Z. Song, E. McGrath, I. McNiece, R. Shadduck, A. Waheed, G. Baber, E. Kleeman, and D. Kaiser. 1987. 69: 827-835.
38. Merchav S., A. Nagler, and I. Tatarsky. 1987. Human stem cell colony-stimulating activity (CFU-GEMM) in medium conditioned by leukemic B-lymphocytes. Exp. Hematol. 15: 115-118.
39. Abe, T., M. Sugimoto, J.P. Fuhrer, M.D. Bregman, A. Kuramoto, M.J.

Murphy, Jr. 1987. Partially purified extract of urine from patients with aplastic anemia has interleukin 3 (IL-3) or burst promoting activity to murine hematopoietic progenitor cells in vitro and in vivo. Exp. Hematol. 15: 529a.
40. Abe T., J.P. Fuhrer, M.D. Bregman, M. Sugimoto, A. Kuramoto, and M. J. Murphy, Jr. 1987. Naturally produced multi-hematopoietic colony stimulating factor in the urine of patients with aplastic anemia. Blood. 70(Suppl 1); 128.
41. Evatt, B.L., J.L. Spivak, and J. Levin. 1976. Relationships between thrombopoiesis and erythropoiesis: With studies of the effects of preparations of thrombopoietin and erythropoietin. Blood 48: 547-558.
42. Levin, J., F.C. Levin, D.F. Hull, III, and D.G. Pennington. 1982. The effects of thrombopoietin on megakaryocyte-CFC, megakaryocytes, and thrombopoiesis: With studies of ploidy and platelet size. Blood. 60: 989-998.
43. Geissler, D., G. Konwalinka, C. Peschel, and H. Braunsteiner. 1987. The role of erythropoietin, megakaryocyte colony-stimulating factor, and T-cell-derived factors on human megakaryocyte colony formation: Evidence for T-cell-mediated and T-cell-independent stem cell proliferation. Exp. Hematol. 15: 845-853.
44. Tsukada, J., M. Misago, T. Sato, M. Kikuchi, S. Oda, S. Chiba, and S. Eto. 1987. The effect of recombinant erythropoietin on murine megakaryocyte colony formation. Int. J. Cell Cloning 5: 401-411.
45. Ishibashi, T., J.A. Koziol, and S.A. Burstein. 1987. Human recombinant erythropoietin promotes differentiation of murine megakaryocytes in vitro. J. Clin Invest. 79: 286-289.
46. McDonald, T.P., M.B. Cottrell, R.E. Clift, W.C. Cullen, and F.K. Lin. 1987. High doses of recombinant erythropoietin stimulate platelet production in mice. Exp. Hematol. 15: 719-721.
47. Mazur, E.M., J.L. Cohen, G.G. Wong, and S.C. Clark. 1987. Modest stimulatory effect of recombinant human GM-CSF on colony growth from peripheral blood human megakaryocyte progenitor cells. Exp. Hematol. 15:1128-1133.
48. Shrader, J.W., I. Clark-Lewis, R.M. Crapper, and G.W. Wong. 1983. P-cell stimulating factor: Characterization, action on multiple lineages of bone marrow derived cells and role in oncogenesis. Immunol. Rev. 76: 79-104.
49. Clark-Lewis, I., S.B.H. Kent, and J.W. Schrader. 1984. Purification to apparent homogeneity of a factor stimulating the growth of multiple lineages of hemopoietic cells. J. Biol. Chem. 259: 7488-7494.
50. Greenberger, J.S., A.M. Krensky, H. Messner, S.J. Burakott, U. Wandl., and M.A. Sakakeeny. 1984. Production of colony-stimulating factor(s) for granulocyte-macrophage and multipotential (granulocyte/erythroid/megakaryocyte/macrophage) hematopoietic progenitor cell (CFU-GEMM) by clonal lines of human IL-2-dependent T-lymphocytes. Exp. Hematol. 12: 720-727.
51. Metcalf, D., C.G. Begley, G.R. Johnson, N.A. Nicola, A.F. Lopez, and D.J. Williamson. 1986. Effect of purified bacterially synthesized murine multi-CSF (IL-3) on hematopoiesis in normal adult mice. Blood 68: 46-57.
52. McDonald, T.P., M. Cottrell, R. Clift, J.A. Khouri, and M.D. Long. 1985. Studies on the purification of thrombopoietin from kidney cell culture medium. J. Lab. Clin. Med. 106: 162.
53. Kawakita, M., T. Miyake, S. Kishimoto, and M. Ogawa. 1982. Apparent heterogeneity of human megakaryocyte colony- and thrombopoiesis-stimulating factors: studies on urinary extracts from patients with aplastic anemia and idiopathic thrombocytopenic purpura. Br. J. Haematol. 52: 429-438.

54. Hoffman, R., H.H. Yank, E. Bruno, and J.E. Straneva. 1985. Purification and partial characterization of megakaryocyte colony-stimulating factor from human plasma. J. Clin. Invest. 75: 1174-1182.
55. Huat, O.S., and N. Williams. 1986. Biochemical characterization on an in-vitro murine megakaryocyte growth activity: megakaryocyte potentiator. Leuk. Res. 10: 403-411.
56. Bagnara, G.P., A. Guarini, L. Gaggioli, G. Zauli, L. Catani, L. Valvasori, G. Zunica, L. Gugliotta, and M. Marini. 1987. Human T-lymphocyte-derived megakaryocyte colony-stimulating activity. Exp. Hematol. 15:679-684.

EFFECT OF SUSTAINED HYPERTRANSFUSION ON RAUSCHER LEUKEMIA VIRUS-VARIANT A (RLV-A) INFECTION IN BALB/c MICE

Gary Paul Leonardi, Michael Manthos, Joseph LoBue, Donald Orlic and Jyotirmay Mitra

Department of Biology, New York University, New York, NY 10003
Department of Anatomy, New York Medical College, Valhalla, NY 11960

ABSTRACT

Infection of BALB/c mice with the RLV-A virus normally induces an erythropoietic dysplasia characterized by hepatosplenomegaly, erythroblastosis, erythroblastemia and severe anemia without reticulocytosis. Time to death varies between 20-30 weeks. Mice were inoculated with RLV-A after being hypertransfused with 75% packed red cells for 42 days which has been shown to eliminate erythropoiesis and modify the microenvironment to favor granulopoiesis. Following RLV-A inoculation, one group did not receive further transfusion (short-term) and another group continued with hypertransfusion weekly (long-term). The pathogenesis of RLV-A in the short-term group paralleled the characteristic RLV-A response. In the long-term group however, the characteristic RLV-A response was never observed. Instead, a granulocytic leukemia was developed. Continued hypertransfusion presumably after establishment of an altered microenvironment resulted in a completely different viral pathogenesis and the development of a transplantable myeloid leukemia.

INTRODUCTION

Hypertransfusion has long been used to modify hemopoiesis since this procedure results in the marked curtailment of erythropoiesis[1]. As a result of hypertransfusion, recognizable red cell elements as well as erythropoietin-dependent progenitor cells (e.g. BFU-E, CFU-E) are drastically reduced whereas erythropoietin-independent progenitors (e.g. CFU-S, CFU-GM) have been shown to increase[2].

Recently, Brookoff and Weiss[3] have induced specific bone marrow stomal cell changes by sustained hypertransfusion in mice. The sequence of events accompanying the marrow shift from reduction in erythropoiesis to enhanced granulopoiesis included the loss of medullary macrophages and an increase in adipocytes and reticular adventitial cells[3]. Their suggestion that the marrow stromal changes induced a shift in the microenvironment to curtail erythropoiesis was strengthened when 2-week hypertransfused mice were unable to mount an erythropoietic response when challenged by hemolytic anemia or phlebotomy until a unique type of stromal cell ("dark stromal cells") appeared[3].

In the present study, sustained hypertransfusion was initiated in animals prior to inoculation with a variant form of Rauscher Leukemia Virus (RLV-A)[4]. This disease is characterized by a slowly developing erythroblastosis accompanied by hepatosplenomegaly and a lack of reticulocytosis in spite of severe terminal anemia[4]. How the pathogenesis of this virally-induced erythroblastic disease is altered in animals, that: (1) have had presumably susceptable erythroid elements eliminated and (2) have a microenvironment which may be changed to favor granulopoiesis, is herein reported.

METHODS

Animals

Male BALB/c mice, obtained from a breeding colony maintained at New York University (N.Y.U.) were used in this study. Both female mice and purchased retired breeders (Charles River, Wilmington, MA) were used as blood donors for hypertransfusion due to the shortage of available male mice in the N.Y.U. colony. All mice were housed 4 per cage and maintained on a diet of lab chow and water ad libitum.

RLV-A Plasma Preparation and Inoculation

RLV-A viremic plasma was obtained from mice bearing a transplantable monomyelocytic leukemia (MML)[5]. Briefly, blood from terminal MML mice was collected using heparinized syringes via cardiocentesis. The blood was centrifuged at 2,000 x g for 10 minutes at 2° C. The supernatant was then collected and recentrifuged a second time at the above conditions. The supernatant was again collected and recentrifuged a third time at 10,000 x g for 15 minutes. Following this treatment, the supernatant was collected in plastic vials and stored in liquid nitrogen. For RLV-A infection, mice were inoculated intraperitoneally with 0.2 ml of rapidly thawed plasma. Care was taken so that for each experiment, the viral potency controls and hypertransfused-viral groups received plasma from the same preparation lot.

Hypertransfusion Studies

This study was divided into 2 experiments: short-term and long-term hypertransfusions. In both studies, 5-to 6-week old animals were inoculated with 1 ml of 75% packed washed erythrocytes, administered intraperitoneally every 7 days for 42 days. Hypertransfusion in mice eliminates erythropoiesis and modifies bone marrow stromal elements so as to favor granulopoiesis[3]. Following the initial 42-day hypertransfusion, groups of at least 4 animals were given RLV-A viremic plasma. In the short-term hypertransfusion experiment, no further hypertransfusion of these animals was done. In the long-term experiments however, hypertransfusion every 7 days continued throughout the experimental period. In both experiments, non-hypertransfused littermates were inoculated with viremic plasma to serve as viral potency control groups and normal untreated animal groups were established. In the long-term experiment, a group of littermates hypertransfused throughout the experimental period but not inoculated with virus was established.

In each experiment, at least 4 animals per group were used and over 1,000 donor animals were required to maintain the hypertransfused state. The preparation of blood for hypertransfusion followed a standard protocol[3]. Blood was collected from etherized donor animals via cardiocentesis using heparinized syringes. This blood was pooled into centrifuge tubes cooled in an ice bath. The blood was then centrifuged at 1,600 x g for 5 minutes at 2°C in a refrigerated centrifuge. The supernatant along with the buffy coat was discarded and the pellet was resuspended in sterile, physiologic saline. This procedure was repeated three times in order to wash the pellet. After the final centrifugation, saline was not added but a measure of the pellet packed red cell volume was made using Strumia micro-micro capillary tubes (Sherwood Medical Industries, St. Louis, MO). Packed red cell volumes in the range of 72-77% were used for hypertransfusion.

Hematologic Parameters

Throughout the experimental period, measurements of the packed red cell volume was made using the technique of Strumia et al.,[6]. Peripheral blood reticulocyte counts were made according to the technique of Brecher[7]. Peripheral blood nucleated cell counts were made using a model "ZM" Coulter counter according to prescribed techniques (Coulter Diagnostics, Hialea, FL). Peripheral blood differential counts were also done.

RESULTS

Hypertransfusion Studies

A group of 8 animals was hypertransfused weekly to ascertain whether the findings of Brookoff and Weiss[3] were applicable in BALB/c mice. On days 7, 14, 21 and 42 a pair of mice were sacrificed and prepared for microscopic examination. Hematologic and morphologic observations in these mice suggested that sustained hypertransfusion in BALB/c mice paralleled the events described for Swiss mice by Brookoff and Weiss[3].

Short-Term Hypertransfusion

PRCV values were initially increased as a result of hypertransfusion (Fig. 1). During this period no reticulocytosis was observed. After cessation of hypertransfusion and inoculation with RLV-A virus, PRCV values steadily declined so that by 6 weeks post-inoculation values were below controls (mean value 39%). PRCV values between short-term hypertransfused and RLV-A viral potency groups paralleled each other throughout the remainder of the disease course, with terminal values between 10 and 20%. A benefit of hypertransfusion was seen as this group showed a slightly greater survival time before succumbing to disease. The half--time of survival in the hypertransfusion group (26 weeks) was 6 weeks later that of RLV-A viral potency controls. However, all the RLV-A infected short-term hypertransfused mice showed the "typical" RLV-A disease course characterized by hepatosplenomegaly, erythroblastosis, erythroblastemia and terminal anemia[4]. Peripheral blood nucleated cell counts were above 100,000 cells/mm^3 with greater than 70% erythroblasts seen in terminal animals (Fig. 2).

Long-Term Hypertransfusion

As previously described, PRCV values were generally elevated above control values as a result of hypertransfusion (Fig. 3). Although there was a slight depression after viral inoculation, sustained hypertransfusion resulted in maintenance of elevated PRCV values. At various times following viral inoculation, an abrupt drop in hematocrit was seen in the long-term hypertransfusion group. This decline was accompanied by elevated white blood cells counts as high as 30,000 cells/mm^3 in which the vast majority of nucleated cells were granulocytic and monocytic. Within 2 weeks these mice responded to hypertransfusion with elevated PRCV and white cell counts returned to normal. Eventually, in each case a massive increase of peripheral blood granulocytic elements as high as 250,000 cells/mm^3

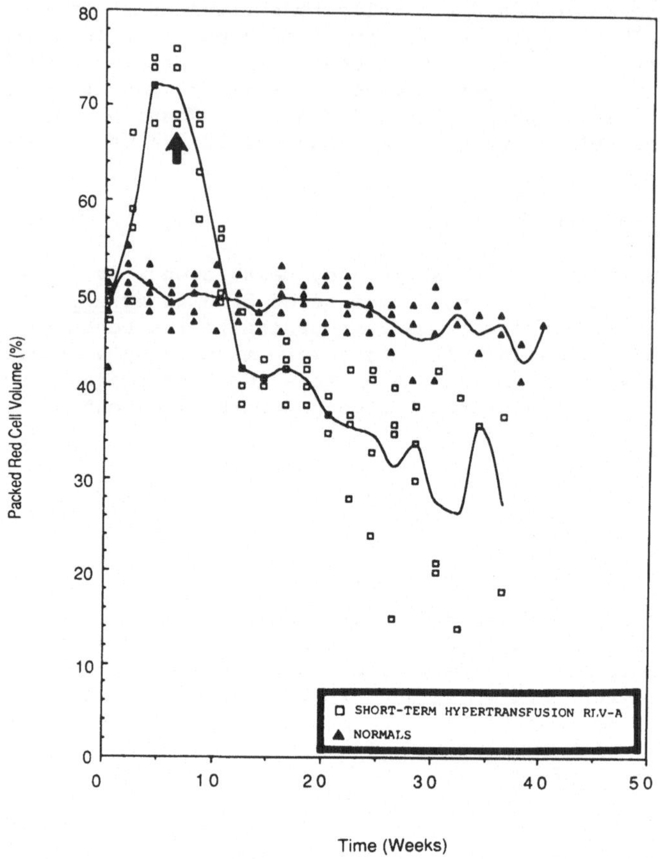

Fig. 1. Packed red cell volume (%) versus time (weeks) for the short-term hypertransfusion RLV-A and normal groups. Graph lines estimate the mean value of scatter points. Arrow ubducated cessation of hypertransfusion and inoculation virus.

with or without an associated drop in PRCV was seen (Fig. 4). Four of the 6 animals in the group responded in this fashion in times ranging from 18-40 weeks post-inoculation. The other 2 died unexpectedly and could not be recovered for transplantation or necropsy. Because of this, we decided to transfer blood as soon as massive increases in leukocyte count was seen in the subsequent four animals. Necropsy showed pronounced splenomegaly, moderate hepatomegaly and 1 animal showed massive increases in kidney size. No members of the long-term hypertransfusion group ever showed the "typical" RLV-A pathogenesis.

Animals that were given blood from the long-term hypertransfusion group all developed a granulocytic leukemia which has since been continually passed. All of the remaining animals in the long-term hypertransfusion group that served as blood donors gave a positive cell transfer, resulting in leukemia. In the passage leukemia, death occurs in 12-14 days, and appears similar in pathogenesis to the donor animals. Inoculation of blood from one of the original male mice was passaged into females. Karyotype analysis of the female mice showed leukemic cells possessing the male karyotype, thus establishing the transplantable nature of the donor cells.[1]

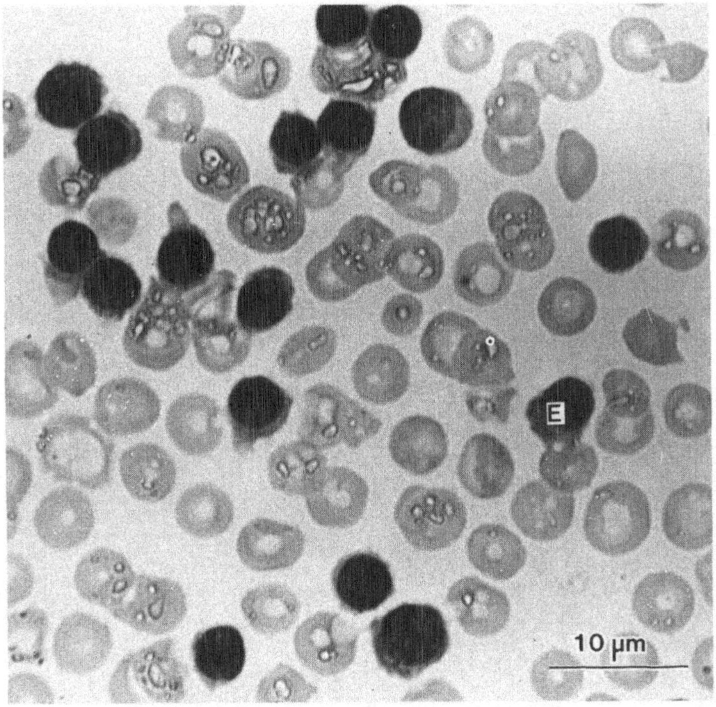

Fig. 2. Light micrograph of peripheral blood from a short-term hypertransfusion RLV-A terminal mouse. Note the erythroplasts (E) present.

DISCUSSION

The use of hypertransfusion to alter the pathogenesis of leukemia has previously been described in mice inoculated with Rauscher Leukemia Virus (RLV)[8]. Long-term hypertransfusion resulted in inhibition of the erythroblastic reaction and increased survival time, however the basic RLV pathogenesis was seen in 5 of the 7 animals surviving past 40 days of hypertransfusion. The other 2 animals developed either a severe erythroblastic response or a severe granulocytic response. In terms of the latter, attempts to establish that these cells were leukemic by transplantation were unsuccessful.

Failure to modify the basic RLV response may have been due to the fact that the virus was administered only 5 days after hypertransfusion was initiated. Thus a significantly altered microenvironment may not have been established prior to viral inoculation. In the present study, short-term hypertransfusion in which an altered microenvironment was at least morphologically established but hypertransfusion not continued after viral inoculation resulted in the characteristic RLV-A disease. Continued hypertransfusion presumably after establishment of an altered microenvironment resulted in a completely different viral pathogenesis and the development of a transplantable myeloid leukemia. Possibly the combined effect of both sustained hypertransfusion and an altered microenvironment in which (1) elimination of virally susceptable "target" cells occurs and (2) modified stromal elements which enhance granulopoiesis are found may be needed to alter viral RLV-A complex expression.

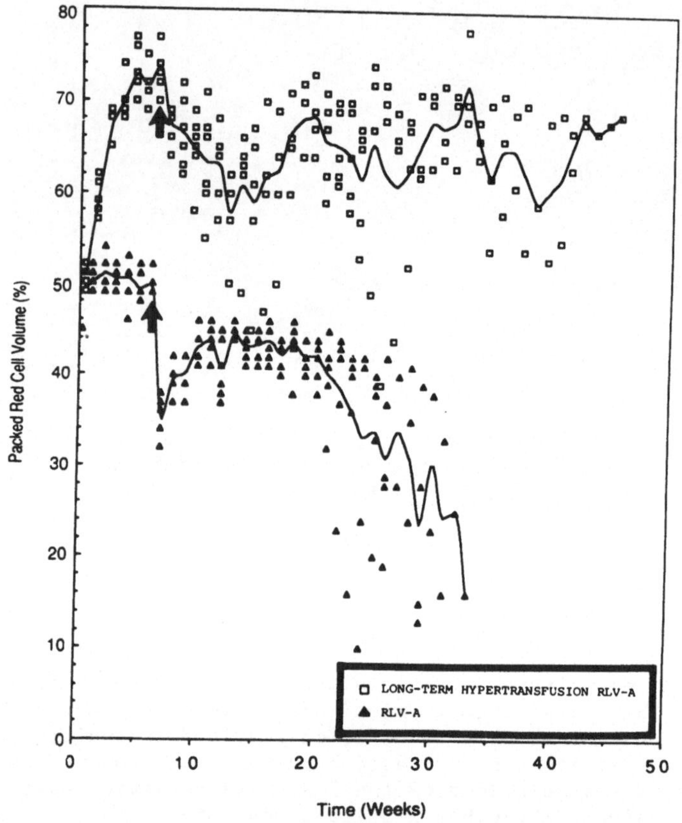

Fig. 3. Packed red cell volume (%) versus time (weeks) for the long-term hypertransfusion RLV-A group and the RLV-A viral potency control group. Graph lines estimate the mean value of scatter points. Arrow indicate inoculation of virus.

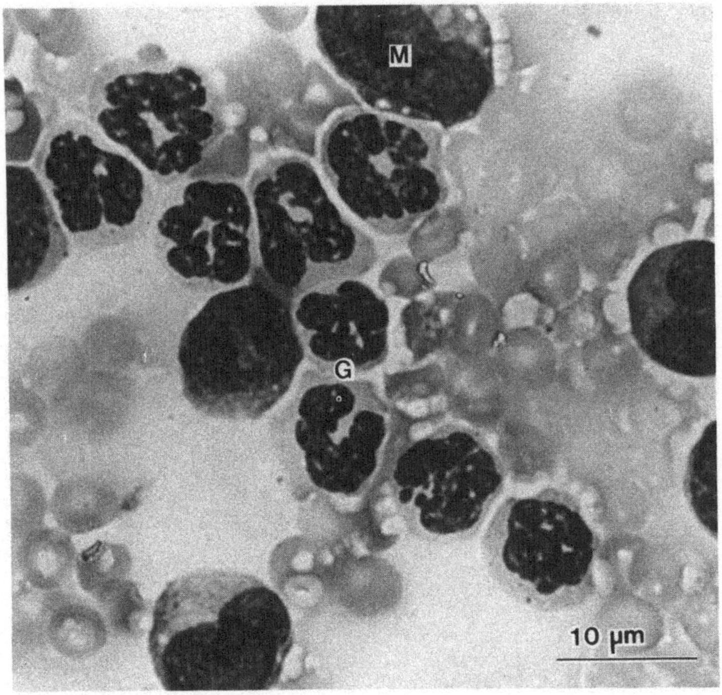

Fig. 4. Light micrograph of peripheral blood from long-term hypertransfusion RLV-A terminal mouse. Note the numerous granulocytic (G) and monocytic (M) elements seen.

REFERENCES

1. Jacobson, L.O., E. Goldwasser and C. W. Gurney. 1960. Transfusion-induced polycythemia as a model for studying factors influencing erythropoiesis. In CIBA Foundation Symposium on Hemopoiesis. G.E.W. Wolstenholme and M.O'Connor, editors. Little Brown and Co., Boston. 423-451.
2. Monette, F.C. 1985. Hypertransfusion-experimental results: Effect on erythropoietic progenitors. In Mathematical Modeling of Cell Proliferation: Stem Cell Regulation in Hemopoiesis. Volume II. Erythropoietic Supression, Combined Stresses, Drug Effects. H.E. Wichmann and M. Loeffler, editors. CRC Press, Inc., Boca Raton. 4-18.
3. Brookoff, D. and L. Weiss. 1982. Adipocyte development and the loss of erythropoietic capacity in the bone marrow of mice after sustained hypertransfusion. Blood 60: 1337-1344.
4. LoBue, J., P.A. Alexander Jr., T.N. Fredrickson, E.F. Schultz, and A.S. Gordon. 1972. Erythrokinetics in normal and diseased states. Virally induced erythroblastosis: A model system. In Regulation of Erythropoiesis. A.S. Gordon, M. Condorelli and C. Peschle, editors. Il Ponte, Milan. 89-101.
5. Fredrickson, T.N., J. LoBue, P.Alexander, E. Schultz, and A.S. Gordon. 1972. A transplantableleukemia from mice inoculated with Rauscher leukemia virus. J. Nat'l. Cancer Inst. 48: 1597-1605.
6. Strumia, M., A. Sample and E. Hart. 1954. An improved microhematocrit

method. Am. J. Clin. Pathol. 24: 1016-1024.
7. Brecher, G. 1949. New methylene blue as a reticulocyte stain. Am. J. Pathol. 19: 895-896.
8. Dunn, T.B., R.A. Malmgren, P.G. Carney and A.W. Green. 1966. Propythiouracil and transfusion modifications of the effects of the Rauscher virus in BALB/c mice. J. Nat'l. Cancer Inst. 36: 1003-1025.

ROLE OF IMMUNOCOMPETENT CELLS IN THE REGULATION OF HUMAN MEGAKARYOCYTOPOIESIS IN VITRO

Robert A. Detrick, Jeanne C. Schulman, Steve W. Mamus, Roger P. McEver, and Esmail D. Zanjani

Department of Medicine
Division of Hematology
Veterans Administration Medical Center
University of Minnesota School of Medicine
Minneapolis, MN 55455

ABSTRACT

T cells and monocytes/macrophages (Mo) have been shown to play important roles in modulating the growth and differentiation of human erythroid and myeloid progenitors and have been implicated in the mechanisms of gamma interferon (γ-IFN) mediated suppression of normal human marrow erythroid progenitors in vitro. In order to assess the importance of T cells and Mo in the growth of human megakaryocytic progenitors (CFU-Mk) in vitro and to investigate γ-IFN effect on human megakaryocytopoiesis, normal human marrow (BM) was cultured in plasma clot in the presence and absence T cells, Mo and γ-IFN under conditions that support the formation of CFU-Mk derived colonies. The removal of T cells from BM (BM-T) caused a significant decrease (71.3 ±3.2 colonies observed vs 231.2 ± 38.5 colonies predicted) in both the number and size of CFU-Mk derived colonies, and no such changes were seen with Mo depletion (BM-Mo); co-culture of autologous T cells with BM depleted of both Mo and T cells (BM-Mo-T) caused a significant increase in CFU-Mk derived colonies and restored colony size. The addition of γ-IFN (<50-10,000 IU/ml) to BM caused a dose dependent inhibition of CFU-Mk (0-90%) as evidenced by decreased colony numbers and reduced colony size. The addition of γ-IFN (50-10,000 IU/ml) to BM-T caused reduced inhibition of CFU-Mk (0-60%); co-culture of T cells (but not Mo) pre-incubated with γ-IFN (10,000 IU/ml; 1 hour, 37 C followed by washing X 3) resulted in supression of CFU-Mk (80% inhibition with the addition of 1:4 T cells:marrow cells).

The results demonstrate that T cells have the ability to modulate the growth of human CFU-Mk in vitro and may, under appropriate conditions, either promote (normal T cells) or inhibit (γ-IFN activated T Cells) human megakaryocytopoiesis.

INTRODUCTION

Cellular interactions have been shown to play an important role in the modulation of human erythropoiesis and myelopoiesis in vitro[1,2], with immunocompetent accessory cells (T cells, monocytes/macrophages:

Mo) having the capacity to support the growth of human erythroid and myeloid progenitors in normal states[3,5], inhibit the growth of these hematopoietic precursors in clinical states of marrow failure[6-8], and produce numerous biologically active proteins (colony stimulating factors, interleukins and interferons) that offer potential mechanisms for many of these effects[9,12].

One of these proteins, γ-interferon (IFN), is a T lymphocyte product[13] that has been shown to cause a dose-dependent inhibition of normal human hematopoietic progenitors in vitro[14-16], to result in a significant incidence of myelosuppressive complications in clinical trials in vivo[17] and to be implicated in the mechanism of at least some cases of aplastic anemia[18]. Furthermore, previous studies in our laboratory have demonstrated that the suppressive effect of human γ-IFN on human erythropoiesis and myelopoiesis in vitro appears to be mediated to a significant degree by T cells and Mo[14,19]. Inhibition by γ-IFN was significantly reduced when the cultured marrow cells were first depleted of T cells and Mo, and the addition of T cells or Mo pre-exposed to γ-IFN to autologous marrow caused a profound suppression of erythroid and myeloid colony development that was not seen with the addition of control T cells or Mo[14,19]. Inhibition by γ-IFN was significantly reduced when the cultured marrow cells were first depleted of T cells and Mo, and the addition of T cells or Mo pre-exposed to γ-IFN to autologous marrow caused a profound suppression of erythroid and myeloid colony development that was not seen with the addition of control T cells or Mo[14,19]. Finally, these studies also suggest that γ-IFN may provide a valuable probe in the investigation of cellular interactions in human hematopoiesis in vitro.

Studies attempting to elucidate the role of T cells and Mo in the regulation of human megakaryocytic progenitors (CFU-Mk) in vitro have been much more limited and the results more inconsistent. Kanz et al[20] described a reduction in CFU-Mk after the depletion of T cells from normal marrow, but Gewirtz et al[21] failed to note a significant change and Geissler et al[22] reported that OKT4+ lymphocytes enhanced while OKT8+ lymphocytes decreased CFU-Mk proliferation. Geissler et al[23] also described an enhanced CFU-Mk cloning efficiency after Mo were depleted from normal marrow cells, but Gewirtz et al[21] did not find Mo depletion to cause a significant change in the growth of CFU-Mk. Gewirtz also described augmented megakaryocyte colony formation with the addition of peripheral blood NK cells. Finally, Nagasawa et al[24] reported T cell mediated supression of megakaryocytopoiesis in a lupus patient with amegakaryocytic thrombocytopenia, and Gewirtz et al[25] found evidence for T cell and Mo mediated suppression of megakaryocytopoiesis in two additional patients with amegakaryocytic thrombocytopenia, suggesting a role for cellular interactions in disease states affecting megakaryocytopoiesis. Thus, T cells and Mo appear capable of modulating megakaryocyte production, but specific roles for T cells, Mo and other marrow accessory cells remain to be clarified.

The studies reported here were undertaken to examine the role of normal T cells and Mo in the regulation of autologous human marrow CFU-Mk activity in vitro and to determine the effect of recombinant human γ=IFN on human CFU-Mk in vitro. As with the erythroid and myeloid progenitor systems, our results demonstrate that T cells are required for the optimal growth of CFU-Mk derived colonies, and that the anti-proliferative effect of γ-IFN on CFU-Mk is mediated to a significant degree through T cells. However, unlike the other progenitor systems, Mo do not appear to play a demonstrable role in either the proliferation of CFU-Mk or the mediation of γ-IFN effect on this hematopoietic progenitor. These studies provide insight into the mechanisms by which γ-IFN affects human megakaryocyte progenitors, contribute to our understanding of cellular interactions in human megakaryocytopoiesis and demonstrate the potential value of biologic response modifiers in elucidating the physiology of human hematopoiesis.

METHODS

Cells and Cell Separation Procedures

All bone marrow aspirations were performed at the posterior iliac crest. Heparinized bone marrow and peripheral blood were obtained from normal, healthy adult donors subsequent to obtaining informed consent. Bone marrow cells (BM) and peripheral blood mononuclear cells (PBM) were isolated by Ficoll-Hypaque density-gradient centrifugation (1.077 sp gr). When required, BM were depleted of monocyte-macrophages (Mo) or T cells by adherence to plastic surfaces and sheep red blood cell (SRBC) rosetting respectively. For Mo depletion, BM (5×10^6 cells/ml) were suspended in Iscove's modified Dulbecco medium (IMDM) with 20% fetal calf serum (FCS), placed in 75 cm^2 plastic tissue culture flasks and allowed to adhere for two h at 37°C in a humidified atmosphere of 5% CO2 in air. Nonadherent cells (BM-Mo) were then collected in the supernatants following gentle agitation of the flasks, spun at 1400 rpm for 10 min, washed once in IMDM, and resuspended in IMDM at desired concentrations. Removal of T lymphocytes from BM (BM-T) and from BM-Mo (BM-Mo-T) was achieved by double-rosetting with SRBC; the second rosette was performed overnight at 4°C. These procedures resulted in complete removal of T cells (0 - 0.2% residual T cells) as determined by FACS analysis with OKT3 and rosetting with SRBC, and near-complete removal of Mo (<0.2% residual Mo) as determined by FACS analysis with LeuM3 and staining with nonspecific esterase. Autologous T cells and Mo for use in co-culture studies were isolated from PBMC. Mo were recovered from the flasks by exposure to EDTA (2% in PBS) for 15 min, washed three times and resuspended in IMDM at desired concentrations; T cells were obtained by lysis of SRBC. The isolated T cell and Mo populations were found to be >90% pure by the phenotypic assessments described above for the depleted populations.

Clonal Assay of CFU-MK, Identification and Enumeration of Human Megakaryocyte Colonies

Initially, CFU-Mk derived megakaryocyte colonies were cultured, fixed, labeled with fluorescent antibody probes and counted, using 1 ml aliquots and maintaining the colonies in 35 mm plates throughout culture, processing and scoring as described by Mazur and Hoffman[26]. However, plates were difficult and time consuming to process, obscured colonies at the periphery of the clots and did not allow for an accurate assessment of megakaryocyte morphology. The plasma clot assay was, therefore, modified as described below to permit rapid processing of clots and consistent high quality morphology. Bone marrow cells (BM: 0.5 - 7.5 x 10^5 cells/1.1 ml; BM-Mo: 0.5 - 7.5 x 10^5 cells/1.1 ml; BM-T: 0.25 - 5 x 10^5 cells/1.1 ml; and BM-Mo-T: 0.25 - 5 x 10^5 cells/1.1 ml) were cultured in plasma clot composed of 0-15% aplastic anemia serum (AAS, see below), 20-25% normal AB serum, 10% citrated bovine plasma, 1% bovine serum albumin, Iscove's Modified Dulbecco Medium (IMDM), 5 x 10^{-5} M mercaptoethanol, 0.02 M minimum essential amino acids, 0.4 M L-glutamine, and 0.2M sodium pyruvate. When required, recombinant human gamma IFN and/or autologous periperal blood (PBL) Mo or T cells were added as described below. This complete mixture was then cultured in duplicate to quadruplicate 0.5 ml aliquots, using 15 mm culture wells. Plates were incubated at 37°C in a humidified atmosphere of 5% in air for 12 days. The clots were then removed, placed on glass slides, fixed with methanol:acetone (1:3) and air dried (approximately 15 min). Occasional slides were then stained with Wrights to assess conventional morphology, but, unless otherwise stated, clots were processed with antibody for phenotypic identification of megakaryocyte colonies as described below. Murine Tab monoclonal antibody to human platelet glycoprotein IIb[27] was diluted 1:1000 in phosphate-buffered saline (PBS) and layered over the fixed plasma clots, and the clots were then incubated for 60 min at 37°C in humidified 5% CO_2 with air. After washing three times with PBS, the specimens were reincubated for an additional 60 min with fluorescein-conjugated goat anti-mouse IgG (Meloy Laboratories, Springfield, VA) diluted 1:200 in PBS. After washing three times with PBS and distilled water, the specimens were counterstained with 0.125% Evans's Blue, washed again with distilled water, wet mounted with barbital:glycerol (1:3), and stored at -20°C until they were scored under a fluorescent microscope. Clusters of three or more brightly fluorescent cells were counted as megakaryocyte colonies, derived from CFU-Mk.

Aplastic Anemia Serum (AAS) Requirements and Plating Efficiency Studies

Initial studies were performed to assess culture requirements for AAS and the effect of increasing concentrations of marrow cells on CFU-Mk plating efficiency. Cultures were established with AAS (0 - 15% v/v, using two lots that had been standardized against a sample kindly supplied by Dr. Ronald Hoffman) and BM (0.5 - 7.5 x 10^5 cells/1.1 ml), BM-Mo (0.5 - 7.5 x 10^5 cells/1.1 ml), BM-T (0.25 - 5 x 10^5 cells/1.1 ml) and BM-Mo-T (0.25 - 5 x 10^5 cells/1.1 ml).

Co-Culture Studies With Autologous Mo and T Cells

Increasing numbers of autologous Mo or T cells (1 x 10^3 - 1 x 10^5) were added to marrow depleted of Mo and T cells (BM-Mo-T; 1 - 2 x 10^5 cells/ml) at the start of culture so that the numbers of Mo or T cells used in these co-culture studies ranged from 1-50% of the total mononuclear cells present in culture.

Studies With Gamma-IFN

The effect of γ-IFN on CFU-Mk cloning efficiency and colony conformation was examined in the following ways. (a) Increasing concentrations of γ-IFN (10 - 10,000 IU/ml) were added directly to the different cell preparations at the start of culture, and the γ-IFN containing mixture was incubated for 12 d as described above. (b) The exact numbers of bone marrow target cells (1 - 5 x 10^5 cells/ml) to be cultured were preincubated for 60 min with either control medium (IMDM with 10% FCS) or medium containing increasing doses of γ-IFN (100 - 10,000 IU/ml) in a 37°C water bath with gentle shaking, washed two to three times to remove residual γ-IFN, resuspended in complete culture medium and then incubated for the 12 d culture period without further addition of γ-IFN. (c) Increasing numbers of autologous Mo or T cells (1 x 10^3 - 1.25 x 10^5/ml) were preincubated for 60 min with either control medium or medium containing increasing doses of γ-IFN (100 - 10,000 IU/ml), washed 3 x, and added to the complete culture medium with BM-Mo-T (1 - 2.5 x 10^5 cells/ml) and incubated for the 12 d culture periods as described above.

Human Recombinant Gamma-IFN

The highly purified (99% pure) recombinant-DNA-derived preparations of γ-IFN used in these studies were provided by Biogen (Cambridge, Massachusetts) and had a specific activity of 7 x 10^7 IU/mg protein. All γ-IFN dilutions were carried out in IMDM.

RESULTS

Megakaryocyte Colony Conformation: Normal BM

With the modification of the procedure described by Mazur and Hoffman[26] it became possible to prepare permanent glass mounted preparations of the culture clot for easier handling and quantitative and morphologic evaluations of colonies. Microscopic evaluation of clots processed with antibody for phenotypic identification of megakaryocytic colonies revealed clusters of brightly fluorescent cells with distinct borders and nuclear lobes (Fig. 1). Cell size and ploidy varied within a given colony, but most cells were large and multinucleate (Fig. 2). Colony size varied with the manipulations described below, but when whole marrow was cultured in the presence of optimal concentrations of AAS, the cellular distribution was as follows: 68% of colonies with 3 -5 cells, 27% with 6-9 cells and 5% with >9 cells (Table 1). Morphologic examination of Wright stained preparation of the clots confirmed the antibody specificity, revealing clusters of large cells with cytoplasmic and nuclear features typical of megakaryocytes (Figs. 3 and 4).

Requirement for AAS and BM Plating Efficiency

Although normal human bone marrow cells formed megakaryocytic colonies in the absence of AAS, the addition of AAS to cultures resulted in significant increases in colony frequency (approximately threefold with 5% AAS), consistency and reproducibility. Subsequent studies were, therefore, performed in the presence of 5 - 10% AAS. Plating efficiency was highly

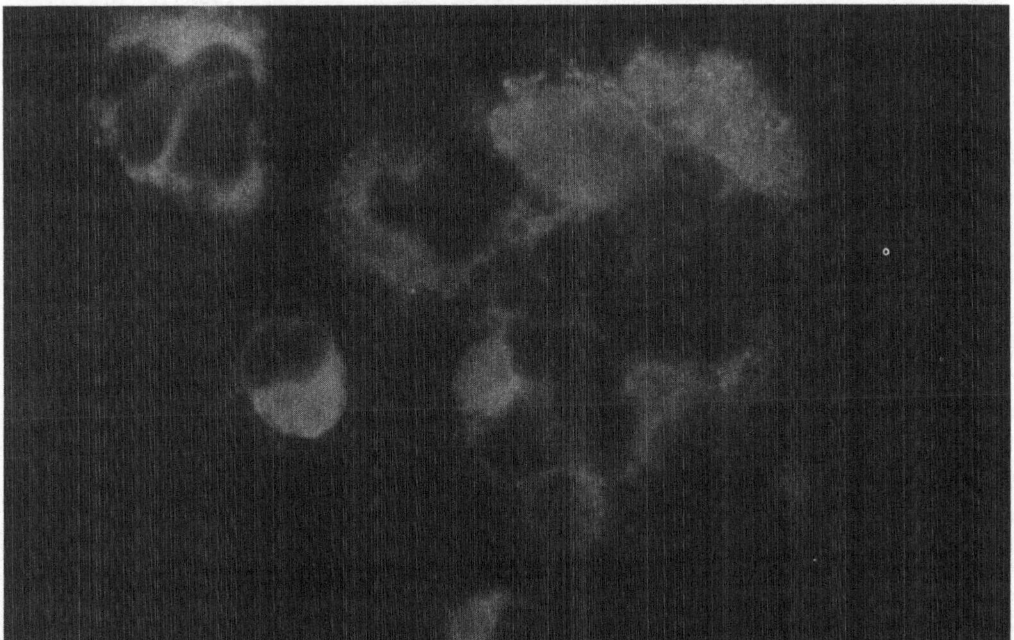

Fig. 1. Photomicrograph of a human megakaryocyte colony at twelve days of culture (40X). The colony was identified by use of a monoclonal Tab primary antibody and a fluorescein-conjugated secondary antibody: all sevel cells in the colony were strongly fluorescent.

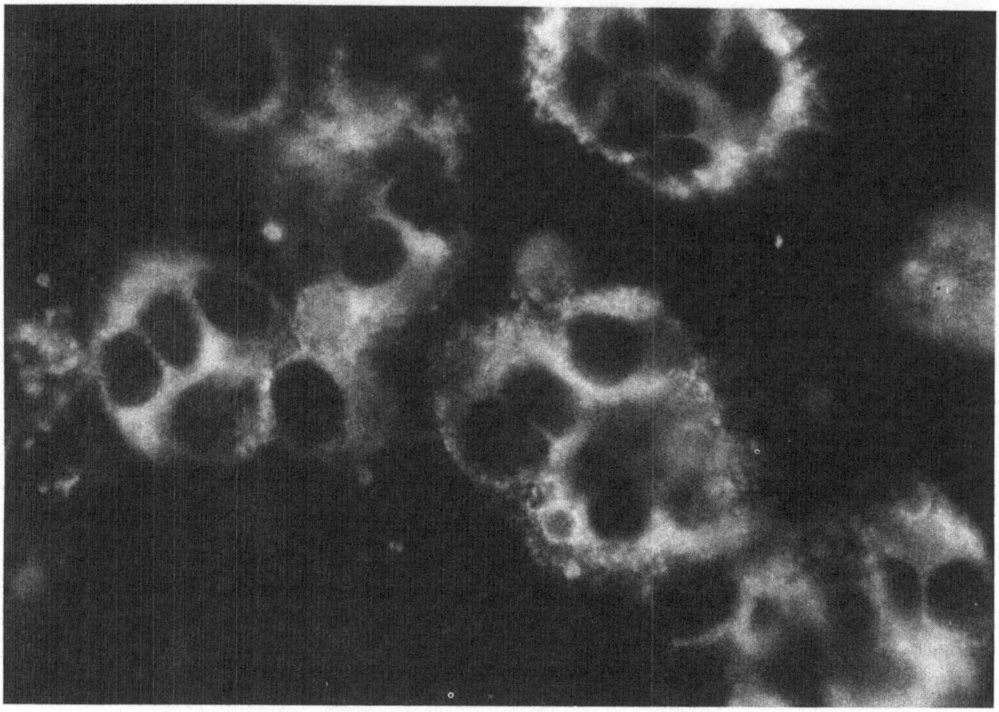

Fig. 2. Photomicrograph of a human megakaryocyte colony at twelve days of culture photographed at higher magnification (100X). The colony was identified with a monoclonal Tab primary antibody and a fluorescein-conjugated secondary antibody. The figure illustrates a range of megakaryocytic size and nuclear ploidy with a predominance of large multinucleate cells.

dependent on the concentration of BM plated. An increase in CFU-Mk cloning efficiency was seen as BM concentrations in culture were increased through a range of $0.5 - 5 \times 10^5$ cells/1.1 ml. However, as is shown in Table 2 this increase was non-linear, hinting at the possibility of accessory cell populations affecting the growth and differentiation of CFU-Mk in vitro.

Studies With T and/or Mo Depleted Bone Marrow Cells

Results presented in Table 3 demonstrate that the removal of Mo from BM prior to culture had no significant effect ($P>0.05$) on the number of CFU-Mk derived colonies. This was in contrast to a significant decrease in erythroid colony formation by the same Mo depleted fraction in vitro (data not shown). The removal of Mo also did not alter the size distribution of CFU-MK colonies (Table 1). However, a highly significant decrease ($P<0.01$) in both the number (Table 3) and size (Table 1) of CFU-Mk derived colonies occurred when bone marrow cells depleted of T lymphocytes were cultured. Table 3 also shows the results of cultures established with BM-Mo-T which exhibited significantly ($p<0.01$) decreased colony formation. This decrease was comparable to the decline in colony formation seen with T cell depletion alone (69.2 ± 4.7% VS 69.7 ± 3.7%) (Table 3); the decreases in colony sizes were also similar for the two experimental groups (Table 1). Furthermore, when increasing concentrations of BM, BM-Mo and BM-T were cultured with AAS and colony counts were corrected for progenitor enrichment entailed in the depletion procedures, only the T cell depleted marrow population showed linear growth properties throughout the concentrations examined (data not shown). No megakaryocytic colonies were detected when isolated populations of marrow Mo and T cells were cultured alone with AAS.

Co-Culture Studies With Autologous Normal T Cells and Mo

To further assess the role of T cells and Mo in the regulation of CFU-Mk activity in vitro, autologous T cells and Mo isolated from peripheral blood were co-cultured with donor marrow that had been depleted of T cells and Mo. Results presented in Figure 5 demonstrate that the ldition of T cells but not Mo to BM-Mo-T caused an impressive rise

Table 1. CFU-Mk derived colony size distribution in vitro[1]

	3-5 Cells/Colony	6-9 Cells/Colony	>9 Cells/Colony
BM	68%	27%	5%
BM + IFN	100%	0%	0%
BM-Mo	73%	27%	0%
BM-Mo + IFN	83%	17%	0%
BM-T	92%	7%	1%
BM-T + IFN	89%	11%	0%
BM-Mo-T	90%	9%	1%
BM-Mo-T + IFN	97%	3%	0%
BM-Mo-T + Mo	92%	7%	1%
BM-Mo-T + T	64%	30%	6%

1. See text for details of cell, interferon (IFN) and culture conditions. Data based on results from three separate studies.

Table 2. Megakaryocytic colony formation by normal human bone marrow cells in vitro[1]

Number of Cells Cultured	CFU-Mk/ml[2]
$5x10^4$ BM/ml	0.6 $\pm$ 0.6
$1x10^5$ BM/ml	2.0 $\pm$ 1.2
$2x10^5$ BM/ml	10.6 $\pm$ 1.4
$5x10^5$ BM/ml	58.0 $\pm$ 8.0

1. See text for details of culture procedure.
2. Mean $\pm$ 1 SEM of results from three separate studies each involving a different normal bone marrow donor.

in CFU-Mk derived colony formation beginning at a point where the added T cells comprised about 25% of the total mononuclear cell population in culture. Figure 5 also shows that a 75% increase in colony formation occurred at a T cell:BM-Mo-T ratio of 1:2. Furthermore, in those donors with low CFU-Mk cloning efficiencies this T cell mediated increase in

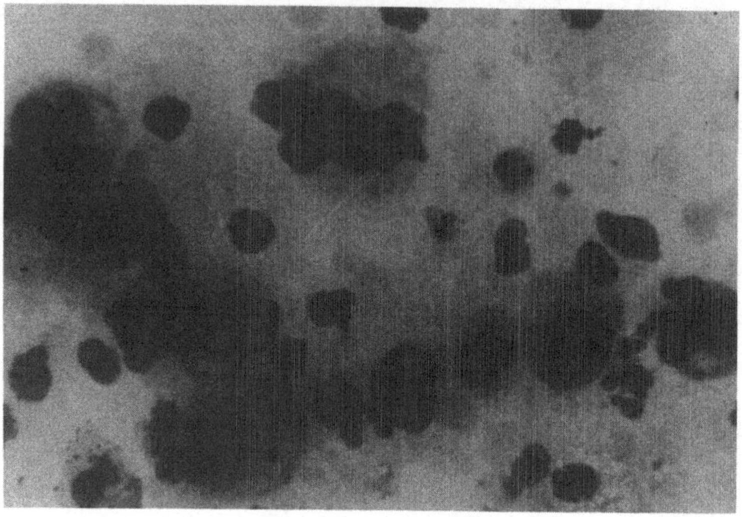

Fig. 3. Photomicrograph of a human megakaryocyte colony at twelve days of culture (40X). The colony was identified by Wright staining.

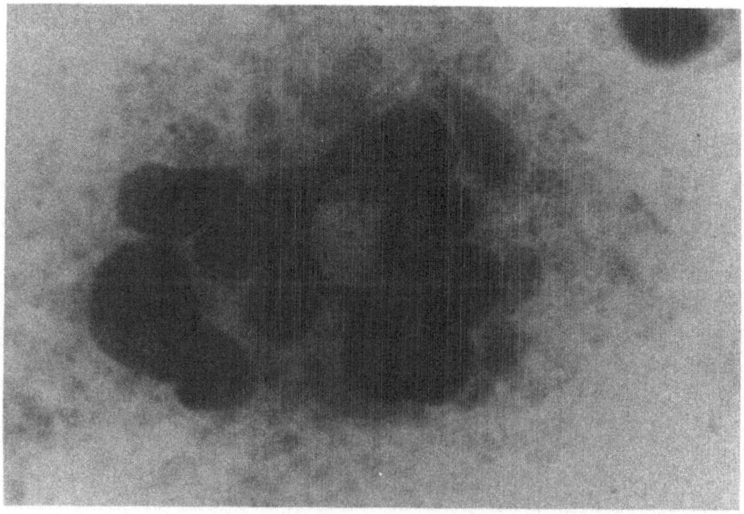

Fig. 4. Photomicrograph of an individual human megakaryocyte at twelve days of culture photographed at higher magnification (100X). The cell was identified by Wright staining and illustrates typical nuclear lobulations and a suggestion of cytoplasmic platelet formation.

megakaryocyte colony formation was significantly more pronounced, and CFU-Mk increases of greater than 100 fold were observed under comparable conditions. By contrast, the addition of Mo either had no significant

Table 3. Megakaryocytic colony formation by monocyte/ macrophage (Mo) and/or T lymphocyte depleted normal human bone marrow cells in vitro[1]

	CFU-Mk / 10^5 Cells[2]		
	Predicted[3]	Observed	% Decrease
BM	-	12.6 $\pm$ 2.1	-
BM-Mo	16.6 $\pm$ 2.8	14.4 $\pm$1.6	13.2
BM-T	231.2 $\pm$ 38.5	71.3 $\pm$ 3.2	69.2
BM-Mo-T	311.1 $\pm$ 51.8	94.4 $\pm$ 3.8	69.7

1. See text for details of Mo and/or T depletions and culture procedures.
2. Mean $\pm$ 1 SEM of results from five separate experiments each invoving a different normal bone marrow donor.
3. Determined with reference to % recovery of cells following Mo and/or T cell depeletions. Isolated Mo or T cell fractions did not produce colonies in vitro.

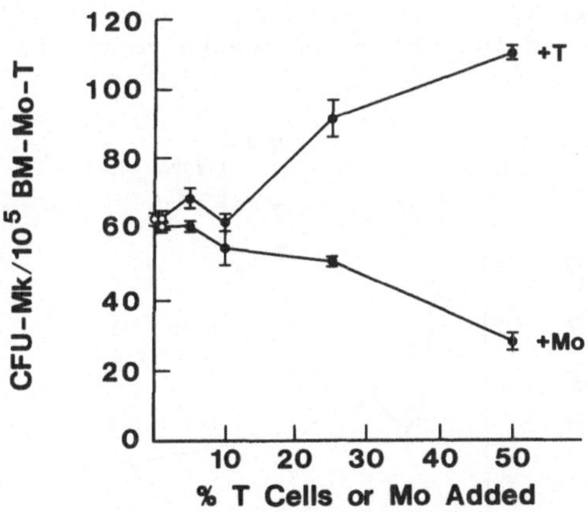

Fig. 5. Stimulation of normal human marrow CFU-MK by autologous T. lymphocytes in vitro. Each data point represents mean ± 1 SEM of results from four separate studies each involving a different normal donor. Monocytes (Mo) or T lymphocytes were added at indicated concentrations to autologous marrow depleted of Mo and T cells (BM-Mo-T). See text for details of cell and culture conditions.

effect on CFU-Mk derived colony formation (donors with low CFU-Mk cloning efficiencies and some donors with high CFU-Mk cloning efficiencies) or caused inhibition at high Mo concentrations (some donors with high cloning efficiencies). The stimulatory effect of T cells on CFU-Mk was accompanied in all cases with a return to larger colony size distribution (Table 1).

Studies With Gamma-IFN

The direct addition of γ-IFN to BM resulted in a dose-dependent inhibition of CFU-Mk (Figure 6). γ-IFN doses of greater than 250 IU/ml resulted in a sharp reduction in colony formation to approximately 80% inhibition at 1500 IU/ml, and higher γ-IFN doses caused a further reduction in CFU-Mk derived colony formation to approximately 90% inhibition at 10,000 IU/ml. γ-IFN doses less than or equal to 250 IU/ml had no significant effect on megakaryocyte colony formation by BM. The suppressive effect of γ-IFN on CFU-Mk was also evident by the decreased size and ploidy of colonies formed (Table 1).

Removing Mo from BM prior to the addition of γ-IFN caused only a modest reduction in the ability of γ-IFN to supress CFU-Mk (Fig. 7), and significant inhibition of megakaryocyte colony formation occurred with nearly all doses of γ-IFN employed. However, when γ-IFN was added to BM-T a different pattern emerged. Figure 8 shows that the abiltiy of γ-IFN to inhibit CFU-Mk colony formation by BM-T was minimal, and only at the highest concentration of γ-IFN used (10,000 IU/ml) a significant inhibition of CFU-Mk (60%) became evident. Lower doses of γ-IFN instead caused an increase in megakaryocyte colony formation; this effect was seen at γ-IFN concentrations of 50-2500 IU/ml, and was most pronounced at the lowest γ-IFN doses (100% with 50 IU/ml) (Fig. 8). Furthermore, the removal of both Mo and T cells from the marrow fraction resulted in a similar pattern of CFU-Mk inhibition to that seen with the removal of T cells alone (Fig. 9). Only very large doses of γ-IFN (10,000 IU/ml) resulted in a pronounced inhibition of megakaryocyte colony formation

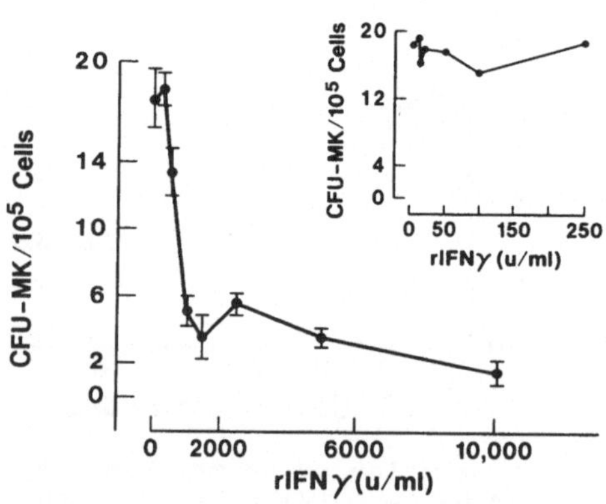

Fig. 6. Effect of direct addition γ-IFN on normal human marrow CFU-Mk in vitro. Each data point represents mean ± 1 SEM of results from five separate experiments each involving a different normal marrow donor. Inset illustrates a lack of inhibition with γ-IFN concentrations $\leq$250 IU/ml. γ-IFN was added at the start of culture. See text for details of culture and γ-IFN conditions.

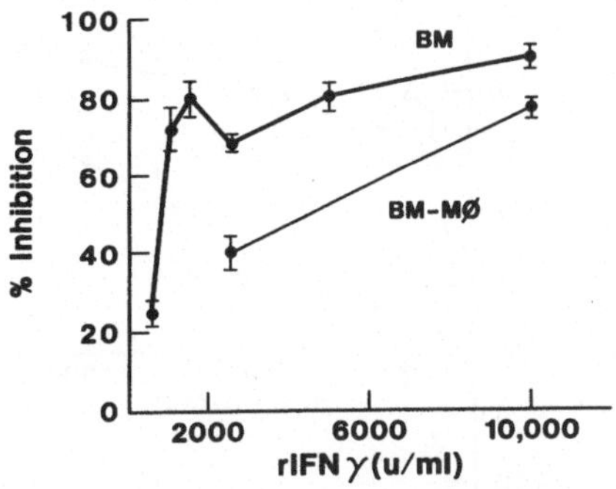

Fig. 7. Effect of direct addition of γ-IFN on megakaryocytic colony formation by monocyte depleted normal human marrow cells (BM-Mo) in vitro. Each data point represents mean ± 1 SEM of results from four separate experiments each involving a different normal marrow donor. γ-IFN was added at the start of culture. See text for details of Mo depletion and culture conditions.

(60%), and low doses of γ-IFN again stimulated rather than inhibited the proliferation of CFU-Mk. Thus T cells also appeared to play an important role in the mediation of the γ-IFN induced suppression of CFU-Mk, and their presence in culture also appeared to mask an additional stimulatory effect by γ-IFN that was easily demonstrated in marrow fractions depleted of T cells.

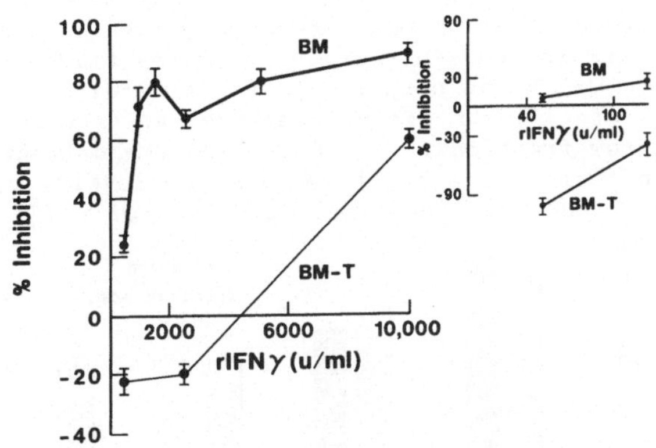

Fig. 8. Effect of direct addition of γ-IFN on megakaryocytic colony formation by T lymphocyte depleted normal human bone marrow cells (BM-T) in vitro. Each data point represents mean ± SEM of results from four separate experiments each involving different normal marrow donor. Points above the horizontal axis reflect inhibition; points below the horizontal axis reflect stimulation. Inset illustrates stimulatory effect of low doses of γ-IFN. γ-IFN was added at the start of culture. See text for details of T cell depletion and culture conditions.

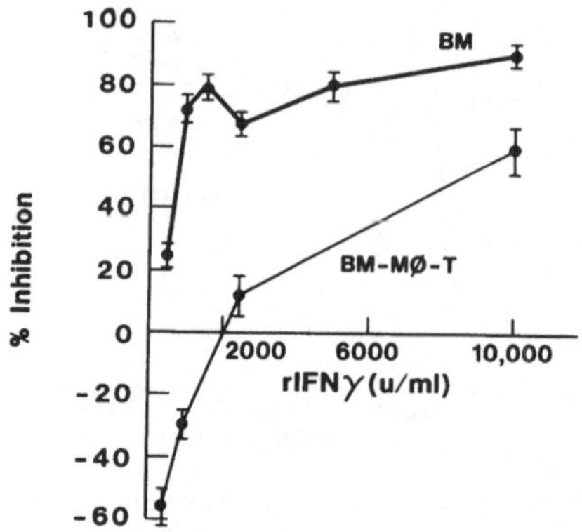

Fig. 9. Effect of direct addition of γ-IFN on megakaryocytic colony formation by Mo and T lymphocyte depleted normal human bone marrow (BM-Mo-T) in vitro. Each data point represents mean ± 1 SEM of results from five separate experiments each involving a different normal marrow donor. Points above the horizontal axis reflect a stimulation. γ-IFN was added at the start of culture. See text for details of Mo and T cell depletions and culture conditions.

Preincubation Studies With Gamma-IFN

To determine whether the suppression of megakaryocytopoiesis by high doses of γ-IFN required continuous γ-IFN exposure, preincubation studies were performed in which the marrow fractions were exposed to 10,000 IU/ml or γ-IFN for 60 min, washed two-three times, cultured without further exposure to γ-IFN and compared to direct addition studies (described above). Again, significant differences emerged between the T replete and T depleted marrow fractions. In the BM and BM-Mo fractions a 60 min. preincubation period with γ-IFN resulted in levels of inhibition comparable to those seen with the direct addition of γ-IFN to culture

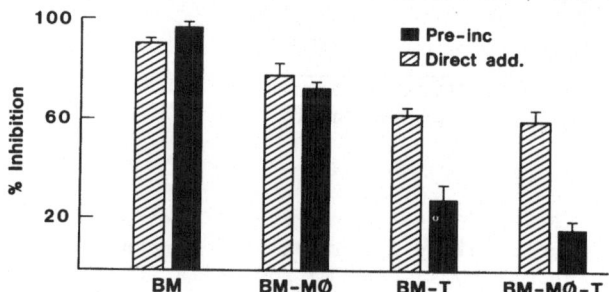

Fig. 10. Effect of γ-IFN on megakaryocytic colony formation by normal (BM), monocyte depleted (BM-Mo), T cell depleted (BM-T) and monocyte and T cell depleted (BM-Mo-T) human bone marrow cells in vitro. Each data point represents mean ± 1 SEM of results from two separate experiments each involving a different normal marrow donor. See text for details of γ-IFN exposure, Mo and T cell depletion and culture conditions.

(Fig. 10), but in the BM-T and BM-Mo-T fractions preincubation studies with γ-IFN resulted in a significantly lower level of inhibition than was seen in direct addition studies. Thus, although pronounced inhibition of CFU-Mk could be demonstrated using only a 60 min preincubation period with γ-IFN, a longer period of γ-IFN contact was required to effect maximal inhibition in the T cell depleted marrow fractions

Studies With Gamma-IFN Activated Autologous Mo and T Cells

To further define the role of immunocompetent cells in human megakaryocytopoiesis and the mediation of γ-IFN effect on human CFU-Mk, autologous Mo and T cells were preincubated with either control medium or γ-IFN (10,000 IU/ml), washed and co-cultured with marrow cells without further addition of γ-IFN. CFU-Mk derived colony counts in the cultures containing Mo or T cells that had been preincubated with γ-IFN were then expressed as a percent of colony counts in the cultures containing Mo or T cells that had been preincubated with only control media (Fig. 11). Autologous T cells that had been preincubated with γ-IFN caused a significant inhibition of CFU-Mk when as few as 5% γ-IFN activated T cells were added, and higher concentrations of γ-IFN treated T cells resulted in more pronounced inhibition, and this contrasts sharply with a modest trend toward CFU-Mk stimulation that occurred with the addition of γ-IFN exposed Mo (Fig. 11).

DISCUSSION

The results presented here demonstrate that T lymphocytes play an important role in the optimal growth and/or differentiation of normal human bone marrow CFU-Mk in vitro. This was evident from our findings that 1) the removal of T cells from normal bone marrow prior to culture resulted in a significant decrease in both the number and size of CFU-Mk colonies, and 2) the addition of autologous T cells to the T cell depleted bone marrow normalized the CFU-Mk colony growth pattern. In this regard, at higher than physiologic concentrations, normal T cells enhanced the growth of megakaryocytic colonies beyond that exhibited by normal marrow in vitro.

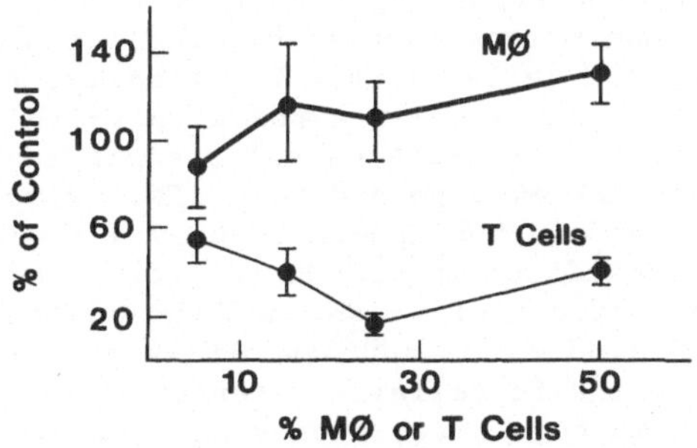

Fig. 11. Effect of γ-IFN activated Mo and T cells on autologous normal CFU-Mk in vitro. Each data point represents mean ± 1 SEM of two separate experiments each involving a different donor. See text for details of Mo and T cell isolations. γ-IFN exposure and culture conditions.

It is now well established that immunocompetent accessory cells (T cells and Mo) can modulate the growth and differentiation of human erythroid (BFU-E and CFU-E[1,3,4], myeloid (CFU-GM)[1,6] and multipotential (CFU-Mix)[28] progenitors in vitro. In this regard, the optimal growth of human BFU-E in vitro has been shown to require the presence of T cells[4] and/or Mo[3]. Furthermore, because optimal BFU-E development in T cell depleted marrow mononuclear cell fractions can also be achieved in the presence of sufficient amounts of burst promoting activity (BPA)[9], the T cell effect on erythropoiesis is believed to be humorally mediated. Similar types of experimental evidence have also been presented for the action of Mo on erythropoiesis in vitro[3,1]. More recent evidence suggests that T cells may also affect the growth of human CFU-Mk in vitro[20,22], and the studies presented here provide strong support for such a modulatory role. And while the mechanism(s) by which normal T cells promote human megakaryocytopoiesis in vitro are not yet known, it is interesting to note that we were unable to abrogate this effect with increased concentrations of aplastic anemia serum, suggesting more complex humoral and/or cell-to-cell contact actions than a single humoral effect. Furthermore, while Mo can greatly influence the growth of BFU-E in vitro[3], they exerted little or no effect on human CFU-Mk in vitro: except for a modest inhibitory effect at very high Mo concentrations, no definitive change in CFU-Mk growth profile was noted when Mo were added to marrow cells depleted of Mo and T cells. Finally, that T cells but not Mo play an important modulatory role in the regulation of human CFU-Mk in vitro was also evident in studies of the effect of human recombinant γ-IFN on CFU-Mk in vitro.

We found that as with the erythroid[14,15] and myeloid[15] progenitors, human γ-IFN is a potent inhibitor of CFU-Mk in vitro. The addition of γ-IFN to cultures of normal human bone marrow cells resulted in a highly significant dose-dependent suppression of CFU-Mk cloning efficiency and colony size. This inhibition was seen both with the direct addition of γIFN as well as after exposure of marrow cells to γ-IFN for brief periods. Thus the continued presence of the lymphokine was not a necessary factor in the suppression of CFU-Mk in vitro. The inhibitory effect was found to be mediated to a significant degree by T lymphocytes: removal of T cells before the addition of γ-IFN significantly reduced the suppressive effect of IFN. Furthermore, γ-IFN in low doses was found to stimulate the formation of CFU-Mk derived colonies by T depleted bone marrow cells, while no such effect was seen in whole marrow samples. It is possible that the absence of a CFU-Mk enhancing effect by low concentrations of γ-IFN in the whole bone marrow samples may reflect a masking inhibitory effect of γ-IFN activated T cells on CFU-Mk. Finally, the results presented here also show that autologous T cells exposed to γ-IFN for short periods and subsequently washed free of γ-IFN caused a significant inhibition of CFU-Mk without direct marrow exposure to γ-IFN. However, unlike the erythroid system[14], Mo did not appear to play a significant role in the mediation of γ-IFN effect on CFU-Mk in vitro: removal of Mo did not alter the inhibitory effect of γ-IFN, and the addition of autologous Mo pre-exposed to γ-IFN failed to suppress CFU-Mk growth in vitro. Such lineage specific differences in cellular interactions may in part reflect fundamental differences in cell size, ploidy and compartmentalization that distinguish megakaryocyte and platelet differentiation from their erythroid and myeloid counterparts.

The results presented here provide evidence that the same accessory cell type (e.g. T lymphocytes) may possess the ability to both promote and, under appropriate conditions, inhibit hematopoietic progenitor cell activity in vitro: the addition of normal T cells increased the size and number of CFU-Mk derived colonies, and the addition of γ-IFN stimulated T cells decreased the size and number of these colonies. Furthermore, while the data indicate that the activation of T cells with γ-IFN renders these cells inhibitory to autologous CFU-Mk, the possibility that these cells may also exert (masked) stimulatory influences on hematopoiesis should also be considered. In this regard, there is considerable evidence to suggest that the tumorcidal activity of IFN activated effector cells is mediated by direct cell-cell contact: in the absence of direct and prolonged contact between the γ-IFN activated cells and the target cell population considerably reduced levels of target cell killing have been noted[29]. Furthermore, γ-IFN exposed marrow accessory cells are capable of producing multilineage hematopoietic growth factors that can promote the proliferation of hematopoietic progenitors in vitro[11]. Thus, it is possible that, depending on their state of activation and their duration of exposure and proximity to (hematopoietic) target cells, accessory T cells may either suppress or enhance hemotopoietic processes. Hence a critical balance of opposing activities could distinguish endpoints of stimulation and inhibition, dependent on the composition of the immediate cellular environment and local concentrations of growth promoting and inhibiting activities.

It should be emphasized that our results do not exclude significant direct effects by γ-IFN on CFU-Mk, and accessory cells other than T cells and Mo may also be important in mediating some of the cellular effects described above. They do, however, demonstrate that in vitro, T cells have the potential to modulate human megakaryocytopoiesis and appear to play an important role in γ-IFN's suppression of megakaryocytic progenitors. These interactions may be important in clinical states of thrombocytopenia characterized by a reduction in marrow megakaryocytes, such as cell-mediated amegakaryocytic thrombocytopenia, viral-associated thrombocytopenia and some cases of prolonged thrombocytopenia following bone marrow transplantation.

REFERENCES

1. Lipton, J. and D. Nathan. 1985. Interaction between lymphocytes and macrophages in hematopoiesis. In Hematopoietic Stem Cells. D. Golde and F. Takaku, editors. Marcel Dekker, New York. 145-202.
2. Cline, M. and D. Golde. 1979. Cellular interactions in haematopoiesis. Nature. 277: 177-181.
3. Zuckerman, K. 1981. Human erythroid burst forming units. Growth in vitro is dependent upon monocyte but not T-lymphocytes. J. Clin. Invest. 67: 702-710.
4. Nathan, D., L. Chess, D. Hillman, B. Clarke, J. Breard, E. Merler, D. Housman. 1978. Human erythroid burst-forming unit: T-cell requirement for proliferation in vitro. J. Exp. Med. 147: 324-339.
5. Ascensao, J., N. Kay, M. Banisadre, and E. Zanjani. 1981. Cell-Cell interaction in human granulopoiesis: role of T lymphocytes. Exp. Hematology 9: 473-478.
6. Bagby, G. and J. Gabourel. 1979. Neutropenia in three patients with rheumatic disorders. Supression of granulopoiesis by cortisol-sensitive thymus-dependent lymphocytes. J. Clin. Invest. 64: 73-82.
7. Zoubos, N., P. Gascon, J. Djeu, S. Trost, and N. Young. 1985. Circulating activated suppressor T-lymphocytes in aplastic anemia. N. Engl. J. Med. 312: 257-262.
8. Zanjani, E., P. McGlave, S. Davies, M. Banisadre, M. Kaplan, and G. Sarosi. 1982. In vitro supression of erythropoiesis by bone marrow adherent cells from some patients with fungal infection. Brit. J. Hematol. 50: 479-490.

9. Aye, M. 1977. Erythroid colony formation in cultures of human marrow: effect of leukocyte conditioned mediu. J. Cell. Physiol. 91: 69-78.
10. Metcalf, D. 1986. The molecular biology and functions of the granulocyte-macrophage colony-stimulating factors. Blood 67: 257-267.
11. Piacibello, W., L. Lu, B. Rubin, and H. Broxmeyer. 1984. Human gamma interferon enhances the release of granulocyte-macrophage colony stimulating factors (GM-CSF) from a sorted population of OKT4+ lymphocytes stimulated by phytohemagglutinin (PHA). Blood 64 (Suppl):455.
12. Herrmann, F. and J. Griffin. 1986. T cell-monocyte interactions in the production of humoral factors regulating human granulopoiesis in vitro. J. Immunol. 136: 2856-2861.
13. Revel, M. 1984. The interferon system in man: nature of the interferon molecules and mode of action. In Antiviral Drugs and Interferon. Becker Y., editor. Marinus Nijhoff Publishing, Boston. 357-433.
14. Mamus, S., S. Beck-Schroeder, and E. Zanjani. 1985. Supression of normal human erythropoiesis by gamma interferon in vitro: role of monocytes and T lymphocytes. J. Clin. Invest. 75: 1496-1503.
15. Broxmeyer, H., L. Lu, E. Platzer, C. Feit, L. Juliano, and R. Rubin. 1983. Comparative analysis of the influences of human gamma, alpha and beta interferons on human multipotential (CFU-GEMM), erythroid (BFU-E) and granulocyte-macrophage (CFU-GM) progenitor cells. J. Immunol. 131: 1300-1305.
16. Lu, L., K. Welte, J. Gabrilove, G. Hangoc, E. Bruno, R. Hoffman, and H. Broxmeyer. 1986. Effects of recombinant human tumor necrosis factor alpha, recombinant human gamma-interferon, and prostaglandin E on colony formation of human hematopoietic progenitor cells stimulated by natural human pluripotent colony--stimulating factor, pluripoietin alpha, and recombinant erythropoietin in serum-free cultures. Cancer Research 46: 4357-4361.
17. Quesada, J. et al. 1986. Clinical toxicity of interferons in cancer patients: a review. J. Clin. Oncol. 4: 234-243.
18. Zoumbos, N., P. Gascon, J. Djeu, and N. Young. 1985. Interferon is a mediator of hematopoietic suppression in aplastic anemia in vitro and possibly in vivo. Proc. Natl. Acad. Sci. USA 82: 188-192.
19. Mamus, S., M. Oken, and E. Zanjani. 1986. Suppression of normal human erythropoiesis by human recombinant DNA-produced alpha-2--interferon in vitro. Exp. Hematol. 14: 1015-1022.
20. Kanz, L., G. Lohr, A. Fauser. 1983. Modulation of human megakaryocytic colony formation. In Normal and Neoplastic Hematopoiesis. D. Golde and P. Marks, editors. Alan R. Liss, Inc., New York 359-368.
21. Gewirtz, A., K. Mangan, and Y. Wen. 1986. Cellular regulation of in vitro human megakaryocytopoiesis. Clin. Res. 34: 458A.
22. Geissler, D., L.Lu, E. Bruno, H. Yank, H. Broxmeyer, and R. Hoffman. 1985. The influence of T-lymphocyte subsets and humoral factors on colony formation by human bone marrow and blood megakaryocyte progenitor cells in vitro. Blood 66. (Suppl):452.
23. Geissler, D., G. Konwalinka, and C. Peschel, et al. 1985. A regulatory role of activated T-lymphocytes on human megakaryocytopoiesis in vitro. Br. J. Haemal. 60: 233-238.
24. Nagasawa, T., T. Sakurai, H. Kashiwagi, and T. Abe. 1986. Cell-mediated amegakaryocytic thrombocytopenia associated with systemic lupus erythematosis. Blood 67: 479-483.
25. Gewirtz, A., M. Sacchetti, R. Bien, and W. Barry. 1986. Cell-mediated suppression of megakaryocytopoiesis in acquired amegakaryocytic

thrombocytopenic purpura. Blood 68: 619-626.

26. Mazur, E., R. Hoffman, J. Chasis, S. Marchesi, and E. Bruno. 1981. Immunofluorescent identification of human megakaryocyte colonies using an antiplatelet glycoprotein antiserum. Blood. 57: 277-286.

27. McEver, R., E. Bennett, and M. Martin. 1983. Identification of two structurally and functionally distinct sites on human platelet membrane glycoprotein IIb-IIIa using monoclonal antibodies. J. Biol. Chem. 258: 5269-5275.

28. Neumann, H. and A. Fauser. 1982. Effect of interferon on pluripotential hemopoietic progenitors (CFU-GEMM) derived from human bone marrow. Exp. Hematol. 10:587-590.

29. Herrmann, F., R. Schmidt, J. Ritz, and J. Griffin. 1987. In vitro refulation of human hematopoiesis by natural killer cells: analysis at a clonal level. Blood 69: 246-254.

THE EFFECTS OF PROSTAGLANDIN E_1 ON MEGAKARYOCYTE PROLIFERATION IN VITRO

George W. Cooper and Xiao Ping Hou

Department of Biology
The City College of the
City University of New York
New York, NY 10031, U.S.A.

ABSTRACT

A concentration of 560 nM Prostaglandin E_1 (PGE_1), in a medium containing either 10% serum and 10% plasma, or 20% plasma and a platelet extract prepared from 24 x 10^6 platelets/ml, caused after four days of incubation in a plasma clot culture system, highly significant increases over the controls in both the number of megakaryocyte colonies/clot and megakaryocytes/clot. A linear dose-response relationship over the entire 14-560 nM range was also demonstrated in cultures containing serum, between the concentration of PGE_1 in each culture, and the number of megakaryocyte colonies/clot and megakaryocytes/clot. However, PGE_1 did not appear to stimulate megakaryocytes in cultures containing only 20% rat plasma.

INTRODUCTION

Williams[1] reported that prostaglandin $E_2(PGE_2)$, in concentrations as low as 3 nM, inhibited differentiation of macrophage colony precursors in agar cultures of mouse bone marrow containing fetal calf serum. Although PGE^2 was reported in this same study to have had no effect on the differentiation of megakaryocyte colonies, when added at levels ranging from 10 µM to 1 pM, the published data implied that PGE_2 did affect megakaryocyte differentiation. The author reported that when the PGE_2 concentrations were kept between 10 µM and 10 nM 2-8% of the colonies contained only megakaryocytes. There were also mixed colonies containing megakaryocytes and other cell types. Neither megakaryocyte colonies nor mixed colonies containing megakaryocytes were found in any cultures containing 100 pM to 1 pM PGE_2. They were also not found in control cultures containing no PGE. These data implied that prostaglandin E might have stimulated megakaryocyte colony differentiation. We studied the effects of prostaglandin E_1 (PGE_1) on the proliferation of megakaryocyte colonies and megakaryocytes in plasma clot cultures of rat bone marrow. The effects of PGE^1 alone and also in combination with either platelet extract or serum on the numbers of megakaryocyte colonies and megakaryocytes were investigated.

MATERIALS AND METHODS

Bone marrow cells, spleen cells, plasma and serum were obtained for each experiment from untreated Wistar rats that had been raised in our laboratory for over ten years. Bone marrow was flushed from both femurs of 2-4 month-old male or female rats into plastic tubes with a mixture of 9 parts Leibovitz' L-15 medium (Gibco, Grand Island, N.Y., U.S.A.) and one part of rat serum. Blood was drawn by cardiac puncture from 6-12 month-old male rats into sterile plastic syringes containing 3.8% (w/v) sodium citrate (1:9 citrate: whole blood). The blood was sedimented at 1100 x g at 4°C for 15 minutes in an anglehead centrifuge to obtain the plasma. Serum was collected by a similar procedure, except that the anticoagulant was not used, and the blood was allowed to clot for over two hours in the refrigerator before centrifugation. Both plasma and serum were sterile-filtered through 0.45 µm pore size Millipore filters after separation.

Pokeweed mitogen (Gibco)-stimulated rat spleen cell conditioned medium (PWM-SCM) was prepared as follows: The spleens of 2-4 month old female rats were minced in a culture dish and homogenized in a plastic test tube with a 10 ml plastic syringe and a 22 gauge, 3.8 cm needle. Nucleated cells were counted and cultured at concentrations of 2×10^5/ml in 35 mm culture dishes for two days. Each dish contained 70% Leibovitz' L-15 medium; 10% bovine embryo extract diluted 1:4 (v/v) with L-15; 10% rat serum; and 10% pokeweed mitogen (1/300 final dilution) in L-15. The medium was collected and filtered through 0.22 µm Millipore filters. Each PWM-SCM was tested for potency before its final use and was stored at -70°C for no longer than two months. Nakeff and Daniels-McQueen's procedure (1976), modified for rat bone marrow, was used to culture the cells in 96 well, round bottom plastic plates. Each 0.3 ml plasma clot culture contained 3×10^5 nucleated cells suspended in 40-50% Leibovitz' L-15 medium; 10% beef embryo extract, diluted 1:4 in L-15; 10% PWM-SCM; 10% rat plasma; either 10% rat serum or an additional 10% rat plasma; 10% isotonic saline containing PGE_1 or a control; and, in one group of experiments, 10% platelet extract dissolved in isotonic saline. The PGE_1 was dissolved in 50% ethanol. The stock solutions were diluted with isotonic saline and, for the lower PGE_1 concentrations, also diluted with 0.25% ethanol in saline, to a final ethanol concentration in the culture medium of 0.025%. Prostaglandin E_1 was not added to the cultures of either of the two control groups. An isotonic saline control containing no ethanol, and also a second control containing 0.025% ethanol in isotonic saline were used.

Rat platelet concentrates containing no visible red blood cells, when examined under the microscope, were prepared from blood anticoagulated with 3.8% sodium citrate. Platelets were separated from the blood by differential centrifugation, redissolved in isotonic saline, and resedimented. The supernatent solutions were removed with plastic Pasteur pipets, removing the platelets from the tubes, and leaving most of the red blood cells behind. The latter step was repeated four more times. The platelets were disrupted by four cycles of freezing and thawing[2] and finally sedimented at 3500 x g for 30 min. The supernatant platelet extract was added to the cultures at the beginning of each incubation at a final concentration equivalent to 24×10^6 platelets/ml.

The cultures were incubated at 37°C in an atmosphere of 6% carbon dioxide in air and 97% relative humidity. Plasma clots were removed daily from the 96-well plates, and examined for megakaryocyte colonies and megakaryocytes.

Acetylcholinesterase positive (AChE+) megakaryocytes and megakaryocyte colonies were identified after staining by use of the "direct coloring" thiocholine method of acetylcholinesterase activity[3]. The cells were counterstained in Harris' hematoxyline #2, and were clarified in running tap water. Final preparations were examined under the light microscope at magnifications of 200 X and 430 X for AChE+ cells and colonies. An AChE+ cell colony was defined as a group of three or more cells in close proximity.

The Student t test was used for the determination of statistical significance.

RESULTS

Prostaglandin E_1 (PGE_1), dissolved in 0.25% ethanol and isotonic saline (final ethanol concentration: 0.025%), was added on day 0 to the wells to final concentrations of 14, 70, 140, 280, or 560 nM. These cultures also contained 10% rat serum, and 10% rat plasma. Isotonic saline, or ethanol dissolved in isotonic saline at a final ethanol concentration of 0.025%, was added, instead of the PGE_1, to the two control groups. There were highly significant increases in the numbers of megakaryocyte colonies/clot, in the cultures containing 560 nM PGE_1 of 98% and 71%, respectively over the saline and 0.025% ethanol control groups ($P < 0.001$ for both groups) (Figure 1). There were twelve to twenty four plasma clots per point. There were also significant increases after four days of incubation in the number of megakaryocytes/clot in these same cultures over the saline and 0.025% ethanol control groups of 92% and 60%, respectively ($P < 0.001$ for both groups) (Figure 2). Each point represented twelve to twenty four plasma clots. There was a linear dose-response relationship between the concentration of PGE_1 in each culture, and both the number of megakaryocyte colonies/clot and the number of megakaryocytes/clot over the entire range of 14-560 nM (Figure 3).

Prostaglandin E_1 was also added in concentrations of 280 or 560 nM to cultures containing platelet extract equivalent to a concentration of 24 x 10^6 platelets/ml, and 20% rat plasma, but no added serum. There

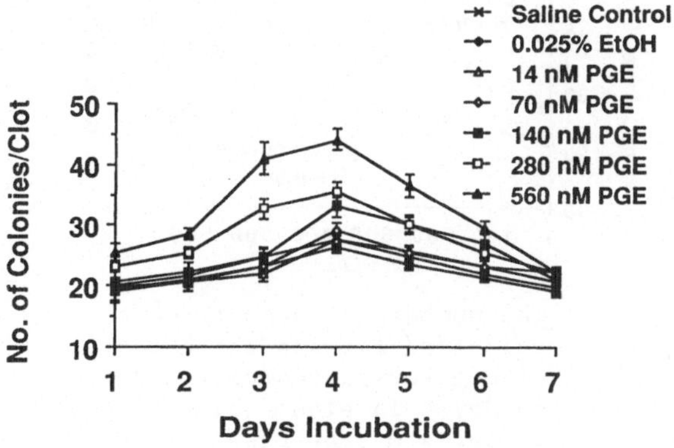

Fig. 1. The number of megakaryocyte colonies/clot identified in plasma clot tissue cultures of rat bone marrow treated with prostaglandin E_1 on day 0. These cultures contained 10% rat plasma. Each point in this and the following figures represented the mean and ± 1 SEM.

were highly significant increases in the number of megakaryocyte colonies/-clot after four days of incubation, over the saline and ethanol controls, of 113% and 85%, respectively, in the cultures containing 560 nM of PGE_1 (P <0.001 for both groups) (Figure 4). There were eight plasma clots per point. A highly significant increase was also noted in megakaryocytes/-clot over the two control groups after four days of 96%, and 75%, respectively (P <0.001 for both groups) (Figure 5).

Although PGE_1 stimulated the growth of megakaryocytic cells and colonies in plasma clot cultures of rat bone marrow containing 10% rat serum or a platelet extract, the hormone did not stimulate the growth of megakaryocyte colonies or megakaryocytes in rat bone marrow cultures that contained 20% rat plasma, but neither serum nor platelet extract (Figures 6 and 7). There were twelve plasma clots for each point.

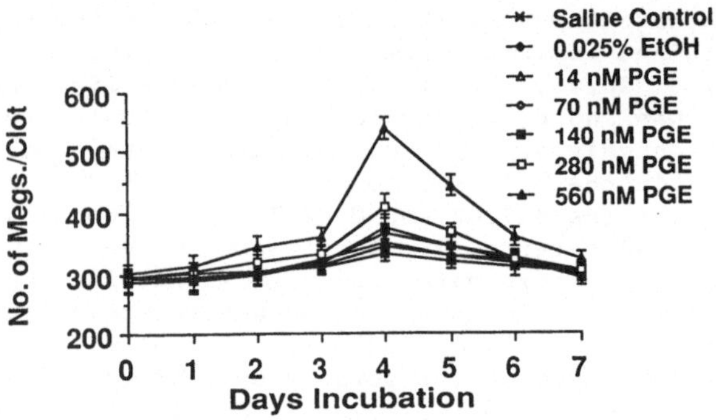

Fig. 2. The numbers of megakaryocytes/clot counted in plasma clot tissue cultures of rat bone marrow containing 10% rat serum and 10% rat plasma. The cultures were treated with prostaglandin E_1 on day 0.

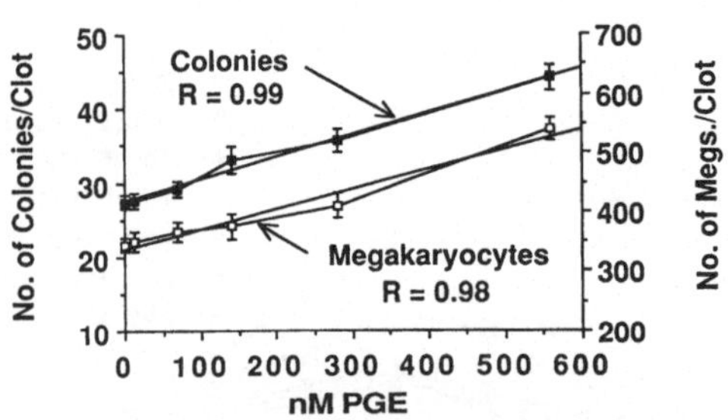

Fig. 3. The numbers of megakaryocyte colonies/clot and the numbers of megakaryocytes/clot counted on day 4 in plasma clot tissue cultures of rat bone marrow containing 10% rat serum and 10% rat plasma. The cultures were treated with prostaglandin E_1 on day 0. The data were from the same experiments as Figure 1 and 2. R is the coefficient of correlation.

DISCUSSION

Ethanol was used as the solvent for the prostaglandin E_1. Since we had showed in a previous study[4] that the administration of a 15% ethanol solution as drinking water to mice for 9-11 weeks directly inhibited megakaryocyte and platelet development in the bone marrow, the concentration of ethanol added to the cultures was kept low, and was also kept

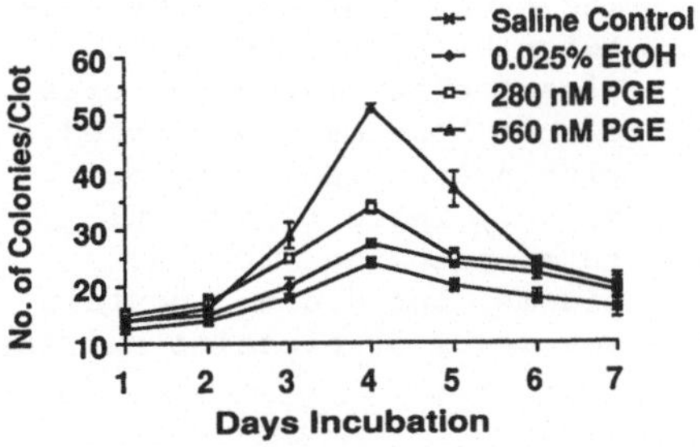

Fig. 4. The numbers of megakaryocyte colonies/clot counted in plasma clot tissue cultures of rat bone marrow treated with prostaglandin E_1 on day 0. These tissue cultures contained an extract of freeze-thawed platelets having a concentration equivalent to 24 x 10^6 platelets/ml, and 20% rat plasma, but did not contain any serum.

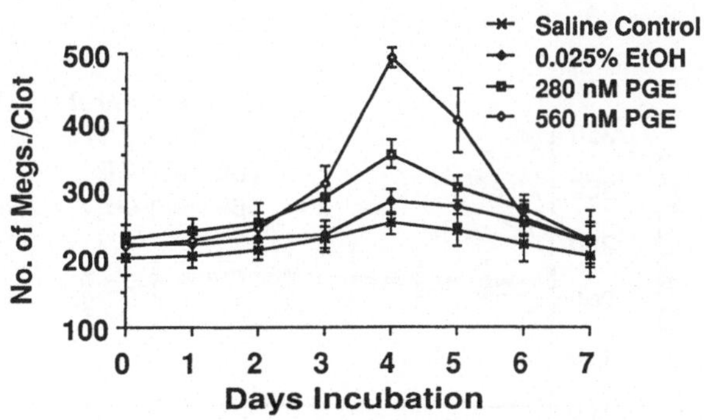

Fig. 5. The numbers of megakaryocytes/clot counted in plasma clot tissue cultures of rat bone marrow treated with prostaglandin E_1 on day 0. These tissue cultures contained an extract of freeze-thawed platelets having a concentration equivalent to 24 x 10^6 platelets/ml, and 20% rat plasma, but did not contain any serum.

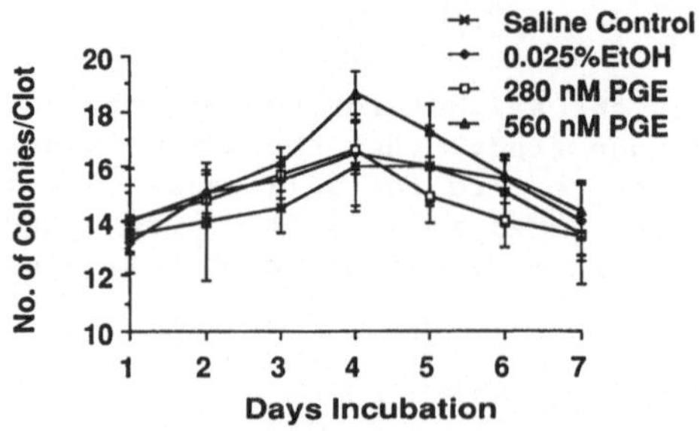

Fig. 6. The numbers of megakaryocyte colonies/clot counted in plasma clot tissue cultures of rat bone marrow treated with prostaglandin E_1 on day 0. These tissue cultures contained 20% rat plasma, but did not contain any serum or platelet extract.

constant. Two control groups were also used. One control group received the vehicle (0.025% ethanol in isotonic saline), and the second group of control cultures received isotonic saline alone. There were no significant differences between the two control groups.

We have shown that prostaglandin E_1 significantly increased the numbers of megakaryocyte colonies and megakaryocytes in plasma clot cultures of rat bone marrow. However, this increase in colony and cell numbers only occurred when either serum or platelet extract was added to the cultures.

Serum was prepared from the clear supernatant of whole blood that was allowed to clot for several hours in the cold. The serum used in this culture system differed from plasma only in that it contained platelet products that were liberated by the platelet release reaction that accompanied blood coagulation, and that it contained no fibrinogen. Platelet-

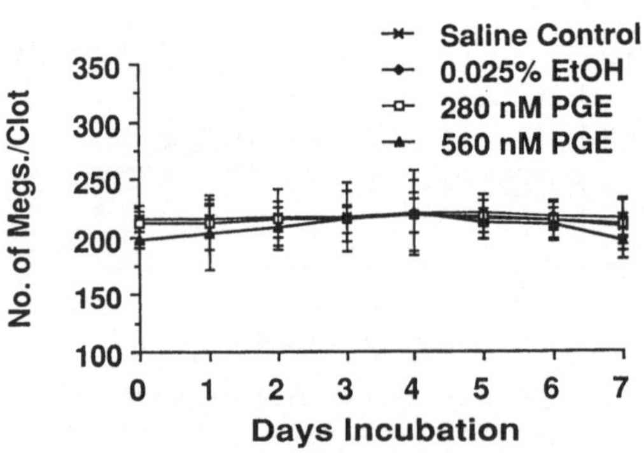

Fig. 7. The numbers of megakaryocytes/clot counted in plasma clot tissue cultures of rat bone marrow treated with prostaglandin E_1 on day 0. These tissue cultures contained 20% rat plasma, but did not contain any serum or platelet extract.

poor plasma was added to all of these cultures, and was allowed to clot in order to form a support for growth of the cells. The substances released from the platelets included several prostaglandins, adenosine diphosphate, transforming growth factor-beta (TGF-β), and platelet-derived growth factor (PDGF). Since a platelet extract substituted for serum in this system, it was inferred that one or more platelet products were also required for the stimulation of megakaryocyte differentiation and proliferation by prostaglandin E_1. The platelet products most strongly suspected to be the stimulating agent(s) were platelet-derived growth factor[5], and transforming growth factor-beta[6,7].

Our results demonstrated that increased megakaryocyte cell and colony numbers were induced by prostaglandin E_1, when platelet extract or serum were added. Very little platelet extract was needed to produce this effect. The amount added to the cultures was equivalent to platelet levels in the circulation of 24×10^9/liter. The normal rat platelet count is approximately 800×10^9/liter.

Stimulation of prostaglandin synthesis by PDGF has been reported in cultures of fibroblasts[8], vascular cells[9,10], and neonatal mouse calvaria[11]. Dukes[12] reported the ability of prostaglandins to enhance heme production in rat bone marrow cells, and he postulated that prostaglandins stimulated erythropoiesis by acting through an adenylcyclase/cAMP mechanism.

Serum contained platelet-derived growth factor, insulin, insulin-like substances, transforming growth factor-beta, epidermal growth factor and numerous other hormones. Prostaglandins might modulate the action of these hormones, rather than act as hormones themselves. The stimulation of differentiation and proliferation of megakaryocyte colonies and megakaryocytes by PGE_1 was demonstrated. This might have been directly regulated by either PDGF or TGF-β. Further investigations of the effects of prostaglandins, purified PDGF, and purified TGF-β on megakaryocyte differentiation and proliferation will be undertaken.

ACKNOWLEDGEMENTS

This research was supported by grant number 2S07RR7132-14 from the National Institutes of Health and grant number 6-63184 from the PSC-CUNY Research Award Program of the City University of New York.

REFERENCES

1. Williams, N. 1979. Preferential inhibition of murine macrophage colony formation by prostaglandin E. Blood 53: 1089-1094.
2. Antoniades, H.N. 1981. Human platelet-derived growth factor (PDGF). Purification of PDGF-I and PDGF-II and separation of their reduced subunits. Proc. Natl. Acad. Sci. USA 78: 7314-7317.
3. Karnovsky, M.J., and L. Roots. 1964. A "direct-coloring" thiocholine method for cholinesterase. J. Histochem. Cytochem. 12: 219-221.
4. Cooper, G.W., H. Dinowitz, and B. Cooper. 1984. The effects of administration of ethyl alcohol to mice on megakaryocyte and platelet development. Throm. Haemostas. 52: 11-14.
5. Stiles, C.D. 1983. The molecular biology of platelet-derived growth factor. Cell 33: 653-655.
6. Sporn, M.B., and A.B. Roberts. 1986. Peptide growth factors and inflammation, tissue repair, and cancer. J. Clin. Invest. 78: 329-332.
7. Sporn, M.B., A.B. Roberts, L.M. Wakefield, and R.K. Assoian. 1986. Transforming growth factor-β. Biological function and chemical

structure. Science 233: 532-534.
8. Shier, W.T. 1980. Serum stimulation of phospholipase A and prostaglandin release in 3T3 cells associated with platelet-derived growth-promoting activity. Proc. Natl. Acad. Sci. USA. 77: 137-141.
9. Coughlin, S.R., M.A. Moskowitz, B.R. Zetter, H.N. Antoniades, and L. Levine. 1980. Platelet-dependent stimulation of prostacyclin synthesis by platelet-derived growth factor. Nature (London). 288: 600-602.
10. Coughlin, S.R., M.A. Moskowitz, H.N. Antoniades, and L. Levine. 1981. Serotonin receptor-mediated stimulation of bovine smooth muscle cell prostacyclin synthesis and its modulation by platelet-derived growth factor. Proc. Natl. Acad. Sci. USA. 78: 7134-7138.
11. Tasjian, A., Jr., E.L. Hohmann, H.N. Antoniades, and L. Levine. 1982. Platelet-derived growth factor stimulates bone resorption via a prostaglandin-mediated mechanism. Endocrinol. 111: 118--124.
12. Dukes, P.P. 1972. Modulating effects of erythropoietin, prostaglandins and cyclic 3'5' nuclotides on erythroid differentiation in marrow cell culture. Fed. Proc. 31: 487.(abstr.).

PARABIOTIC DEMONSTRATION OF A SUBSTANCE RELEASED FROM BURNED TISSUE AFFECTING MARROW MEGAKARYOCYTE DIAMETER AND NUMBER IN MICE

G.D. Kalmaz*, M.M. Guest**, and E.E. Kalmaz***

*University of Texas Medical Branch, Department of Internal Medicine, Division of Hematology/Oncology
Galveston, Texas 77550

**Shriners Burns Institute, Galveston Unit
Division of Hematology, Galveston, TX 77550

***National Aeronautics and Space Administration, Johnson Space Center, Biomedical Research Division SD-4
Houston, TX 77058

ABSTRACT

Four groups of C3H mice were studied to determine effects of substances released into blood from burned tissue on platelet counts and megakaryocyte numbers and sizes. Burns were full thickness scald injuries over 20% of body. Groups were: 1) Single, non-burned (controls); 2) Single, burned; 3) Parabiosed non-burned; 4) Parabiosed, one burned in each pair. Platelet counts and megakaryocyte counts and sizing were done at sacrifice. On 2nd postburn day platelet counts of group 4 were significantly lower ($P<0.0005$) than in group 1, but higher than in group 2. On day 5 platelet counts of group 4, returned to normal, then rose above normal. Two to five days post-burn, megakaryocyte diameters were significantly larger in groups 2 and 4. Fifty megakaryocytes from each mouse of treatment groups were measured. A blood-borne substance(s) derived from burned tissue appears responsible for waning and waxing of megakaryocyte activity.

INTRODUCTION

Parabiosis, the union of two living individuals to produce a continuous blood exchange has been used extensively in biological research since the technique of uniting rats was first described by Paul Bert in 1864[1]. The total blood volume of each animal is exchanged approximately 10 times every 24 hr; blood components are continuously being transferred between partners[2].

It is generally assumed that the thrombocytopenia in burns is caused by factors released from the burned tissue. It appears probable that the released factors are carried by the circulating blood. Surgical parabiosis can be used to detect circulating factors.

MATERIALS AND METHODS

ANIMALS: Ten-week old male C3H mice were used. Four groups were studied: (1) single (non-parabiosed) non-burned controls; (2) single burned mice; (3) non-burned parabiosed mice; (4) burned mice parabiosed with non-burned mice. After 10-14 days in parabiosis mice were sacrificed or separated for continued study. Each mouse of a parabiotic pair was sampled only one time for platelet count, megakaryocyte number and size.

PARABIOSIS: All parabiotic pairs were united in an identical fashion using a method similar to that described by Finerty[3]. In brief, 24-25 g mice were pared at as near as possible to identical weights. Mice were anesthetized with pentobarbital, 6.5 mg/100g body weight. After clipping and depilating the area to be involved in the surgical union, it was cleaned with 70% ethyl alcohol. Midlateral incisions were made from behind the ears to points approximately 1 cm anterior to the tails along the projected adjacent sides of each animal. The adjoining muscles on the lateral side of the peritoneum of the two animals were then divided and the abdominal muscles on the right side of one animal were sutured to the abdominal muscles on the left side of the other animal, so that the peritoneal surfaces were in contact. To prevent separation of the animals, ligatures were placed around the right and left clavicles and braided 4.0 silk was used to suture gluteal muscles. The partners were then sutured skin-to-skin with ventral and dorsal skin opposed. A single 9 mm round clip was placed at the posterior end of the surgical connection to prevent twisting. Following the surgery parabiotic pairs were weighed, then placed in individual cages and allowed to recover. Body weight was measured weekly until the time of the experiment and 2 times weekly for the remainder of the study. At death, the body weights were obtained before and after separation.

BURN: After confirming cross-circulation between the parabionts - visible excretion of phenolsulfonphalein (PSP) in the urine 20-30 min after injection of 1 ml of 0.8% (W/v) PSP into the gluteal muscle of the other animal - the back of one of each pair in group 4 was burned (full thickness over 20% of its body surface).

The surface area of each mouse to be burned was estimated from Meeh's formula[4].

$$A = KW\ 2/3$$

where: A = surface area in cm

w = body weight in g

$k = 6.5$

(The proportionality constant, k, used in this study is a mean of previously reported values for the C3H mice).

Following pentobarbital anesthesia, the scald burn was produced by immersing the depilated area in water at 70°C for 6 seconds while the mouse was held in a device that limited the area of exposure and protected the other mouse of the pair from burn injury. This procedure produces a uniform burn with sharp margins. The burned animals were dried upon removal from the water, and were resuscitated immediately postburn with saline intraperitoneally (I.P.). Normal mice were given the same single I.P. injection of saline. After the burn, the parabiotic pairs recovered from anesthesia and were active, apparently without pain. The parabiosis only slightly interfered with mobility; the animals ate and drank normally. Histologic evaluation of the burn area showed complete destruction of the epidermis and some subcutaneous damage. Although wound swabs showed occasional skin contaminants, mostly staphylococcus aereus, no invasive or systematic infection was observed.

PLATELET COUNTS: Platelet counts were obtained from a single drop of blood taken from the retro-orbital sinus at sacrifice. The phase microscope was used to determine platelet numbers.

SIZING MEGAKARYOCYTE DIAMETER: Mice were killed at intervals following infliction of the burn. After killing the mice the femurs were removed and fixed in Zenkers' formol solution, decalcified, and stained with hematoxylin-eosin. The average number of megakaryocytes per high powered field (HPF) was determined in one entire longitudinal section from each femur as previously described[5]; in brief, six HPF per femur were counted and the average was recorded as a single observation. Megakaryocyte diameters were measured using an eyepiece micrometer at a magnification of 420X. The eyepiece was calibrated with a disk-micrometer; each calibration represented 0.01 mm and measurements were made to the nearest µm. The diameter of a megakaryocyte was calculated as the average of two measurements made at right angles from each megakaryocyte[6]. From each treatment group, fifty megakaryocytes were measured and numbers were corrected for multiple counting according to cell diameter measurements using the formula described by Harker[7].

$$N = N'/(d/t)+1$$

where N is the actual number of cells per section

N' is the number of cells counted

d is the mean cell diameter

t is the thickness of the section

CONDUCT OF RESEARCH

This research was conducted according to the "Guide for the Care and Use of Laboratory Animals" as promulgated by the Guide for Grants and Contracts of the NIH. In all cases data were managed and analyzed using the CLINFO Data Analysis System of the General Clinical Research Center (supported in part by grant from General Clinical Resources Center Program, DRR, NIH-RR0073).

RESULTS

Full-thickness scald burns of 20% of the body surface area led to marked changes in platelet counts in single burned mice and in non-burned mice parabiosed with burned mice (Fig. 1). By day 2 postburn the platelet count fell to below $3 \times 10^5/mm^3$ in the single burned mice and to $3.9 \times 10^5/mm^3$ in the burned mice parabiosed with non-burned mice. Platelet counts in the non-burned mice parabiosed with burned mice ($5.2 \times 10^5/mm^3$) were significantly lower ($P<0.0005$) than in the non-burned parabiosed mice ($11.2 \times 10^5/mm^3$). Platelet counts in the non-burned parabiosed mice were significantly above the counts in normal mice ($9.2 \times 10^5/mm^3$). The cause for the increase in platelet counts of non-burned parabionts has not yet been ascertained. The platelet counts of non-burned parabionts has not yet been ascertained. The platelet counts in single non-burned mice and in non-burned mice parabiosed with burned mice had essentially returned to the normal level by day 5. From day 5 postburn to day 10 (last day of observation) the platelet counts in the single burned mice and in the parabiosed mice with a burned

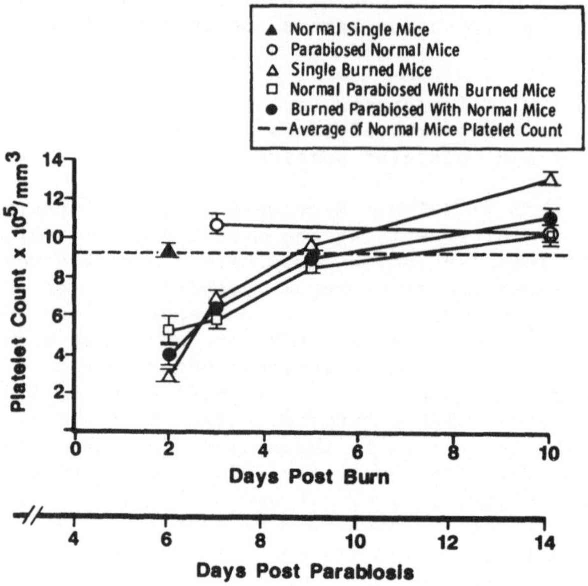

Fig. 1. Platelet counts in mice following full thickness scald burns over 20% of body surface area. Standard errors are indicated by vertical bars.

partner slowly increased above the normal level. However, the platelet count in the single non-burned mice continued to rise above the normal level while the counts in the parabiosed pairs with one member burned reached only the approximate level of the non-burned parabiosed pairs.

Effects of 20% full thickness scald burns, on the number of megakaryocytes in mice in the 4 experimental groups is graphed in Fig. 2. Compared to normal mice, burn injury caused a significant increase in the number of marrow megakaryocytes/HPF on days 2,3 and 5 postburn. Megakaryocytes/HPF of burned mice parabiosed with non-burned mice were increased markedly but the numbers were not as high as in single burned mice. In both cases the number of megakaryocytes/HPF plateaued by day 5 postburn. The number of megakaryocytes/HPF were higher in non-burned parabiosed mice compared to non-burned single mice. Megakaryocyte sizes (expressed as μm diameter) were significantly increased ($P<0.0005$) in burned mice, non-burned mice parabiosed with burned mice, and burned mice parabiosed with non-burned mice on days 2-5 postburn (Fig. 3). However, megakaryocyte size was larger in single burned mice than in the non-burned mice parabiosed with burned mice and in burned mice parabiosed with non-burned mice. Megakaryocyte sizes in non-burned parabiotic mice were significantly decreased when compared to their sizes in non-burned single mice.

DISCUSSION

Previous studies[8-11] have shown changes in platelet counts following thermal injury in man and experimental animals. The present study indicates that full thickness burn injury over 20% of the body surface causes thrombocytopenia in single burned mice, in burned mice parabiosed to non-burned mice and in non-burned mice parabiosed to burned mice during the first 4 days postburn with a nadir at 2 days. There is then a progressive increase in platelet counts that exceeds control levels by day 10 postburn in these groups. It was also observed that parabiosis of burned mice with non-burned mice moderates the thrombocytosis when compared to single burned mice, but thrombocytosis is not eliminated (Fig. 1). Wallner et al[12] also reported in mice an increase in platelet counts above normal values by day 8. A study by Eurenius et al[13] showed that rat platelet counts following a 30% full thickness burn decreased during the first 2 days postburn and increased above control levels by day 3. Our earlier study in mice[14] is consistent with the observations of Eurenius et al[13] in burned rats with the exception that the platelet count did not exceed the control level in our study until day 8. The rapid return of platelet counts above control levels by day 3 in the study of Eurenius and co-workers could be the result of the larger burn injury or their use of a different species of rodents.

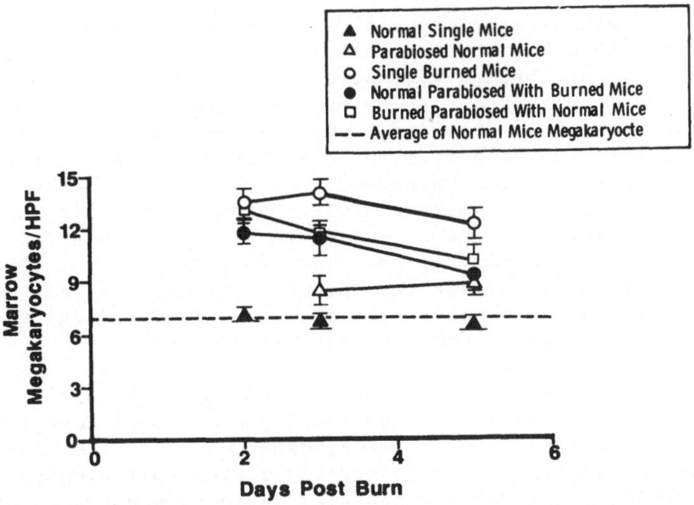

Fig. 2. The number of megakaryocytes per high powered field following 20 full thickness scald burns. The horizontal dashed line represents the average number of megakaryocytes/HPF for control mice. Standard errors are indicated by vertical bars.

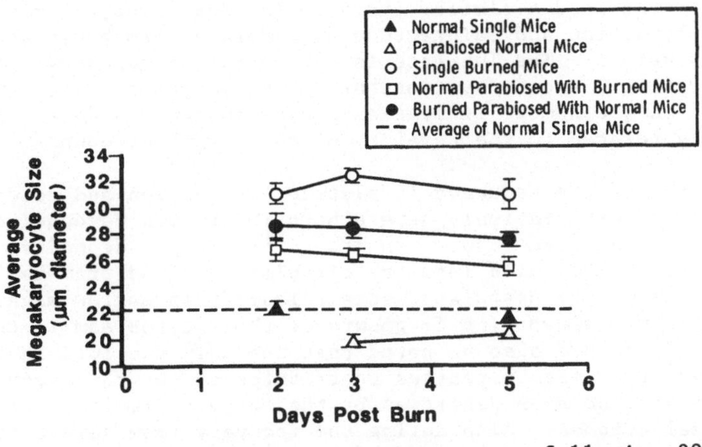

Fig. 3. Diameters of marrow megakaryocytes following 20% scald burns. Average diameter of megakaryocytes in control mice is represented by horizontal dashed line. Vertical lines indicate standard errors.

The mechanism of burn-induced thrombocytopenia is controversial. Some authors[15-17] have suggested that surgery is the cause for the observed thrombocytopenia in patients with burns while others believe that thermal injury produces toxic hypersplenism, which in turn causes a reduction in platelet production[18-19]. However, in one of our previous studies[20] thrombocytopenia occured in burned mice even though the blood was collected from animals that had not undergone surgery; thus it is unlikely that surgical procedures are the primary cause for platelet suppression following burn trauma. Our data indicate that release of inhibitory factors into the circulation from the heat injured tissue plays the major role. Furthermore, the effect of a "toxic hypersplenism" remains an unproven postulate. In our experiments we have only observed splenic enlargement in two mice (9th and 10th day) at a time when the platelet counts were above normal.

To study the mechanisms responsible for thrombocytopenia in the early postburn phase and thrombocytosis in the later phase, mice with deep thermal injury parabiosed to non-burned mice were used to examine the effects of factor(s) released from burned tissues on the number and diameter of bone marrow megakaryocytes. Our data indicate that 20% full thickness burns resulted in increases in the number and diameter of megakaryocytes in the marrows of single burned mice, burned mice parabiosed to non-burned mice, and non-burned mice that were partners with burned mice. Although marked increases in the number and size of megakaryocytes were observed 2 days postburn, the complete recovery of the platelet count was not observed before 5 days postburn. This may be due to continued release of toxic materials from the burned tissues, prolonging thrombocytopenia by damaging mature megakaryocytes and thereby preventing platelet release.

Previous data[20] from our laboratory showed that 20% full thickness burn injury caused an increase in the proportion of small acetylcholinesterase positive cells in the marrow of mice, continuing until 4 days postburn. Increases in the average size of the megakaryocytes following induction of thrombocytopenia by antisera or platelet depleted donor sera have been reported to occur by other investigators in bone marrow sections[5,21-23] and in smears[24]; thus the present study which shows that full thickness burns (20% of surface) resulted in significant increases in the concentration and diameter of megakaryocytes in the marrow of mice is in essential agreement with the above referenced studies. Also our observations are in agreement with Odell et al[25]; they observed increases in the number of megakaryocytes/HPF at 3 days after injection of sera from platelet-depleted donor rats. Harker[7] described a similar increase in megakaryocyte number (3 times normal) after induction and maintenance of moderate thrombocytopenia by exchange transfusions. The increase in the number of megakaryocytes was found to be present 2 and 5 days postburn. The length of time before onset of an increase in the number of megakaryocytes in these studies appears to be inversely related to the severity of the thrombocytopenia.

The etiology of the sequence in platelet production and release from the bone marrow after relatively severe burn injury can be concluded from our studies and other burn injury reports to be largely dependent upon a toxic factor that is released into the circulating blood from the burn-injured tissue. This factor, we postulate, acts primarily on marrow megakaryocytes to prevent them from disgorging fragments of their cytoplasm in the form of platelets. It should also be noted that not only the lull in formation of new platelets from megakaryocytes contributes to the thrombocytopenia, but also platelets had been destroyed by the temperature rise during the abnormal thermal episode. Also during the recovery from severe thrombocytopenia increases of megakaryocyte size and number were observed. Thus far we have been unable to determine which response occurs first or whether both responses occur simultaneously. As a result of the reduction in the number of circulating platelets, a feedback mechanism is postulated to cause an increase in small acetylcholinesterase positive cells that are precursors of megakaryocytes. Within a day, or perhaps two or three, after the burn injury occurs, the release of toxic burn factor(s) declines; consequently the megakaryocytes again become capable of releasing platelets into the circulating blood. Since both the number and size of megakaryocytes are then greater than normal and the feedback stimulus for the production and release of platelets is present, thrombocytosis occurs.

In brief summary of our explanation of the results based on studies in this laboratory and from other laboratories it may be stated: that immediately following severe burn a toxic factor (or factors) is released from burned tissues into the circulating blood; the toxic factor blocks formation and extrusion of platelets from megakaryocytes; the resulting thrombocytopenia then triggers a feedback mechanism which stimulates production of megakaryocytes; the increased size and number of megakaryocytes contribute to the thrombocytosis when the block to the release of platelets is removed.

ACKNOWLEDGEMENTS

Supported partly by an Intramural Grant (DHHS.2S07RR07205-06). The authors wish to thank Moretta Phillips for preparation of this manuscript and Sandra Baxter for the art work.

REFERENCES

1. Bert, P. 1864. Sur La Greffe Animale. J. de L'Anatomie et de La Physiologic 1: 69-87.
2. Harris, R.B.S., and R.J. Martin. 1984. Recovery of body weight from below "set point" in mature female rats. J. Nutr. 114: 1143-1150.
3. Finerty, J.C. 1952. Parabiosis in physiological studies. Physiol. Rev. 32: 277-302.
4. Walker, H.L., and A.D. Mason. 1968. A standard animal burn. J. Trauma. 8: 1049-1051.
5. McDonald, T.P. and G.D. Kalmaz. 1983. Effects of thrombopoietin on the number and diameter of marrow megakaryocytes of mice. Exp. Hematol. 11: 91-97.
6. Jackson, C.W. and C.C. Edwards. 1977. Evidence that stimulation of megakaryocytopoiesis by low dose vincristine results from an effect on platelets. Br. J. Haematol. 36: 97-105.
7. Harker, L.A. 1968. Megakaryocyte quantitation. Clin. Invest. 47: 452-457.
8. Arturson, G., and G. Wallenius. 1964. Hypercoagulability of blood after burn trauma in rats: A preliminary report. Acta Chir. Scand. 128: 340-345.
9. Hergt, K. 1972. Blood levels of thrombocytes in burned patients: observation on their behavior in relation to clinical condition of the patient. J. Trauma 12: 599-607.
10. Gabilando, F.Y., E.H. Huerta, and M.C. Jimenez. 1976. Hemostasis in burns: A preliminary report. Burns 3: 24-28.
11. Coctow, W.T. 1981. Comparative coagulation abnormality patterns in moderately and severely burned patients. Burns 7: 434-437.
12. Wallner, S.F., R. Vautrin, J. Murphy, S. Anderson, and V. Peterson. 1984. Haematopoietic response to burning. Burns 10: 236-251.

13. Eurenius, K., R.F. Mertensen, P.M. Meserol, and P.W. Curreri. 1972. Platelet and megakaryocyte kinetics following thermal injury. J. Lab. Clin. Med. 79: 247-257.
14. Kalmaz, G.D., J.D. Bessman, D.N. Herndon, and E.E. Kalmaz. 1985. The effect of early excision of burn tissue on megakaryocyte proliferation and maturation in vitro. Blood 66(5): 457a (Abstr).
15. Warren, R., J. Lauridsen, and J.S. Belko. 1953. Alterations in numbers of circulating platelets following surgical operation and administration of adrenocorticotropic hormone. Circulation 7: 481-486.
16. Peppe, H., and S. Lindsay. 1960. Responses of platelets, eosinophils and total leukocytes during and following surgical procedures. Surg. Gynec. Obstet. 110: 319-326.
17. Epeberg, O. 1962. Changes in the coagulation system following major surgical procedures. Acta Med. Scand. 171: 679-685.
18. Monteil, P., C. Reynier, and J. Vigne. 1963. Etude de variations des plaquettes sanquines jans la cadre des troubles de la obagulation cheg les brules. Hemostase 3: 239-249.
19. Asko-Serjavaara, S. 1974. Inhibition of bone marrow cell proliferation in burned mice. Scand. J. Reconstr. Surg. 8: 192-197.
20. Kalmaz, G.D., and M.M. Guest. 1987. Burn injury causes suppression of megakaryocyte progenitors detected by small acetycholinesterase positive cell assay. Proc. Amer. Burn. Assoc. 19: 5a (Abstr.)
21. Penington, D.G., and T.E. Olsen. 1970. Megakaryocytes in states of altered platelet production: cell numbers, size and DNA content. Br. J. Haematol. 18: 447-463.
22. McPherson, G.G. 1974. Changes in megakaryocyte development following thrombocytopenia. Br. J. Haematol. 26: 105-115.
23. Cullen, W.C., and T.P. McDonald. 1986. Comparison of stereologic techniques for the quantification of megakaryocyte size and numbers. Exp. Hematol. 14: 782-788.
24. Ebbe, S., J. Stohlman, Jr., J. Overcash, J. Donovan and D. Howard. 1968. Megakaryocyte size in thrombocytopenic and normal rats. Blood 32: 383-392.
25. Odell, T.T., Jr., C.W. Jackson, T.J. Friday, and D.E. Charsha. 1969. Effect of thrombocytopenia on megakaryocytopoiesis. Brit. J. Haematol. 17: 91-101.

PRECLINICAL AND CLINICAL EFFECTS OF THE HEMATOPOIETIC

COLONY STIMULATING FACTORS[a]

Hal E. Broxmeyer,[1-3] Saroj Vadhan-Raj,[4] Giao Hangoc,[1]
Li Lu,[1] Jordan U. Gutterman,[4] and Douglas E. Williams.[1]

Departments of[1] Medicine (Hematology/Oncology), [2]Microbiology
and Immunology, and the
[3]Walther Oncology Center
Indiana University School of Medicine
Indianapolis, IN

and

Department of [4]Clinical Immunology and Biological Therapy
University of Texas
M.D. Anderson Hospital and Tumor Institute
Houston, TX

ABSTRACT

Hematopoietic colony stimulating factors are available in purified recombinant form and have used for assessment of hematopoietic activities in mice, monkeys and in phase I clinical trials with humans. This report reviews the preclinical and clinical studies involving these factors.

INTRODUCTION

Hematopoietic colony stimulating factors (CSFs) were operationally defined based on their capacity to stimulate cells to proliferate/differentiate into groups or colonies of blood cells in culture in a semi-solid support medium such as agar, methylcellulose or plasma clot containing a source of serum[1-3]. Serum can be replaced by defined ingredients such as transferrin[4]. The different CSFs that have now been defined based on biological action, protein purification and gene cloning include interleukin-3 (IL-3, multi CSF), granulocyte-macrophage (GM)-CSF, granulocyte (G)-CSF and macrophage (M)-CSF (also termed CSF-1)[1-3]. Other hematopoietically-active factors include erythropoietin (Epo, in some cases also considered under the broad title of CSFs), and factors which act in concert with CSFs to enhance, in an additive or synergistic manner, the actions of the CSFs. These factors include IL-1, IL-2 (T cell growth factor), IL-4 (B-cell Growth Factor-1), IL-5 (Eosinophil Growth factor), IL-6 (B-Cell Growth factor-1, Interferon-Beta 2, Hybridoma Growth factor), the Inferferons (IFN)-Gamma, Alpha and Beta, and the Tumor Necrosis factors-Alpha and Beta (Lymphotoxin)[1-3].

The interesting component of the system of blood cell regulation is the complexity and redundancy of positive and negative feedback loops[1-3].

Thus, one cell-type can respond in a similar phenotypic manner to more than one factor, although the actual mechanisms of action may be different. Moreover, a single molecule can have more than one action, although it is not always clear that the multiple actions of the single molecules are a result of the direct action of that molecule. Interpretations are difficult since absolutely pure populations of cells are rarely used, and it is clear that one molecule can trigger the gene expression, production and release of other biologically active molecules. To elucidate these cascading events, scientists have used a reductionalist approach by isolating pure populations and subpopulations of cells and assessing the effects of purified natural and recombinant molecules on these pure cells in the absence of serum, but in the presence of better defined ingredients. This approach has been successful not only with studies on accessory cells, but also with studies on hematopoietic progenitor cells[5-8]. Thus a great deal of insight into the complexity of cell-regulation has been gained. Yet, our understanding of the regulatory processes which keep blood cell production in vivo within its narrow steady-state flux is not as well understood. It may be even more complex than what is seen in vitro based on the probability that we have not yet identified all the molecules and cell-types that can impinge on these regulatory processes.

Without necessarily knowing exactly how a molecule may work, it was of importance to determine if actions of the molecules in vivo at least mimicked what would be expected of that molecule based on the direct as well as indirect actions manifested in vitro. With the availability of relatively large amounts of purified molecules, made possible by recombinant technologies of gene-cloning and- expression, the hematopoietic CSFs have been evaluated in animal models. More recently, phase I trials have been done in the clinic with some of these factors. The rest of this report describes our own studies, as well as studies of others, on the in vivo actions of the CSFs.

ANIMAL STUDIES IN VIVO

IL-3[9-19] and CSF-1[12,13,15,16,17,20-24] have thus far been studied in mice, while GM-CSF[12,15,17,25-28] and G-CSF[21,29-35] have been evaluated in both mice and monkeys. Increased circulating white blood cell counts have been noted in mice or monkeys given multiple injections or continuous infusions of high concentrations of CSFs. Maintenance of elevated white cell counts requires the continuous administration of the CSFs.

Our own studies[12-17,20,21,34] have demonstrated that effects of CSFs on cycling rates and absolute numbers of progenitors are dose related. Single injections of low dosages of either recombinant murine IL-3, recombinant murine GM-CSF, recombinant human G-CSF, recombinant human CSF-1, or natural murine CSF-1 increase the cycling rates (percent of cells in S-phase of the cell-cycle) of multipotential (CFU-GEMM), granulocyte-macrophage (CFU--GM), erythroid (BFU-E), and high proliferative potential (HPP-CFC) progenitor cells. However, increases in absolute numbers of these progenitor cells require larger dosages of single injections of the CSF. While the effects were apparent in previously untreated mice, they were much more obvious in mice in which myelopoiesis was first suppressed by administration of iron-saturated purified human lactoferrin[12,13] or during the overbound phase after recovery from sublethal dosages of cyclophosphamide[14,20]. In the lactoferrin-pretreated mice, serum levels of CSF were depressed by >90%[36] which may relate to a direct decrease of CSF from mononuclear phagocytes[37], or to a direct decrease of a factor, such as IL-1, from mononuclear phagocytes which then triggers other cells such as fibroblasts, endothelial cells or T-lymphocytes to release CSFs[38-40]. In fact lactoferrin decreases release of IL-1 from human monocytes in vitro[41].

The administration of single dosages of a single CSF, even at concentrations that increased absolute numbers of progenitors, did not result in increases in nucleated cellularity in the marrow, spleen or blood[12-17,20,21,34]. This suggested that continuous presence of much higher dosages of CSFs are required for manifestation of increased nucleated cellularity.

Of the four CSFs, CSF-1[42] has the interesting property of being able to induce the release from monocytes/macrophages in vitro of molecules such as acidic ferritin[43], prostaglandin E[44,45], tumor necrosis factor-alpha[46], and interferon-alpha[45,46] all of which are suppressive to hematopoietic progenitor cells. These in vitro phenomena were consistent with our recent observation that while a single injection of 20,000 units natural murine or recombinant human CSF-1 to previously untreated mice increased the cycling rates of CFU-GEMM, BFU-E and CFU-GM as measured 36 hours later, three injections of 20,000 units of CSF-1 given at times 0, 12 and 24 hours, caused a suppression in the cycling rates of the progenitors[47]. Others have shown that multiple additions of 25,000 units CSF-1 per day for many days does not result in increased numbers of progenitors or nucleated cells[22], a fact consistent with our studies. However, more recent studies by us have shown that even CSF-1 can result in a moderate (two-fold) increase in circulating leukocytes, but only if multiple additions of extremely high concentrations of CSF-1 ($\sim 8 \times 10^5$ units/injection) are used. This reflects a large increase in splenic progenitors with a concomitant decrease in marrow progenitors[48]. It is of interest that in mice the major organ of hyperthrophy in response to high dose CSFs is the spleen[11,25,33,48].

Animal studies have demonstrated that CSFs may act as a protective agent, if given after chemotherapy or immunosuppressive therapy in association with transplantation, to influence an accelerated repopulation of leukocytes such as neutrophils[24,28,29,31]. In this context it is of interest that pretreatment of mouse marrow cells in vitro with recombinant murine IL-3 enhances the reconstituting ability of these cells when injected into irradiated mice[49].

SYNERGISTIC EFFECTS OF CSFs IN VIVO IN MICE

Of considerable importance to the actions of the CSFs in vivo is that they can synergise in vivo[16,17,21,34] as well as in vitro[1,2]. In mice pretreated with lactoferrin[17] the concentrations of murine CSF-1, IL-3 and GM-CSF needed to increase the cycling rates of CFU-GM, BFU-E and CFU-GEMM in vivo were reduced by factors of 40-200, 10-50 and 40 to >400, respectively for the CSFs, and the concentrations needed to increase the progenitor cell numbers were reduced by factors of 40-500, 20-80 and >40 to >200, respectively for CSF-1, IL-3 and GM-CSF, when these CSFs were given in combination with one other CSF at concentrations of each which were not effective by themselves. Similar synergistic effects have been noted with recombinant human CSF-1 and recombinant human G-CSF given to lactoferrin-pretreated mice[34]. Recombinant murine IL-3 synergises with natural murine CSF-1 when given to mice to cause the increased proliferation of HPP-CFC in vivo[16].

Many of the effects noted in vivo, especially those involving the synergistic action of multiple forms of CSFs, are probably a result of the cascading effects of molecules acting indirectly on accessory cells as noted above for in vitro studies. It is possible however that some of the synergistic effects in vivo are due to direct molecule action at the level of the progenitor cell. Using a purified population of progenitors for the granulocyte-macrophage lineage (CFU-GM)[5], additive and synergistic effects of different CSFs have been detected[6]. Thus, these effects were apparent in the absence of accessory cells. This population of CFU-GM has also been used to characterize the receptor for murine GM-CSF[50], to show that the proportion of cells expressing the c-fms oncogene (CSF-1 receptor) is equal to those forming a colony or cluster of macrophages in vitro in response to CSF-1[51], and to demonstrate that IL-1 alpha can promote the survival of CFU-GM in vitro in the absence of accessory cells[52].

CLINICAL TRIALS WITH THE CSFs

Recombinant human Epo has been shown to be capable of correcting the anemia of end stage renal failure in phase I and phase II clinical trials[53,54] and recombinant human GM-CSF when used in phase I trials has been successful in increasing circulating white blood cell counts in patients with acquired immunodeficiency syndrome (AIDS)[55], in patients with myelodysplastic syndromes (MDS)[56], and in other patients with bone marrow failure and malignancy[57]. The peripheral responses appeared to be dose-related and reflected the continuous administration of GM-CSF. Increases were seen in circulating neutrophils, eosinophils and monocytes. In some patients with one or more cytopenia that received at least two 14 day cycles of treatment, multilineage responses were noted which were characterized by two-fold or greater increases in platelet count to a level above 100,000, two fold or greater increases in corrected reticulocyte count, and a reduced requirement for red cell transfusions[57].

Bone marrow cells from some of the patients with leukemia, MDS, cancer and other cytopenic disorders on the phase I clinical trial with recombinant human GM-CSF[56,57], were assessed in vitro for CFU-GM growth patterns, cycling rates, and responsiveness in vitro to CSFs and colony inhibiting factors[58]. The growth patterns reflecting the ratio of CFU-GM colonies to clusters, which have been used as a diagnostic and prognostic indicator[59], remained unchanged during the treatment when compared to pretreatment values. This was especially relevant for the marrow cells from patients with MDS since these patients responded in vivo with increases in numbers of polymorphonuclear neutrophils. Thus, abnormal growth patterns of the CFU-GM did not change even though mature cells were seen circulating in the blood after treatment. This suggests that the mature cells in the patient may have derived from an abnormal clone, a fact consistent with studies using the technique of premature chromosomal condensation[57,58]. The dose response curves of CFU-GM to stimulation by GM-CSF in vitro also did not change during the clinical trial[58] a fact consistent with the increases in peripheral leukocyte counts which remained high as long as the GM-CSF was administered to the patients. Both the in vitro and the clinical results suggested that there was no down-modulation of GM-CSF receptors on CFU-GM in response to the continuous administration of GM-CSF. The percentage of CFU-GM in DNA synthesis, which is a measure of the proliferative rates of these cells, determined by high specific activity tritiated thymidine kill technique in vitro, was markedly enhanced in a reversible fashion after administration in vivo of GM-CSF[58]. CFU-GM in a slow- or non-cycling state prior to administration of GM-CSF were placed into rapid cycle, but returned to a slow- or non-cycle state after cessation of GM-CSF administration. These would appear to be the first evidence that administration of GM-CSF to patients is having an effect at the level of the hematopoietic progenitors, since colony numbers themselves cannot be used as an indication of activity due to the fact that marrow cellularity cannot be accurately quantitated and thus absolute numbers of marrow progenitors can not be determined.

Combinations of biomolecule therapy will no doubt be used in the future, and to this end we were interested in whether the responsiveness of patient CFU-GM to other factors might be changed after the clinical administration of GM-CSF[58]. It was found that marrow CFU-GM, from patients given GM-CSF *in vivo*, were increased in sensitivity to inhibition *in vitro* by recombinant human H-subunit (acidic) ferritin in 2 of 8 cases, and were increased in sensitivity to inhibition by lower dosages of recombinant human tumor necrosis factor-alpha in all patients evaluated. The sensitivity of CFU-GM to inhibition *in vitro* by recombinant human interferon-gamma and prostaglandin E_1 did not change during the clinical trial. It is possible that the above information may be of use in planning phase II/III clinical trials combining GM-CSF with chemotherapy and/or other biotherapy.

Recombinant human G-CSF is also being evaluated in clinical trials and it is likely that trials will soon begin with recombinant human IL-3 and recombinant human CSF-1. What is probable is that the successful use of these factors will depend on the specific diseases that they are used in, the stage of the disease and how and when they are administered.

ACKNOWLEDGEMENTS

We wish to thank Stephanie Moore for typing the manuscript.

REFERENCES

1. Broxmeyer, H.E. 1986. Biomolecule-cell interactions and the regulation of myelopoiesis. Int. J. Cell Cloning 4: 378-405.
2. Broxmeyer, H.E. and D.E. Williams. 1988. The production of myeloid blood cells and their regulation during health and disease. CRC Crit. Rev. Oncol./Hematol. In Press.
3. Metcalf, D. 1985. The granulocyte-macrophage colony stimulating factors. Science (Wash. DC) 229: 16-22.
4. Lu, L., K. Welte, J.L. Gabrilove, G. Hangoc, E. Bruno, R. Hoffman, and H.E. Broxmeyer. 1986. Effects of recombinant human tumor necrosis factor-alpha, recombinant human interferon-gamma and prostaglandin E on colony formation of human hematopoietic progenitor cells stimulated by natural human pluripotent colony stimulating factor, pluripoietin-alpha and recombinant erythropoietin in serum-free cultures. Cancer Res. 46: 4357-4361.
5. Williams, D.E., J.E. Straneva, R-N. Shen, and H.E. Broxmeyer. 1987. Purification of murine bone marrow-derived granulocyte-macrophage colony-forming cells and partial characterization of their growth and regulation *in vitro*. Exp. Hematol. 15: 243-250.
6. Williams, D.E., J.E. Straneva, S. Cooper, R.K. Shadduck, A. Waheed, S. Gillis, D. Urdal, and H.E. Broxmeyer. 1987. Interactions between purified murine colony stimulating factors (natural CSF-1, purified recombinant GM-CSF, and purified recombinant IL-3) on the *in vitro* proliferation of purified murine granulocyte-macrophage progenitor cells. Exp. Hematol. 15: 1007-1012.
7. Lu, L., D. Walker, H.E. Broxmeyer, R. Hoffman, W. Hu, and E. Walker. 1987. Characterization of adult human marrow hematopoietic progenitors highly enriched by two-color sorting with My10 and major histocompatability (MHC) class II monoclonal antibodies. J. Immunol. 139: 1823-1829.
8. Williams, D.E., S. Cooper, and H.E. Broxmeyer. 1988. The effects of hematopoietic suppressor molecules on the *in vitro* proliferation of purified murine granulocyte-macrophage progenitor cells (CFU-GM). Cancer Res. 48: 1548-1550.
9. Kindler, V., B. Thorens, S. deKossodo, B. Allet, J.F. Eliason, D. Thatcher, N. Farber, and P. Vassalli. 1986. Stimulation of hematopoiesis *in vivo* by recombinant bacterial murine interleukin 3. Proc. Natl. Acad. Sci. USA 83: 1001-1005.
10. Lord, B.I., G. Malineux, N.G. Testa, M. Keely, E. Spooncer, and T.M.

Dexter. 1986. The kinetic response of haemopoietic precursors cells *in vivo* to highly purified recombinant interleukin 3. *Lymphokine Res.* 5: 97-104.

11. Metcalf, D., C.G. Begley, G.R. Johnson, N.A. Nicola, A.F. Lopez, D.J. Williamson. 1986. Effects of purified bacterially synthesized murine multi-CSF (IL-3) on hematopoiesis in normal adult mice. *Blood* 68: 46-57.
12. Broxmeyer, H.E., D.E. Williams, S. Cooper, R.K. Shadduck, S. Gillis, A. Waheed, D.L. Urdal and D.C. Bicknell. 1986. Comparative effects *in vivo* of recombinant murine interleukin-3, natural murine colony-stimulating factor-1 and recombinant murine granulocyte-macrophage colony stimulating factor on myelopoiesis in mice. *J. Clin. Invest.* 79: 721-730.
13. Broxmeyer, H.E., D.E. Williams, G. Hangoc, S. Cooper, P. Gentile, R-N. Shen, P. Ralph, S. Gillis and D.C. Bicknell. 1987. The opposing actions *in vivo* on murine myelopoiesis of purified preparations of lactoferrin and the colony stimulating factors. *Blood Cells* 13: 31-48.
14. Broxmeyer, H.E., D.E. Williams, and S. Cooper. 1987. The influence *in vivo* of natural murine interluekin 3 on the proliferation of myeloid progenitor cells in mice recovering from sublethal dosages of cyclophosphamide. *Leuk. Res.* 11: 201-205.
15. Broxmeyer, H.E., D.E. Williams, H.S. Boswell, S. Cooper, R.K. Shadduck, S. Gillis, A. Waheed, and D.L. Urdal. 1986. The effects *in vivo* of purified preparations of murine macrophage colony-stimulating factor-1, recombinant murine granulocyte-macrophage colony-stimulating factor and natural and recombinant murine interleukin 3 without and with pretreatment of mice with purified iron-saturated human lactoferrin. *Immunobiology* 172: 168-174.
16. Williams, D.E., G. Hangoc, S. Cooper, H.S. Boswell, R.K. Shadduck, S. Gillis, A. Waheed, D. Urdal, and H.E. Broxmeyer. 1987. The effects of purified recombinant murine IL-3 and/or purified natural murine CSF-1 *in vivo* on the proliferation of murine high (HPP-CFC) and low (LPP-CFC) proliferative potential colony forming cells: demonstration of *in vivo* synergism. *Blood* 70: 401-403.
17. Broxmeyer, H.E., D.E. Williams, G. Hangoc, S. Cooper, S. Gillis, R.K. Shadduck, and D.C. Bicknell. 1987. Synergistic myelopoietic action *in vivo* after administration to mice of combinations of purified murine colony-stimulating factor 1, recombinant murine interleukin 3, and recombinant murine granulocyte-macrophage colony-stimulating factor. *Proc. Natl. Acad. Sci. USA* 84: 3871--3875.
18. Metcalf, D., C.G. Begley, N.A. Nicola, and G.R. Johnson. 1987. Quantitative responses of murine hemopoietic populations *in vitro* and *in vivo* to recombinant multi-CSF (IL-3). *Exp. Hematol.* 15: 288-295.
19. Hangoc, G., L. Lu, A. Oliff, S. Gillis, W. Hu, D.C. Bicknell, D. Williams, and H.E. Broxmeyer. 1987. Modulation of Friend virus infectivity *in vivo* by administration of purified preparations of human lactoferrin and recombinant murine interleukin-3 to mice. *Leukemia* 1: 762-764.
20. Broxmeyer, H.E., D.E. Williams, S. Cooper, A. Waheed, and R.K. Shadduck. 1987. The influence *in vivo* of murine colony stimulating factor-1 on myeloid progenitor cells in mice recovering from sublethal dosages of cyclophosphamide. *Blood* 69: 913-918.
21. Broxmeyer, H.E., D.E. Williams, S. Cooper, P. Ralph, S. Gillis, D.C. Bicknell, G. Hangoc, R. Drummond, and L. Lu. 1988. Synergistic interactions of hematopoietic colony stimulating and growth factors in the regulation of myelopoiesis. *In*: Behring Institute Mitteilungen (BIM), Behring, Marburg, Germany, In Press.
22. Shadduck, R.K., A. Waheed, F. Boegel, G. Pigoli, A. Porcellini, and

V. Rizzoli. 1987. The effect of colony stimulating factor-1 in vivo. Blood Cells. 13: 49-63.

23. Yanoi, N., M. Yamada, Y. Watanabe, M. Saito, M. Kuboyama, K. Motoyoshi, F. Takaku, S. Funakoshi, and M. Watanabe. 1983. The granulopoietic effect of human urinary colony-stimulating factor on normal and cyclophosphamide treated mice. Exp. Hematol. 11: 1027-1036.

24. Motoyoshi, K., F. Takaku, T. Maekawa, Y. Miura, K. Kimura, S. Furusawa, M. Hattori, T. Nomura, H. Mizoguchi, M. Ogawa, K. Kinugasa, T. Tominaga, M. Shimoyama, K. Deura, K. Ohta, T. Taguchi, T. Masaoka, and I. Kimura. Protective effect of partially purified human urinary colony-stimulating factor on granulocytopenia after antitumor chemotherapy. Exp. Hematol. 14: 1069-1075.

25. Metcalf, D., C.G. Begley, D.J. Williamson, E.C. Nice, J. deLamarter, J. Mermod, D. Thatcher, and A. Schmidz. 1987. Hemopoietic responses in mice injected with purified recombinant murine GM-CSF. Exp. Hematol. 15: 1-9.

26. Donahue, R.E., E. Wang, D.K. Stone, R. Kamen, G.G. Wong, P.K. Seghal, D.C. Nathan, and S.C. Clark. 1986. Stimulation of hematopoiesis in primates by continuous infusion of recombinant human GM-CSF. Nature 321: 872-875.

27. Mayer, P., C. Lam, H. Obenaus, E. Liehl, and J. Besemer. 1987. Recombinant human GM-CSF induces leukocytosis and activates peripheral blood polymorphonuclear neutrophils in nonhuman primates. Blood 70: 206-213.

28. Monroy, R.L., R.R. Skelly, T.J. MacVittie, T.A. Davis, J.J. Sauber, S.C. Clark, and R.E. Donahue. 1987. The effect of recombinant GM-CSF on the recovery of monkeys transplanted with autologous bone marrow. Blood 70: 1696-1999.

29. Welte, K., M.A. Bonilla, A.P. Gillio, T.C. Boone, G.K. Potter, J.L. Gabrilove, M.A.S. Moore, R.J. O'Reilly, and L.M. Souza. 1987. Recombinant human granulocyte colony-stimulating factor. Effects on hematopoiesis in normal and cyclosphosphamide-treated primates. J. Exp. Med. 165: 941-948.

30. Tsuchuja, M., H. Nomura, S. Asano, Y. Kaziro, and S. Nagata. 1987. Characterization of recombinant human granulocyte colony-stimulating factor produced in mouse cells. EMBO J. 6: 611-616.

31. Shimamura, M., Y. Kobayashi, Y., A. Yuo, A. Urabe, T. Okabe, Y. Komatsu, S. Itoh, and F. Takaku. 1987. Effect of human recombinant granulocyte colony-stimulating factor on hematopoietic injury in mice induced by 5 fluorouracil. Blood 69: 353-355.

32. Cohen, A.M., K.M. Zsebo, H. Inove, D. Hines, T.C. Boone, V.R. Chazin, L. Tsai, T. Ritch, and L.M. Souza. 1987. In vivo stimulation of granulopoiesis by recombinant human granulocyte colony-stimulating factor. Proc. Natl. Acad. Sci. USA 84: 2484-2488.

33. Tsuchiya, M., H. Nomura, S. Asano, Y. Kaziro, and S. Nagata. 1987. Characterization of recombinant human granulocyte colony-stimulating factor produced in mouse cells. EMBO J. 6: 611-616.

34. Broxmeyer, H.E., D.E. Wiliams, S. Cooper, G. Hangoc, and P. Ralph. 1987. Purified recombinant human-granulocyte colony-stimulating factor and- macrophage colony stimulating factor synergise in vivo to enhance proliferation of granulocyte-macrophage, erythroid and multipotential progenitor cells in mice. Blood 70: 168a (abstract).

35. Welte, K., M.A. Bonilla, J.L. Gabrilove, A.P. Gillio, G.K. Potter, M.A.S. Moore, R.J. O'Reilly, T.C. Boone, and L.M. Souza. Recombinant human granulocyte colony-stimulating factor: In Vitro and In Vivo effects on myelopoiesis. Blood Cells 13: 17-30.

36. Broxmeyer, H.E., A. Smithyman, R.R. Eger, P.A. Meyers, and M. DeSousa. 1978. Identification of lactoferrin as the granulocyte-derived inhibitor of colony-stimulating activity production. J. Exp. Med. 148: 1052-1067.

37. Broxmeyer, H.E. and E. Platzer. 1984. Lactoferrin acts on I-A and I-E/C antigen subpopulations of mouse peritoneal macrophages in the absence of T lymphocytes and other cell types to inhibit production of granulocyte-macrophage colony-stimulating factors in vitro. J. Immunol. 133: 306-314.
38. Zucali, J.R., C.A. Dinarello, D.J. Oblon, M.A. Gross, L. Anderson, and R.S. Weiner. 1986. Interleukin-1 stimulates fibroblasts to produce granulocyte-macrophage colony-stimulating activity and prostaglandin E_2. J. Clin. Invest. 77: 1857-1863.
39. Zucali, J.R., H.E. Broxmeyer, C.A. Dinarello, M.A. Gross, R.S. Weiner. 1987. Regulation of early human hematopoietic (BFU-E and CFU-GEMM) progenitor cells in vitro by interleukin-1 induced fibroblast conditioned medium. Blood 69: 33-37.
40. Bagby, G.C., C.A. Dinerello, P. Wallace, C. Wagner, S. Hefeneider, and E. McCall. 1986. Interleukin 1 stimulates GM-CSA release by vascular endothelial cells. J. Clin Invest. 78: 1316-1323.
41. Zucali, J.R., D.A. Levy, H.E. Broxmeyer, and C.A. Morse. 1987. Lactoferrin acts on human monocytes to inhibit production of interleukin 1 and the ability of monocyte-conditioned medium to induce fibroblasts to produce granulocyte-macrophage colony stimulating factors. Blood 70: 191a (abstract).
42. Ralph, P., M.B. Ladner, A.M. Wang, E.S. Kawasaki, L. McConlogue, J.F. Weaver, S.A. Weiss, P. Shadle, K. Koths, M.K. Warren, E.R. Stanley, and H.E. Broxmeyer. 1987. The molecular and biological properties of human and murine members of the CSF-1 family. In: Molecular Basis of Lymphokine Action, Webb, D.R., C.W. Pierce, and S. Cohen, eds., The Humana Press Inc., 295-311.
43. Broxmeyer, H.E., L. Juliano, A. Waheed, and R.K. Shadduck. 1985. Release from mouse macrophages of acidic isoferritins that suppress hematopoietic progenitor cells is induced by purified L cell colony-stimulating factor and suppressed by human lactoferrin. J. Immunol. 135: 3224-3231.
44. Pelus, L.M., H.E. Broxmeyer, J.I. Kurland, and M.A.S. Moore. 1979. Regulation of macrophage and granulocyte proliferation: specificities of prostaglandin E and lactoferrin. J. Exp. Med. 150: 277-292.
45. Moore, R.N., F.J. Pitruzzelo, H.S. Larsen, and R.T. Rouse. 1984. Feedback regulation of colony-stimulating factor CSF-1 induced macrophage proliferation by endogenous E prostaglandins and interferon. J. Immunol. 133: 541-543.
46. Warren, M.K., and P. Ralph. 1986. Macrophage growth factor CSF-1 stimulates human monocyte production of interferon, tumor necrosis factor and myeloid CSF. J. Immunol. 137: 2281-2285.
47. Chikkappa, G., H.E. Broxmeyer, S. Cooper, D.E. Williams, H. Hangoc, and R.K. Shadduck. 1987. Feedback suppression of hematopoietic progenitor cell proliferation in vivo after administration of multiple injections of pure natural murine and recombinant human macrophage colony stimulating factor (CSF-1) to mice. Blood 70: 168a (abstract).
48. Broxmeyer, H.E. and P. Ralph. 1987. Unpublished observations.
49. Fabian, I., I. Bleiberg, I. Riklis, and Y. Kletter. 1987. Enhanced reconstitution of hematopoietic organs in irradiated mice, following their transplantation with bone marrow cells pretreated with recombinant interleukin 3. Exp. Hematol. 15: 1140-1144.
50. Williams, D.E., D.C. Bicknell, L.S. Park, J.E. Straneva, S. Cooper, and H.E. Broxmeyer. 1988. Purified murine granulocyte-macrophage progenitor granulocyte-macrophage colony stimulating factor (GM-CSF). Proc. Natl. Acad. Sci. USA 85: 487-491.
51. Bicknell, D.C., D.E. Williams, and H.E. Broxmeyer. 1988. Correlation between CSF-1 responsiveness and expression of (CSF-1 receptor)

c-fms in purified murine granulocyte-macrophage progenitor cells (CFU-GM). Exp. Hematol. 16: 240-243.

52. Williams, D.E. and H.E. Broxmeyer. 1988. Interleukin-1-alpha enhances the in vitro survival of purified murine granulocyte-macrophage progenitor cells in the absence of colony-stimulating factors. Blood, In Press.

53. Wineolis, C.G., D.O. Oliver, M.J. Pippard, C. Reid, M.R. Downing, and P.M. Cotes. 1986. Effect of human erythropoietin derived from recombinant DNA on the anemias of patients maintained by chronic haemodialysis. Lancet ii: 1175-1178.

54. Eschbach, J.W., J.C. Egrie, M.R. Downing, J.K. Browne, and J.W. Adamson. 1987. Correction of the anemia of end stage renal disease with recombinant human erythropoietin. New Engl. J. Med. 316: 73-78.

55. Groopman, J.E., R.T. Mitsyasu, M.J. DeLeo, D.H. Oette, and D.W. Golde. 1987. Effect of recombinant human granulocyte-macrophage colony--stimulating factor on myelopoiesis in the acquired immunodeficiency syndrome. N. Engl. J. Med. 317: 593-598.

56. Vadhan-Raj, S., M. Keating, A. LeMaistre, W.H. Hittelman, K. McCredie, J.M. Trujillo, H.E. Broxmeyer, C. Henny, and J.U. Gutterman. 1987. Effects of recombinant human granulocyte-macrophage colony-stimulating factor in patients with myelodysplastic syndromes. N. Engl. J. Med. 317: 1546-1552.

57. Vadhan-Raj, S., S. Buescher, A. LeMaistre, M. Keating, R. Walters, H. Kantarjian, W.N. Hittelman, H.E. Broxmeyer, and J.U. Gutterman. 1988. Stimulation of hematopoiesis in patients with bone marrow failure and in patients with malignancy by recombinant human granulocyte-macrophage colony-stimulating factor. Blood, In Press.

58. Broxmeyer, H.E., S. Cooper, D.E. Williams, G. Hangoc, J.U. Gutterman, and S. Vadhan-Raj. 1988. Growth characteristics of marrow hematopoietic progenitor cells in patients on a phase I clinical trial with purified recombinant human granulocyte-macrophage colony stimulating factor. Exp. Hematol., In Press.

59. Mertelsmann, R., M.A.S. Moore, H.E. Broxmeyer, C. Cirrincione, and B. Clarkson. 1981. The diagnostic and prognostic significance of the CFU-C assay in acute non-lymphoblastic leukemia (ANLL). Cancer Res. 41: 4844-4848.

CURRENT STATUS OF THROMBOPOIETIN

T.P. McDonald

University of Tennessee
College of Veterinary Medicine
Knoxville, Tennessee

INTRODUCTION

This review outlines the effects of a thrombocytopoiesis-stimulating factor (TSF or thrombopoietin) on thrombocytopoiesis and megakaryocytopoiesis. Thrombopoietin is a hormone that is now known to be a major controlling factor of blood platelet production. The background, including a model for megakaryocytopoiesis, controlling factors, sources of the hormone, the role of the kidney in thrombopoietin production, a possible gene location for the stimulating factor, assay procedures, antibodies, and tests for other stimulators in TSF-rich preparations will be discussed briefly. In addition, new studies on purification and characterization of the hormone will be presented.

BACKGROUND

Model for Megakaryocytopoiesis

A model for megakaryocytopoiesis has been presented previously by Ebbe[1]. The model shows that megakaryocytes originate from an uncommitted pluripotent stem cell, and the first step is commitment of this hematopoietic progenitor cell to megakaryocyte differentiation. Once the cells are committed, cellular proliferation occurs and this step is followed by nuclear proliferation. Following mitotic amplification of the megakaryocyte precursor pool, cells undergo nuclear proliferation. Endomitotic nuclear division occurs following this proliferation step, leading to cytoplasmic growth and maturation with the acquisition of platelet-specific proteins and cell organelles. Finally, platelets are released into the circulation.

Controlling Factors

A current hypothesis of the controlling mechanism of megakaryocytopoiesis and thrombocytopoiesis, derived from *in vitro* studies, indicates that there may be two types of factors involved, i.e., the megakaryocyte-colony stimulating factor (Meg-CSF) and a potentiator, thrombopoietin[2,3]. Several previous studies have shown that *in vitro* Meg-CSF stimulates the early proliferative events of megakaryocytopoiesis and TSF acts synergistically to enhance colony formation by stimulating cytoplasmic maturation[3]. Meg-CSF has been found in various preparations, including culture media from spleen cells[4,5], WEHI-3-conditioned media[6], urine from patients with aplastic anemia[7,8], and sera or plasma from irradiated animals[9,10]. Also, Interleukin-3 and GM-CSF have been shown to stimulate megakaryocyte colony formation in vitro[11,12].

Sources of Thrombopoietin

Several previous studies have utilized plasma, serum and urine from thrombocytopenic animals and patients[13-22] as a source of thrombopoietin for several studies. Thrombopoietin-like materials have also been reported to be present in body fluids of patients with thrombocytosis and poor platelet function[23,24]. However, another source of TSF that appears to be more stable and potent than that from in vivo sources is medium from human embryonic kidney (HEK) cell cultures[14,25-39]. Because of its potency and stability, this preparation will continue to be a valuable source of TSF for a variety of studies.

Role of the Kidney in Thrombopoietin Production

In an early study, it was concluded that the kidney was necessary for the appearance of a "post-hemorrhagic thrombopoietic factor"[40]. In rats made thrombocytopenic by bleeding, thrombopoietin activity was not found in serum of rats after nephrectomy, but the factor was present in the serum of rats with intact kidneys[40]. The thrombocytopoietic-stimulating factor could be produced in rats after other organs had been removed, i.e., spleen, adrenal glands and pituitary.

In a follow-up study, it was reported that the kidney was necessary for the appearance of TSF in plasma after production of acute thrombocytopenia[41]. Extracts of plasma from normal rats, rats injected with rabbit anti-rat platelet serum, nephrectomized rats, and nephrectomized rats injected with antisera were tested in the immunothrombocythemic mouse assay for the presence of TSF. Only an extract of plasma from unoperated rats, made thrombocytopenic by an injection of antisera, gave positive results for TSF. These results indicate that, in rats, the kidney is necessary for TSF production.

In agreement with this work, Klener et al.[42] showed that vinblastine treatment caused a significant increase in serum thrombocytopoietic activity in animals. The vinblastine-induced increase in thrombopoietic activity was abolished by bilateral nephrectomy but was unaffected by bilateral ligation of the ureters. These workers in agreement with previous studies, concluded that, in rodents, the kidney is probably a major source of the serum thrombocytopoietic factor.

The hypothesis that the kidney is a site of thrombopoietin production received additional support when McDonald et al.[25] reported the detection of TSF in medium of HEK cell cultures. TSF production seems to be localized in the kidney since cultures of fetal mouse liver did not have demonstrable thrombopoietin levels[43]; other studies using adult livers from other species[44] confirmed this finding.

Another study showed that bilaterally nephrectomized rats not only failed to show increased numbers of a megakaryocyte precursor cell [the small acetylcholinesterase-positive (SAChE+) cell] after being made thrombocytopenic, but other anephric rats that were not given anti-platelet serum had decreased percentages of SAChE+ cells when compared to untreated control rats[45]. These findings indicate the importance of the kidney in the day-to-day production of thrombopoietin for the maintenance of the number of SAChE+ cells. Interestingly, unilateral nephrectomy and ligation of the ureters had little effect on altering the percentages of SAChE+ cells, suggesting that surgical stress and uremia do not interfere with the release and action of thrombopoietin.

In a recent study[46], it was shown that platelet counts were reduced and mild thrombocytopenia was frequent in patients with chronic renal failure. Thrombopoietic activity in plasma from 7 thrombocytopenic patients was measured using ^{75}Se-selenomethionine incorporation into platelets of mice. TSF activity was found to be reduced. It was concluded that a possible cause for the platelet count reduction was insufficient thrombopoietic activity[46].

Since anephric man is not usually markedly thrombocytopenic, it seems possible that, in humans, extrarenal sites may play a significant role in production of the hormone. However, the data outlined above and elsewhere[47] show that in rodents, the intact kidney is necessary for TSF production, indicating the possible absence of extrarenal sites in these animals for TSF production. Clarification of the site of production of TSF will require additional work.

Thrombopoietin Gene Location

A patient with acute megakaryoblastic leukemia who had a paracentric inversion of the 3q chromosome with increased thrombopoietin titers was described[48]. Based on these findings it was postulated that a gene on chromosome 3q has a role in thrombopoietin production and led to the suggestion that either 3q21 or 3q26 contains the locus for a gene that regulates thrombocytopoiesis[48]. In additional work two patients were described with acute blastic transformation of chronic myeloid leukemia associated with strikingly elevated platelet counts, abnormalities of chromosome 3q, and the standard Philadelphia chromosome translocation[49]. This latter work suggested that 3q21 is the relevant site for thrombopoietin production, but other studies are needed to establish conclusively the location for the TSF gene.

Assays For Thrombopoietin

Previous assays for thrombopoietin have utilized platelet counting[15,50,51], measurement of alterations in platelet labeling with radioisotopes[21,52,53] changes in platelet sizes[20,29,54,55], an immunoassay[56], increases and alterations of megakaryocytes *in vivo*[31,32] and *in vitro*[6], and determination of the number of SAChE+ cells in the marrow of rodents[57]. In some of these studies, the thrombopoietin recipient animals have been pretreated, i.e., splenectomized[15], transfused with platelets[21,30], or injected with platelet specific antisera[22,52] in an attempt to increase their sensitivity to TSF. Of these several attempts to find the optimum conditions for the assay of thrombopoietin, mice in rebound-thrombocytosis appear to be more sensitive to exogenous TSF preparations than are normal mice[47]. Other studies have improved these early assay techniques and have shown that the route of administration of TSF did not seem to be important since TSF injected either subcutaneously or intraperitoneally gave essentially the same result; however, multiple injections were more effective than single injections[58]. The time of measurement of the ^{35}S incorporation into platelets and selection of the mouse strain determines, to a large degree, the sensitivity of the TSF assay[59]. Sex of the mice does not appear to be important; similar responses were found in both male and female mice[59]. In dose-response experiments ^{35}S-sodium sulfate gave significantly greater values at the higher TSF doses than did ^{75}Se-selenomethionine, but either isotope can be used to measure platelet production rates in mice stimulated with TSF[60]. A sensitive assay for thrombopoietin that utilizes measurement of the SAChE+ cell in marrow of mice was previously described[57]. This assay technique was able to detect smaller doses of TSF than the rebound-thrombocytotic mouse assay and should prove to be helpful in future experiments in assaying for thrombopoietin.

Antibodies to Thrombopoietin

Both polyclonal[56,61,62] and monoclonal[63] antisera have been raised against TSF utilizing, as antigens, thrombopoietin from thrombocytopenic sheep[56,61] and humans[62,63] and TSF purified from HEK cell culture medium[62]. The results indicate that antibodies made against thrombopoietin from sheep plasma or human urinary thrombopoietin can neutralize the biological activity of both HEK-cell TSF[62,63] and endogenous thrombopoietin of mice[61]. Specifically, thrombopoietin, when assayed in the immunothrombocythemic mouse assay, was neutralized *in vitro* by addition of serum obtained from a rabbit previously immunized with a thrombopoietin-rich preparation from sheep[61]. The anti-TSF serum had no effect on platelets; however, a depression in thrombocytopoiesis was observed after injection of anti-TSF serum into normal mice[61]. It was concluded that the injections of an antibody into normal mice developed against thrombopoietin interrupt normal thrombocytopoiesis by reducing endogenous thrombopoietin levels[61]. Moreover, a monoclonal antibody made against human urinary thrombopoietin was shown to reduce the bioactivity of TSF extracted from HEK cell culture media[63] when tested in the immunothrombocythemic mouse assay. These data indicate that thrombopoietin is required for the day-to-day maintenance of platelet counts[61] and that antibodies made against one source of thrombopoietin will neutralize the biological activity of others[61-63].

Tests for Other Hematopoietic Stimulators in HEK Cell Culture Media

The supernatant fluid of HEK cell culture medium is free of erythropoietin and *in vivo* WBC stimulating factors[25]. Although the crude medium contains a granulocyte colony stimulating factor (G-CSF), TSF and G-CSF are clearly separate entities[64].

In previous studies[25] it was shown, that WBC counts were not altered in mice after injection of HEK cell culture media, when compared to other mice injected with control production media, indicating that the TSF-rich media did not contain *in vivo* WBC-stimulating factors.

Moreover, when the culture media were subjected to erythropoietin and erythrogenin assays, negative results were found[25]. Production media or growth media with or without kidney cell growth did not increase the %^{59}Fe-RBC incorporation into exhypoxic mice when compared to values from other mice injected with saline. In other work (Garcia J.F., unpublished), it was found that the HEK cell culture media did not show positive results when tested in a radioimmunoassay for erythropoietin, indicating that HEK cell culture medium is essentially free of RBC-stimulating factors.

In another study[64], both TSF-rich culture media and CSF-rich preparations were used in three separate experiments; the results indicate a lack of correlation ($r = 0.003$) between CSF activity and the ability of the preparations to stimulate %^{35}S incorporation into platelets of assay mice. Moreover, purified L cell CSF was inactive in the TSF assay. When the colony stimulating activity of preparations rich in TSF was neutralized with anti-CSF serum, the material's ability to stimulate platelet production in mice was not affected. Based on these results it was concluded that TSF and CSF are separate entities.

PURIFICATION OF THROMBOPOIETIN

Chemical Characteristics

Previously employed purification procedures have helped to clarify some of the chemical characteristics of thrombopoietin. For example, the fact that thrombopoietin precipitates at ethanol concentrations of >40%[67] and ammonium sulfate centrations of 60-80%[16,17,19,26-28,37] indicates that the molecule is relatively hydrophilic and/or compact enough to limit surface hydrophobic interactions. Affinity for the lectins (both wheat germ agglutinin and concanavalin A) indicates that thrombopoietin is a glycoprotein[19,38] with N-acetyl glucosamine or sialic acid and mannose residues in its carbohydrate side chains. Binding to both anion and cation exchange resins suggests that the charge characteristics of thrombopoietin are appropriate for a glycoprotein.

Previously a five-step purification procedure was used[26,28] to purify to homogeneity TSF from HEK cell culture media[25]. The pure material was obtained by a series of column chromatographic and precipitation procedures to include: ammonium sulfate precipitation, Sephadex G-75 column chromatography, diethylaminoethyl (DEAE) ion exchange chromatography, both DEAE and size exclusion-high performance liquid chromatography (HPLC), and sodium dodecyl sulfate-polyacrylamide gel electrophoresis (SDS-PAGE)[26]. Table 1 shows that the DEAE column chromatographic step resulted in significant losses of the hormone. The final product was potent (approximately 11,000 U/mg of protein), but it was obtained in very small amounts. New studies to characterize the TSF molecule are outlined below.

Reaction of TSF to Endo F and H

Enzyme digestion was accomplished by diluting endoglycosidases F (*F. meningosepticum*, endo-β-N-acetylglucosaminidase F) and H (*S. plicatus*, endo-β-N-acetylglucosaminidase H) into a buffer of 75 mM Na_2HPO_4-NaH_2PO_4, 15 mM EDTA, 0.015% SDS, 0.75% ethanol, 0.15 Nonidet P-40^R^ at pH 6.2[65]. Enzymes were added at a concentration of 1 U/mg protein to a partially purified TSF preparation (Step I, refer to Table 1 for definition of steps) or human serum albumin (HSA). The mixtures were incubated for 18 hours at 37°C with mixing and dialyzed against 3 changes of saline using 1,000 molecular weight (MW) cutoff membranes. After dialysis the mixtures were freeze-dried and assayed for TSF content by use of the immunothrombocythemic mouse assay[52]. As shown in Table 2, HSA diluted with buffer and mixed with either Endo F or Endo H did not stimulate platelet production in mice. However, when TSF was incubated with diluting

Table 1. Purification of Thrombopoietin from HEK Cell Culture Media

Step	Procedure	% Recovery from starting material: Protein	% Recovery from starting material: Units	Specific activity (units/mg Protein)
Crude	Culture Media	100	100	0.11
I	Sephadex	2.2	71	3.6
II	DEAE	0.5	9.8	2.3
III	HPLC-SE	0.02	4.4	47.4
IV	HPLC-DEAE	0.0009	1.3	100
V	SDS-PAGE	0.00003	1.1	11,000

Table 2. Effect of Endoglycosidases F and H on Biological Activity of TSF

Material	$\%^{35}S$ Incorporation $x10^3$ ± SE
Saline	2.70 ± 0.18
TSF Lab Standard	6.25 ± 0.31***
HSA + Buffer	3.06 ± 0.30
HSA + Endo F	2.17 ± 0.15
HSA + Endo H	2.56 ± 0.36
TSF + Buffer	3.62 ± 0.39*
TSF + Endo F	3.75 ± 0.14**
TSF + Endo H	3.99 ± 0.56*

TSF was used at 0.9 mg/mouse; human serum albumin (HSA) was tested at 1.0 mg/mouse; Both Endo F and Endo H were used at 1 μg/mg protein. 4-5 mice were used in each treatment group. Values were significantly elevated over saline-control, *P < 0.05; ** P < 0.005; *** P < 0.0005. TSF-Lab standard was a potent TSF source of HEK-cell culture media.

buffers or either Endo F or Endo H, the $\%^{35}S$ incorporation in all cases was significantly increased over saline controls, indicating that the enzymes did not alter TSF bioactivity.

Digestion of TSF With Cyanogen Bromide

Cyanogen bromide (CNBr) cleavage was accomplished by dissolving Step I TSF in 70% formic acid to a concentration of 5 mg/ml. A solution of CNBr was made up in 70% formic acid so that when mixed with TSF, CNBr was in a 50% molar excess over methionine residues[66]. After mixing, nitrogen was bubbled through the solutions and the mixtures were incubated at 22°C for 24 hours in the dark. After incubation the mixtures were diluted 15-20 fold with water, freeze-dried, and assayed for TSF content[52]. Table 3 shows that CNBr destroyed the bioactivity of TSF. It should also be noted that the CNBr-control TSF was subjected to the same procedures as the experimental group but without CNBr, i.e., dilution into formic acid, flushing with nitrogen, incubation at room temperature, and lyophilization; as shown, the control TSF retained bioactivity ($P < 0.005$).

Additional Chemical Characteristics

Thrombopoietin is not dialyzable[16-19,26-28,37,38,56,67]. The factor from HEK cell culture media does not deteriorate in both ammonium sulfate and Sephadex fractions during four to five weeks of storage at -76°C[37] and it is heat stable[38]. TSF will withstand lyophilization[26-28,37,38,56,67] and does not deteriorate at pHs of 1-8[38]. TSF was shown to have an isoelectric pH of 4.7[26] and it exists in trace amounts in both plasma of thrombocytopenic animals[19] and crude HEK cell culture medium[26-28]. Several of these studies indicate that TSF is a relatively stable molecule supporting the conclusion that it is a glycoprotein. The fact that TSF is stable to mercaptoethanol[26] indicates that disulfide bridges are probably not necessary for biological activity. It was shown in this report that TSF activity was unaffected by Endoglycosidases F and H (Table 2). Therefore, glycosylation of TSF by "high mannose" or "complex carbohydrates" linked through asparagine to the protein backbone is probably not necessary for biological activity. However, the loss of TSF activity after breakage of the primary structure with either trypsin[38] or cyanogen bromide (Table

Table 3. Effects of Cyanogen Bromide Digestion on Biological Activity of TSF

Material	$\%^{35}S$ Incorporation x10^3 ± SE
Saline	2.72 ± 0.23
TSF Lab Standard	4.48 ± 0.34**
CNBr-control TSF	4.57 ± 0.41**
CNBr-treated TSF	3.51 ± 0.46^{NS}

CNBr: Cyanogen bromide; 4-5 mice were used in each treatment group. Values were significantly elevated over saline-control ** $P < 0.005$; ^{NS} not significantly different from saline-control.

3) indicates that TSF requires at least some finite protein structure involving methionine for activity.

Molecular Weight Determinations of Thrombopoietin

In recent experiments, it was found that TSF could be purified by subjecting the crude materials to Sephadex G-75 chromatography, ethanol precipitation, SDS-PAGE and reverse phase (RP)-HPLC. If the TSF-rich post SDS-PAGE material that was used for the RP-HPLC purification step was subjected to additional electrophoresis after dialysis and lyophilization using denaturing conditions, a broad diffuse protein band from 14-20 Kd was found. Further purification on RP-HPLC showed two major bands and four minor bands; one of the minor bands (15Kd) appeared to be thrombopoietin. It was also found that if this 15 Kd MW TSF was dialyzed in a weak phosphate buffer, freeze-dried and re-electrophoresed on SDS-PAGE without boiling or without the use of strong denaturing conditions, the material that was previously shown to be 15 Kd MW protein migrated to 30 Kd. This finding agrees with our previous work[26], showing that TSF may exist at a MW of 30 Kd. Also, this finding shows that TSF has a strong ability to self-associate[68]. TSF probably exists as a dimer in nondenaturing conditions (30 KdMW).

SIMILARITIES OF TSF AND MSF

Recently, a megakaryocyte stimulating factor (MSF) that has a MW of 15 Kd was isolated from both HEK cell culture media and plasma from thrombocytopenic rabbits[68,69]. The factor was shown to stimulate the synthesis of platelet factor-4 in alpha granules of megakaryocytes and to elevate the rate of cytoplasmic maturation during megakaryocyte development. There appears to be many similarities between this factor and TSF. For example: 1) both factors have the same approximate MW, 2) both stimulate megakaryocytopoiesis, 3) they both bind wheat germ agglutinin, 4) both have a remarkable tendency to self-associate under nondenaturing conditions, 5) they have similar total purification factors, 6) the procedures that have been used for their purification are very similar, and 7) the presence of both these factors in plasma of animals depends on the level of circulating platelets. It is not known if MSF and TSF are the same or different factors, since MSF has not yet been tested *in vivo*; however, the data that exist so far indicate that these two factors are probably identical.

ACKNOWLEDGEMENTS

The author is grateful to Marilyn Cottrel and Rose Clift for expert technical assistance and to his many collaborators, both present and past, for their contributions to the data outlined in this review. This work was supported by a grant (HL14637) from the Heart, Lung and Blood Institute.

REFERENCES

1. Ebbe, S. 1970. Responses of the thrombocytopoietic system to platelet depletion and irradiation. _In_: Stohlman F (ed) Hemopoietic cellular proliferation, New York, Grune and Stratton, p 285.
2. McDonald, T.P. 1988. Thrombopoietin: its biology, purification and characterization. _Exp Hematol_. 16: 201.
3. Mazur, E.M. 1987. Megakaryocytopoiesis and platelet production. _Exp Hematol_ 15: 340.
4. Nakeff A., S. Daniels-McQueen. 1976. In vitro colony assay for a new class of megakaryocyte precursors: colony-forming unit megakaryocyte (CFU-M). _Proc Soc Exp Biol Med_ 151: 587.
5. Levin, J., F. C. Levin, D.G. Penington, and D. Metcalf. 1981. Measurement of ploidy distribution in megakaryocyte colonies obtained from culture: with studies of the effects of thrombocytopenia. _Blood_ 57: 287.
6. Williams, N., R.R. Eger, H.M. Jackson, and D.J. Nelson. 1982. Two-factor requirement for murine megakaryocyte colony formation. _J Cell Physiol_ 110: 101.
7. Enomoto, K., M. Kawakita, S. Kishimoto, N. Katayama, and T. Miyake 1980. Thrombopoiesis and megakaryocyte colony stimulating factor in the urine of patients with aplastic anemia. _Br J Haematol_ 45: 551.
8. Miyake, T., M. Kawakita, K. Enomoto, and M.J. Murphy, Jr. 1982. Partial purification and biological properties of thrombopoietin extracted from the urine of aplastic anemia patients. _Stem Cells_ 2: 129.
9. Mazur, E.M., and K. South. 1985. Human megakaryocyte colony-stimulating factor in sera from aplastic dogs: partial purification, characterization and determination of hematopoietic cell lineage specificity. _Exp Hematol_ 13: 1164.
10. Miura, M., C.W. Jackson, and S.A. Lyles. 1984. Increases in circulating megakaryocyte growth-promoting activity in the plasma of rats following whole body irradiation. _Blood_ 63: 1060.
11. Burstein, S.A. 1986. Interleukin-3 promotes maturation of murine megakaryocytes _in vitro_. _Blood Cells_ 11: 469.
12. Quesenberry, P.J., J.N. Ihle, and E. McGrath. 1985. The effect of interleukin 3 and GM-CSA-2 on megakaryocyte and myeloid clonal colony formation. _Blood_ 65: 214.
13. Levin, J., and B.L. Evatt. 1979. Humoral control of thrombopoiesis. _Blood Cells_ 5: 105.
14. McDonald, T.P. 1981. Thrombopoietin and its control of thrombocytopoiesis and megakaryocytopoiesis. _In_: Evatt B.L., Levine R., Williams, N. (eds) Megakaryocyte biology and precursors: in vitro cloning and cellular properties. New York: Elsevier North Holland, p 39.
15. O'Dell, T.T., Jr., T.P. McDonald, and T.C. Detwiler. 1961. Stimulation of platelet production by serum of platelet-depleted rats. _Proc Soc Exp Biol Med_ 108: 428.
16. Evatt, B.L., D.P. Shreiner, and J. Levin. 1974. Thrombopoietic activity of fractions of rabbit plasma: studies in rabbits and mice. _J Lab Clin Med_ 83: 364.

17. Evatt, B.L., J. Levin, and K.M. Algazy. 1979. Partial purification of thrombopoietin from the plasma of thrombocytopenic rabbits. Blood 54: 377.
18. Dassin, E., J. Bourebia, Y. Najean, and A.M. Rosset. 1983. Partial purification of a thrombocytopoiesis stimulating factor present in the serum of thrombocytopenic rats. Acta Haematol 69: 249.
19. Hill, R., and J. Levin. 1986. Partial purification of thrombopoietin using lectin chromatography. Exp Hematol 14: 752.
20. Levin, J., F.C. Levin, D.F. Hull, III, and D.G. Penington. 1982. The effects of thrombopoietin on megakaryocyte-CFC, megakaryocytes, and thrombopoiesis: with studies of ploidy and platelet size. Blood 60: 989.
21. Evatt, B.L., and J. Levin. 1969. Measurement of thrombopoiesis in rabbits using ^{75}selenomethionine. J Clin Invest 48: 1615.
22. Penington, D.G. 1970. Isotope bioassay for "thrombopoietin". Br Med J 1: 606.
23. McDonald, T.P. 1975. Assay of thrombopoietin utilizing human sera and urine fractions. Biochem Med 13: 101.
24. Shreiner, D.P., J. Weinberg, and D. Enoch. 1980. Plasma thrombopoietic activity in humans with normal and abnormal platelet counts. Blood 56: 183.
25. McDonald, T.P., R. Clift, R.D. Lange, C. Nolan, I.I.E. Tribby, and G.H. Barlow. 1975. Thrombopoietin production by human embryonic kidney cells in culture. J Lab Clin Med 85: 59.
26. McDonald, T.P., M. Cottrell, R. Clift, J.A. Khouri, and M.D. Long. 1985. Studies on the purification of thrombopoietin from kidney cell culture medium. J Lab Clin Med 106: 162.
27. McDonald, T.P., R. Clift, M. Cottrell, and M.D. Long. 1986. Further studies on the purification and assay of thrombopoietin. In: Levine, R.F., Williams, N., Levin J., Evatt, B.L. (eds) Megakaryocyte development and function, New York: Alan R. Liss, p 215.
28. McDonald, T.P., R. Clift, M. Cottrell, and M.D. Long. 1987. Recovery of thrombopoietin during purification. Biochem Med Metab Biol 37: 335.
29. McDonald, T.P. 1980. Effect of thrombopoietin on platelet size of mice. Exp Hematol 8: 527.
30. McDonald, T.P., R. Clift, C. Nolan, and I.I.E. Tribby. 1976. A comparison of mice in rebound-thrombocytosis with platelet-hypertransfused mice for the assay of thrombopoietin. Scand J Haematol 16: 326.
31. McDonald, T.P., and G.D. Kalmaz. 1983. Effects of thrombopoietin on the number and diameter of marrow megakaryocytes of mice. Exp Hematol 11: 91.
32. Cullen, W.C., and T.P. McDonald. 1986. A comparison of stereologic techniques for the quantification of megakaryocyte size and number. Exp Hematol 14: 782.
33. Kalmaz, G.D., and T.P. McDonald. 1981. Effects of antiplatelet serum and thrombopoietin on the percentage of small acetylcholinesterase-positive cells in bone marrow of mice. Exp Hematol 9: 1002.
34. Raha, S.W., W. Wesemann, and T.P. McDonald. 1985. Isolation of mouse megakaryocytes:I.Separation of two fractions enriched in different maturational stages. Eur J Cell Biol 37: 111.
35. McDonald, T.P., M. Cottrell, C.C Congdon, O. Walasek, and G.H. Barlow. 1978. Stimulation of megakaryocytic spleen colonies in mice by thrombopoietin. Life Sci 22: 1853.
36. McDonald, T.P., M. Cottrell, and R. Clift. 1977. Hematologic changes and thrombopoietin production in mice after x-irradiation and platelet-specific antisera. Exp Hematol 5: 291.
37. McDonald, T.P., and C. Nolan. 1979. Partial purification of a

thrombocytopoietic-stimulating factor from kidney cell culture medium. Biochem Med 21: 146.
38. McDonald, T.P., R.B. Andrews, R. Clift, and M. Cottrell. 1981. Characterization of a thrombocytopoietic-stimulating factor from kidney cell culture medium. Exp Hematol 9:288.
39. McDonald, T.P., M. Cottrell, C. Nolan, and O. Walasek. 1977. Immunologic similarities of thrombopoietin from different sources. Scand J Haematol 18: 91.
40. Krizsa, F. 1971. Study on the development of posthaemorrhagic thrombocytosis in rats. Acta Haematol 46: 228.
41. McDonald, T.P. 1976. Role of the kidneys in thrombopoietin production. Exp Hematol 4: 27.
42. Klener, P., O. Marcibal, L. Donner, and F. Kornalik. 1977. Serum thrombopoietic activity following administration of vinblastine. Scand J Haematol 19: 287.
43. Zucali, J.R., T.P. McDonald, D.F. Gruber, and E.A. Mirand. 1977. Erythropoietin, thrombopoietin and colony stimulating factor in fetal mouse liver culture media. Exp Hematol 5: 385.
44. Ogle, J.W., C.D.R. Dunn, T.P. McDonald, and R.D. Lange. 1978. The in vitro production of erythropoietin and thrombopoietin. Scand J Haematol 21: 188.
45. McDonald, T.P., and G.D. Kalmaz. 1983. Nephrectomy abolishes the increase in small acetylcholinesterase-positive immature rat megakaryocytes induced by acute thrombocytopenia. Proc Soc Exp Biol Med 174: 131.
46. Gafter, U., H. Bessler, T. Malachi, D. Zevin, M. Djaldetti, and J. Levi. 1987. Platelet count and thrombopoietic activity in patients with chronic renal failure. Nephron 45: 207.
47. McDonald, T.P. 1981. Annotation: assay and site of production of thrombopoietin. Br J Haematol 49: 493.
48. Pinto, M.R., M.A. King, G.D. Goss, W.R. Bezwoda, F. Fernandes-Costa, B. Mendelow, T.P. McDonald, E. Dowdle, and R. Bernstein. 1985. Acute megakaryoblastic leukemia with 3q inversion and elevated thrombopoietin (TSF): an autocrine role for TSF? Br J Haematol 61: 687.
49. Bernstein, R., A. Bagg, M. Pinto, D. Lewis, and B. Mendelow. 1986. Chromosome 3q21 abnormalities associated with hyperactive thrombopoiesis in acute blastic transformation of chronic myeloid leukemia. Blood 68: 652.
50. Spector, B. 1961. In vivo transfer of a thrombopoietic factor. Proc Soc Exp Biol Med 108: 146.
51. de Gabriele, G., and D.G. Penington. 1967. Regulation of platelet production: thrombopoietin. Br J Haematol 13: 210.
52. McDonald, T.P. 1973. Bioassay for thrombopoietin utilizing mice in rebound-thrombocytosis. Proc Soc Exp Biol Med 144: 1006.
53. McDonald, T.P. 1976. A comparison of platelet size, platelet count, and platelet ^{35}S incorporation as assays for thrombopoietin. Br J Haematol 34: 257.
54. Weintraub, A.H., and S. Karpatkin. 1974. Heterogeneity of rabbit platelets. II. Use of the megathrombocyte to demonstrate a thrombopoietic stimulus. J Lab Clin Med 83: 896.
55. Weiner, M., and S. Karpatkin. 1972. Use of the megathrombocyte to demonstrate thrombopoietin. Thrombos Diathes Haemorrh 28: 24.
56. McDonald, T.P. 1973. The hemagglutination-inhibition assay for thrombopoietin. Blood 41:219.
57. Kalmaz, G.D., and T.P. McDonald. 1982. Assay for thrombopoietin: a new, more sensitive method based on measurement of the small acetylcholinesterase-positive cell. Proc Soc Exp Biol Med 170: 213.
58. McDonald, T.P. 1977. Effects of different routes of administration

and injection schedules of thrombopoietin on ^{35}S incorporation into platelets of assay mice. Proc Soc Exp Biol Med 155: 4.
59. McDonald, T.P., R. Clift, and M. Cottrell. 1979. Assay for thrombopoietin: a comparison of time of isotope incorporation into platelets and the effects of different strains and sexes of mice. Exp Hematol 7: 289.
60. Clift, R., and T.P. McDonald. 1979. A comparison of ^{35}S sodium sulfate and ^{75}Se-selenomethionine as platelet labels for the assay of thrombopoietin. Proc Soc Exp Biol Med 162: 380.
61. McDonald, T.P. 1974. Immunological studies of thrombopoietin. Proc Soc Exp Biol Med 147: 513.
62. McDonald, T.P. 1978. Neutralizing antiserum to thrombopoietin. Proc. Soc. Exp. Biol. Med. 158: 557.
63. McDonald, T.P., R. Clift, and M. Cottrell. 1986. Monoclonal antibodies to human urinary thrombopoietin. Proc Soc Exp Biol Med 182: 151.
64. McDonald, T.P., and R.K. Shadduck. 1982. Comparative effects of thrombopoietin and colony-stimulating factors. Exp Hematol 10: 544.
65. Elder, J.H., and S. Alexander. 1982. Endo-β-N-Acetylglucosaminidase F: Endoglycosidase from Flavobacterium meningosepticum that cleaves both high-mannose and complex glycoproteins. Proc Natl Acad Sci USA 79: 4540.
66. Allen, G. 1981. Sequencing of proteins and peptides. New York, Elsevier North Holland, p 62.
67. McDonald, T.P., M. Cottrell, R. Clift, and K. Lane. 1974. Purification and assay of thrombopoietin. Exp Hematol 2: 355.
68. Tayrien, G., and R.D. Rosenberg. 1987. Purification and properties of a megakaryocyte stimulatory factor present both in the serum-free conditioned medium of human embryonic kidney cells and in thrombocytopenic plasma. J Biol Chem 262: 3262.
69. Greenberg, S.M., D.J. Kuter, and R.D. Rosenberg. 1987. Stimulation of megakaryocyte maturation by megakaryocyte stimulatory factor. J Biol Chem 262: 3269.

ABNORMALITIES OF THE HEMATOPOIETIC REGULATORY NETWORK

Grover C. Bagby Jr., Brenda Wilkinson,
Elaine McCall, and Melinda Lee

Medical Research, VA Medical Center
and the Divisions of Hematology and
Medical Oncology and Geriatrics
Department of Medicine
Oregon Health Sciences University
Portland, Oregon 97201

INTRODUCTION

It is now widely recognized that mononuclear phagocytes play a major role in regulating the production of hematopoietic growth factors by a variety of cell types[1-17]. As shown in Fig. 1, two monokines, interleukin 1 (IL-1) and tumor necrosis factor alpha (TNF) are known to induce the expression of G- and GM-CSF genes in fibroblasts and in endothelial cells[7,9,13,15]. Such observations have supported our notion that mononuclear phagocytes may serve as pivotal regulatory cells in growth factor gene expression[12]. The principal hurdle to be overcome is the uncertainty surrounding the biological relevance of these findings. This is a legitimate concern in view of the highly reductionistic design of the studies used in support of this model. For example, studies in our laboratory begin with the purification and culture, independently of other collaborating cell populations, of peripheral blood monocytes in complete tissue culture medium to generate monocyte conditioned medium (MCM), followed by the introduction of serial dilutions of that MCM to confluent cultures of stromal cells such as fibroblasts or endothelial cells, and the assay of the stromal cell supernatants for hematopoietic growth factors[1-4,6,11,14,15]. In such experiments, IL-1 secreted by mononuclear phagocytes, induces the expression of G- and GM-CSF genes[7,9,15], a phenomenon which can be detected by using the fibroblast or endothelial cell supernatants in specific bioassays with neutralizing antibodies[15], by using western blot analysis, and by measuring CSF gene transcripts by northern or S1 nuclease analysis[7,9,15] (Fig. 2).

While these types of in-vitro approaches are important and informative, and may serve as paradigms of regulatory interactions between macrophages and stromal cells in the setting of the intact hematopoietic microenvironment, there exist important potential pitfalls, one of which reflects the major generic weakness of reductionistic in-vitro approaches. Specifically, data in support of this model are gleaned from experiments in which monocytes and hematopoietic stromal cells were removed from their respective contexts. Consequently, given the growing complexity of this putative network model of hematopoietic control, we are obliged

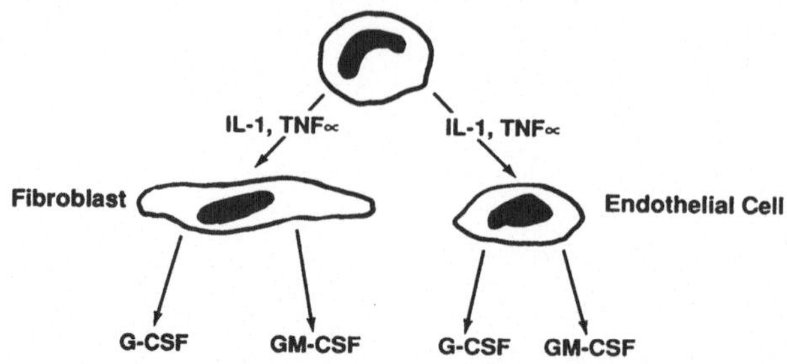

Fig. 1. The monokines IL-1 and TNF induce the expression of GM-CSF and G-CSF genes in endothelial cell and fibroblasts. Mononuclear phagocytes themselves can be induced to express G-CSF and M-CSF genes.

to address it's potential biological relevance. One way to approach this problem is to identify disease states in which the network might be perturbed. In this regard we reason that either abnormally high IL-1 production rates or high specific activity of native IL-1 might lead to granulopoietic hyperplasia. Conversely, we hypothesize that failure of IL-1 to induce CSF gene expression should lead to marrow failure. We summarize below results of studies in our laboratory which, in our view, have supported these notions.

CHRONIC MYELOGENOUS LEUKEMIA OF THE JUVENILE TYPE

Chronic myelogenous leukemia in children consists of heterogeneous clinical subsets. In one variety, paradoxically known as the "adult" type, children exhibit leukocytosis, largely neutrophilia, hepatosplenomegaly, and generally exhibit the Philadelphia chromosome translocation[18,19]. In a second type, "juvenile chronic myelogenous leukemia," children demonstrate variable degrees of organomegaly, and often exhibit neutrophilia, monocytosis, fetal erythropoiesis, and absence of the classic Philadelphia chromosome[18-21]. Over the past 11 years, we have carried out a series of in-vitro studies using bone marrow and blood cells from nine children with JCML and five children with the adult variety. We noted a consistent and striking abnormality of granulocyte macrophage colony growth in these children. As shown in Table 1, when bone marrow from children with JCML were cultured in the absence of exogenous CSFs, a remarkably intense pattern of "spontaneous" colony growth occurred. The addition of a potent crude source of GM and G-CSF, human placental conditioned medium, increased colony growth by only a mean value of 6%. In cultures of normal bone marrow and of marrow cells from the 5 children with the adult variety of CML, spontaneous colony growth was less than 10% of maximum[22]. Colonies were predominantly monocytic but neutrophil and eosinophil colonies were also noted in the spontaneously forming clones.

We reasoned that colony growth in these children was either autonomous (the colony forming units either didn't require CSF at all, or made their own CSF) or resulted from a paracrine mechanism in which auxiliary cells in the marrow stimulated CFU-GM proliferation. To address this concern, we cultured marrow from seven of these children in the presence and absence of a source of CSF (either leukocyte conditioned medium or human placental conditioned medium[15,23]). Moreover, marrow cells were cultured before and after depletion of auxiliary cells. In every

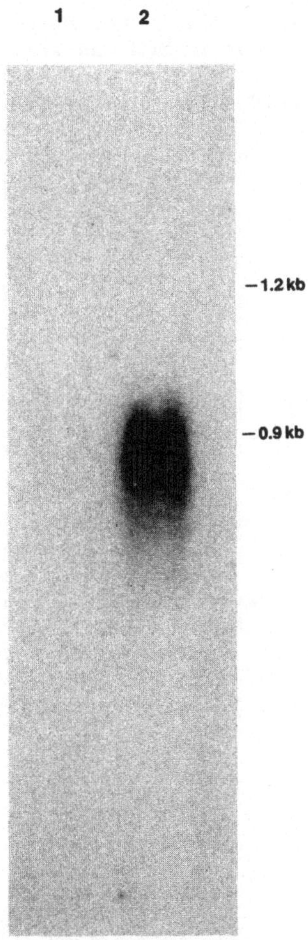

Fig. 2. Northern blot analysis of total cytoplasmic RNA from human umbilical vein endothelial cells stimulated (lane 2) and unstimulated (lane 1) with recombinant IL-1 alpha. A random hexamer labeled GM-CSF cDNA (kindly provided to us by Dr. Steven Clark, Genetic Institute) was used as a probe.

case so tested, the removal of adherent cells, or phagocytes from the marrow cells resulted in a major decline in spontaneous colony growth. When the depleted cell population was cultured in the presence of CSFs, colony formation returned to baseline levels, indicating clearly that the depletion phenomenon failed to remove CFU-GM.

We next isolated T-lymphocytes and monocytes from the marrow and blood of these patients and tested the culture supernatants for GM-CSA in colony growth assays. Unstimulated T lymphocytes failed to produce CSA, but monocyte conditioned medium (MCM) did induce colony growth. Interestingly, as shown in Fig. 3, MCM- induced colony growth depended on the presence of auxiliary cells in the normal target marrow, suggesting strongly that the active factor in MCM was an indirectly acting factor. For this reason, we tested the ability of these supernatants to induce CSF expression in endothelial cells. These studies were positive[22]. In addition, in chromatofocusing experiments, the CSF gene inducing activity in JCML MCM eluted at a pI of approximately 7.0, a fraction which, again, contained no CSA[22]. Finally, we also found that the activity in JCML MCM which induced CSF gene expression was neutralized with a

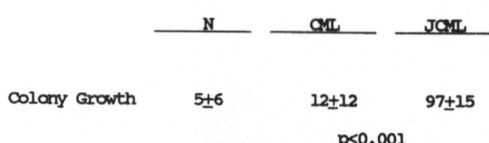

	N	CML	JCML
Colony Growth	5±6	12±12	97±15
			p<0.001

Table 1. High Level Colony Growth in Children with JCML. Shown is granulocyte/macrophage colony growth in three groups of subjects expressed as mean (± S.D.) percent maximum where maximum was defined as the GM colony number induced in 10^5 low density bone marrow cells by 10% HPCM. CML, adult type CML (n=5); N, normal volunteers (n=25), and JCML, juvenile chronic myelogenous leukemia (n=9).

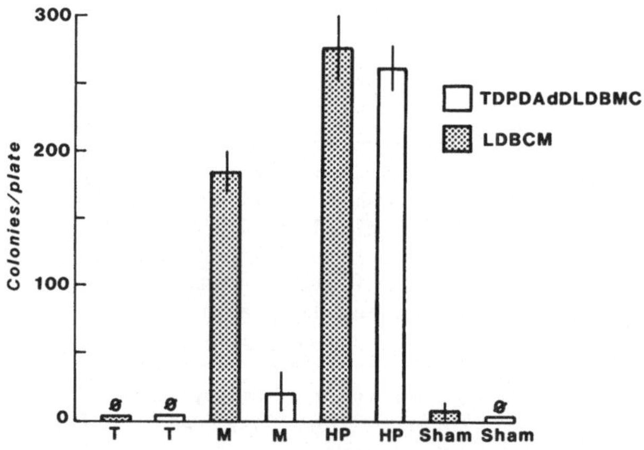

Fig. 3. Unfractionated and auxiliary cell depleted bone marrow cells from a normal volunteer were exposed to T-cell (T) and monocyte (M) supernatants from the peripheral blood of two patients with JCML. Supernatants of monocyte cultures induced substantial colony formation but only in cultures containing auxiliary cells capable of producing CSF. Sham conditioned medium stimulated cultures and HPCM-induced cultures were appropriately negative and positive respectively.

rabbit antiserum against native IL-1 (an antibody which neutralizes both $IL-1_a$ and $IL-1\beta$[2,34]). Moreover, in one experiment, the addition of the antibody to cultured JCML marrow cells significantly inhibited spontaneous colony growth[22].

In summary, spontaneous colony growth is a unique characteristic of JCML and serves to distinguish children with this disease from those with CML of the "adult" type. Secondly, spontaneous colony growth depends on the production of CSFs in the culture medium by auxiliary cells and these cells express CSF genes under the influence of native IL-1 produced by the leukemic monocytes. Is this unique pattern of colony growth reflective of an in-vivo defect of the hematopoietic network? If IL-1 driven CSF accounted for the granulopoietic hyperplasia in these children we might expect to detect elevated CSA titers in plasma or serum. In fact, as shown in Fig. 4, serum CSA titers in two children with JCML were significantly elevated above normal levels. We also reasoned that, if CSA titers were "driving" CFU proliferation in these children, CFU-GM suicide rates should be higher than normal. Indeed, experiments in which marrow cells were exposed, for 15-60 minutes, to 50ug/ml cytosine arabinoside, showed a significantly increased suicide fraction in progenitor cells from JCML patients (90% in the patients, 40% in normal volunteers[22].

We do not yet know the exact mechanism by which spontaneous colony growth occurs in marrow cells from these children. Theoretically, the

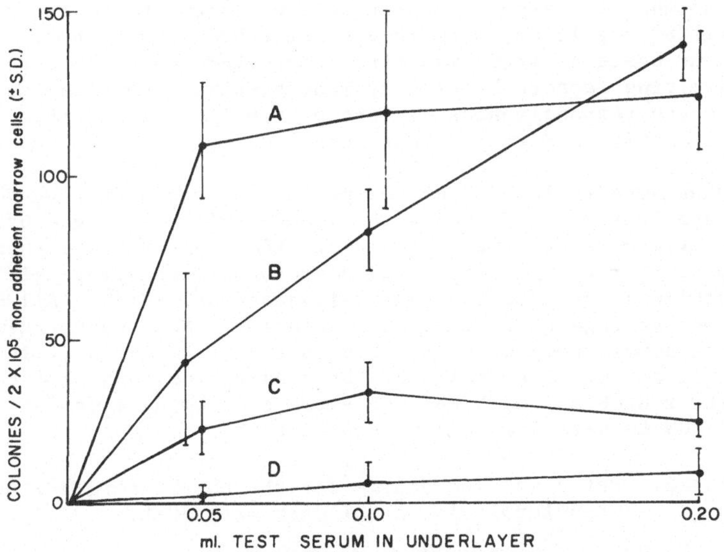

Fig. 4. Serum CSA assays. Serum samples were obtained from three patients with acquired, self limited drug induced neutropenia (A), five samples of umbilical cord blood (C), 5 normal adults (D), and two children with JCML (B). Sera were stored at -20°C until being used in colony growth assay. Multiple dilutions of sera were placed in agar underlayers and overlayered with normal, nonadherent low density bone marrow cells. Colonies were counted after 7 days of incubation. Each paint represents mean ± S.D. colonies/10^5 cells, values which reflect, when read from the linear portion of the dose response curve, an estimate of CSA concentration in the serum samples.

abnormality might be explained in a number of ways, most of which are outlined in Table 2. In view of these myriad possibilities one might well ask what we have learned in our eleven years of study of this disease. Clearly, we have determined the following: that intense spontaneous colony growth was a universal characteristic in children with JCML and served to distinguish it from JCML in children with the adult type of CML.[1] Moreover, this unique pattern of colony growth is driven by either secreted or membrane bound[25-29] IL-1 produced by mononuclear phagocytes because no marrow cell, including monocytes themselves, independently produced high amounts of colony stimulating activity. In addition, although CFU-GM did not form colonies autonomously, they are highly proliferative when freshly obtained, and serum CSA titers are abnormally high, at least using simple bioassays. We argue that these findings leave little doubt that a major abnormality in the hematopoietic network exists in these children and that studies using currently available techniques in molecular biology will resolve many of the remaining questions.

MALFUNCTION OF THE MARROW MICROENVIRONMENT IN THE ELDERLY

Another potential defect in the regulatory network, a defect which, if confirmed in larger studies, may prove to be very common, is that which we have found in studies on normal elderly volunteers[30]. A number of clinical studies on patients with lymphoma have supported the hypothesis that the elderly are at greater risk of complications of aggressive chemotherapy[31-33]. Notwithstanding it's broad acceptance, this notion has been inadequately tested. Accordingly we sought to determine whether the monokine/CSF regulatory network was perturbed in healthy elderly subjects. Our basic strategy was first, to determine whether the production of recruiting factors by elderly monocytes was subnormal and second, to determine the responsiveness of bone marrow fibroblasts from elderly volunteers to native and recombinant interleukin-1.

Monocytes were isolated from the peripheral blood of five volunteers ranging in age from 25 to 35 years and from 5 volunteers aged 65 to 75. These cells were utilized to generate MCM. The MCM was then applied to confluent cultures of third passage autologous and allogeneic bone marrow fibroblasts prepared as previously described[1,3,34]. All of the volunteers met stringent criteria for inclusion. None were taking medication, all had normal complete blood counts and normal renal, liver, and thyroid function. The design of the studies was such that each set of marrow fibroblast cultures was exposed to both "elderly" MCM and "young" MCM in parallel culture experiments

Table 2. Potential Mechanisms for the observed Paracrine Granulopoiesis in children with JCML.

1. Inappropriate production of high levels of IL-1
2. High level sensitivity of CSA producing cells to IL-1
3. Absence of factors which normally repress IL-1 gene expression
4. Qualitative abnormality (eg. mutation) of IL-1 resulting in inordinately high specific activity
5. High specific activity of induced CSF gene product

---

1. In the more recent years of this study we began to accurately predict, based on the pattern of spontaneous colony growth, that the banded chromosome analyses would (or would not) show the Philadelphia chromosome rearrangement.

The results of these studies[30] (summarized in Table 3) indicated that monocyte conditioned media from elderly volunteers contained factors which induced the production of a normal (relative to "young" MCM) amount of GM-CSA but that fibroblasts from elderly volunteers were apparently less responsive to MCM than fibroblasts from young volunteers. Accordingly, because we had previously determined that the CSF inducing factor in MCM is IL-1[3,15], we tested the responsiveness of elderly and young fibroblasts to graded doses of recombinant IL-1β. The results (Fig. 5) indicate that marrow fibroblasts obtained from three of the five elderly volunteers were less responsive to all doses of IL-1 beta tested.

Table 3. Monocyte conditioned media from young and elderly volunteers induce expression of CSA by marrow fibroblasts from 5 young and 5 elderly volunteers.

	Young Fibroblasts	Elderly Fibroblasts
Young MCM	normal	subnormal
Elderly MCM	normal	subnormal

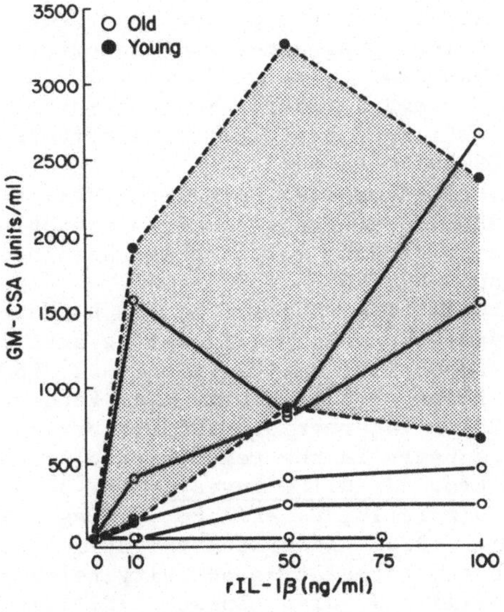

Fig. 5. IL-1 resistance marrow fibroblasts from elderly volunteers. CSA production shown on the ordinate as a function of rIL-1β (kindly provided by Dr. Charles Dinarello) dose in cultured marrow fibroblasts from five elderly volunteers. The shaded area is the range of fibroblast CSA responses in five young volunteers. Fibroblasts from three of the five elderly donors were abnormally insensitive to rIL-1β.

In summary, MCM from elderly volunteers is as effective as MCM from young volunteers in inducing CSF expression by marrow fibroblasts. In contrast, the induced CSF expression by marrow fibroblasts from some elderly donors is significantly decreased compared to fibroblasts from young donors. Although the number of volunteers in our study is small, we speculate that induced expression of hematopoietic growth factor genes is abnormal in cells of the marrow microenvironment in many elderly volunteers and that this defect may account for the intolerance of elderly patients to cytotoxic chemotherapy and radiation.

CONCLUSIONS

Much remains to be done with both of these models of network malfunction. Studies using recombinant growth factors and assays for those factors must be carried out in children with JCML. A larger study must be done in elderly healthy volunteers before we would presume to propose that the observed defects are commonplace or representative of important in-vivo defects. Nonetheless, such studies must be performed, so that we can develop a greater understanding of how the hematopoietic regulatory network is designed to function in vivo, and to develop a meaningful rationale for support of elderly patients with recombinant hematopoietic growth factors at times of hematopoietic stress.

REFERENCES

1. Bagby, G.C., E. McCall, and D.L. Layman. 1983. Regulation of colony stimulating activity production. Interactions of fibroblasts, mononuclear phagocytes and lactoferrin. J. Clin. Invest. 71: 340-344.
2. McCall, E., and G.C. Bagby. 1985. Monocyte-derived recruiting activity. Kinetics of production and effects of endotoxin. Blood 65: 689-695.
3. Broudy, V.C., K.S. Zuckerman, S. Jetmalani, J.H. Fitchen, and Jr. Bagby GC. 1986. Monocytes stimulate fibroblastoid bone marrow stromal cells to produce multilineage hematopoietic growth factors. Blood 68: 530-534.
4. Bagby, G.C., E. McCall, K.A. Bergstrom, and D. Burger. 1983. A monokine regulates colony-stimulating activity production by vascular endothelial cells. Blood 62: 663-668.
5. Bagby, G.C., V.D. Rigas, R.M. Bennett, A.A. Vandenbark, and H.S. Garewall. 1981. Interaction of lactoferrin, monocytes, and T-lymphocyte subsets in the regulation of steady-state granulopoiesis in vitro. J. Clin. Invest. 68: 56-63.
6. Bagby, G.C., C.A. Dinarello, P. Wallace, C. Wagner, S. Hefeneider, and E. McCall. 1986. Interleukin-1 stimulates granulocyte macrophage colony stimulating activity release by vascular endothelial cells. J. Clin. Inest. 78: 1316-1323.
7. Sieff, C.A., S. Tsai, and D.V. Faller. 1987. Interleukin 1 induces cultured human endothelial cell production of granulocyte-macrophage colony-stimulating factor. J. Clin. Invest. 79: 48-51.
8. Gerson, S.L., H.M. Friedman, and D.B. Cines. 1985. Viral infection of vascular endothelial cells alters production of colony-stimulating activity. J. Clin. Invest. 76: 1382-1390.
9. Broudy, V.C., K. Kaushansky, J.M. Harlan, and J.W. Adamson. 1987. Interleukin-1 stimulates human endothelial cells to produce granulocyte macrophage colony-stimulating factors and granulocyte colony-stimulating factor. J. Immunol. 139: 464-468.
10. Zucali, J.R., C.A. Dinarello, D.J. Oblon, M.A. Gross, L. Anderson,

and R.S. Weiner. 1986. Interleukin-1 stimulates fibroblasts to produce granulocyte-macrophage colony stimulating activity and prostaglandin E2. J. Clin. Invest. 77: 1857-1863.

11. Segal, G.M., E. McCall, T. Stueve, and G.C. Bagby. 1987. Interleukin 1 stimulates endothelial cells to release multilineage human colony-stimulating activity. J. Immunol. 138: 1772-1778.

12. Bagby, G.C. 1987. Production of multilineage growth factors by hematopoietic stromal cells: an intercellular regulatory network involving mononuclear phagocytes and interleukin-1. Blood Cells 13: 147-159.

13. Broudy, V.C., K. Kaushansky, G.M. Segal, J.M. Harlan, and J.W. Adamson. 1986. Tumor necrosis factor type α stimulates human endothelial cells to produce granulocyte-macrophage colony-stimulating factor. Proc. Natl. Acad. Sci. USA 83: 7467-7471.

14. Zuckerman, K.S., G.C. Bagby, E. McCall, B. Sparks, J. Wells, and V. Patel. 1985. A monokine stimulates production of human erythroid burst promoting activity by endothelial cells in-vitro. J. Clin. Invest. 75: 722-725.

15. Zsebo, K.M., V. Yuschenkoff, S. Schulter, D. Chong, E. McCall, C.A. Dinarello, B. Altrock, and G.C. Bagby. 1988. Vascular endothelial cells and granulopoiesis Interleukin-1 stimulates release of G-CSF and GM-CSF. Blood (in press):

16. Munker, R., J. Gasson, M. Ogawa, and H.P. Koeffler. 1986. Recombinant human TNF induces production of granulocyte macrophage colony-stimulating factor. Nature 323: 79-82.

17. Fibbe, W.E., J. Van Damme, A. Billiau, P.J. Voogt, N. Dunkerken, P.M.C. Kluck, and J.H.F. Falkenburg. 1986. Interleukin 1 (22K factor) induces release of granulocyte-macrophage colony-stimulating activity from human mononuclear phagocytes. Blood 68: 1316-1321.

18. Mays, J.A., R.C. Neerhout, G.C. Bagby, and R.D. Koler. 1980. Juvenile chronic granulocytic leukemia. Am. J. Dis. Child. 134: 654-658.

19. Barak, Y., S. Levin, R. Vogel, I.J. Cohen, B. Wallach, E. Nir, and R. Zaizov. 1981. Juvenile and adult types of chronic granulocytic leukemia of childhood: growth patterns and characteristics of granulocyte-macrophagee colony forming cells. Am. J. Hematol. 10: 269-275.

20. Altman, A.J., C.G. Palmer, and R.L. Baehner. 1974. Juvenile "chronic granulocytic" leukemia: a panmyelopathy with prominent monocytic involvement and circulating monocyte colony forming cells. Blood 43: 341-347.

21. Brodeur, G.J., L.W. Dow, and D.L. Williams. 1979. Cytogenetic features of juvenile chronic myelogenous leukemia. Blood 53: 812-819.

22. Bagby, G.C., C.A. Dinarello, R.C.Neerhout, D. Ridgway, and E. McCall. 1988. IL-1 dependent paracrine granulopoiesis in chronic granulocytic leukemia of the juvenile type. J. Clin. Invest. (in press).

23. Schlunk, T., and M. Schleyer. 1980. The influence of culture conditions on the production of colony stimulating activity by human placenta. Exp. Hematol. 8: 179-184.

24. Dinarello, C.A., H.A. Bernheim, J.G. Cannon, G. Lopreste, S.J.C. Warner, A.C. Webb, and P.E. Auron. 1985. Purified, ^{35}MET, ^{3}-LEU-labelled human monocyte interleukin-1 (IL-1) with endogenous pyrogen activity. Brit. J. Rheumatol. 24 (Suppl 1): 59-64.

25. Bakouche, O., D.C. Brown, and L.B. Lachman. 1987. Subcellular localization of human monocyte interleukin 1: evidence for an inactive precursor molecule and a possible mechanism for IL 1 release. J. Immunol. 138: 4249-4255.

26. Conlon, P.J., K.H. Grabstein, A. Alpert, K.S. Prickett, T.P. Hopp, and S. Gillis. 1987. Localization of human mononuclear cell interleukin 1. J. Immunol. 139: 98-102.
27. Merluzzi, V.J., R.B. Faanes, M. Czajkowski, K. Last-Barney, P.C. Harrison, J. Kahn, and R. Rothlein. 1987. Membrane-associated interleukin 1 activity on human U937 tumor cells: stimulation of PGE2 production by human chondrasarcoma cells. J. Immunol. 139: 166-168.
28. Le, J.M., D. Weinstein, U. Gubler, and J. Vilcek. 1987. Induction of membrane-associated interleukin 1 by necrosis factor in human fibroblasts. J. Immunol. 138: 2137-2142.
29. Weaver, C.T., and E.R. Unanue. 1986. T cell induction of membrane IL-1 on macrophages. J. Immunol. 137: 3868-3873.
30. Lee, M., and G.C. Bagby. 1988. The hematopoietic microenvironment in the elderly: defects in induced CSF gene expression. (submitted for publication) 0:
31. Armitage, J.O., and J.F. Potter. 1984. Aggressive chemotherapy for diffuse histiocytic lymphoma in the elderly: Increased complications with advancing age. J. Am. Geriatr. Soc. 32: 269-273.
32. Austin-Seymour, M.M., R.T. Hoppe, R.S. Cox, S.A. Rosenberg, and H.S. Kaplan. 1984. Hodgkin's disease in patients over sixty years old. Ann. Int. Med. 100: 13-18.
33. Peterson, B.A., T.F. Pajak, M.R. Cooper, N.I. Nissen, O.J. Glidewell, J.F. Holland, C.D. Bloomfield, and A.J. Gottlieb. 1982. Effect of age on therapeutic response and survival in advanced Hodgkin's disease. Cancer Treat. Rep. 66: 889-898.
34. Lee, M., G.M. Segal, and G.C. Bagby. 1987. Interleukin-1 induces human bone marrow-derived fibroblasts to produce multilineage hematopoietic growth factors. Exp. Hematol. 15: 983-988.

GROWTH FACTOR-RELATED MECHANISMS OF LEUKEMOGENESIS

J.W. Schrader*+, K.B. Leslie*+, I. Clark-Lewis#+, H.J. Ziltener*+, and S. Schrader*+

*The Biomedical Research Centre
University of British Columbia
222 Health Sciences Mall
Vancouver, BC, V6T 1W5, Canada

+The Walter and Eliza Hall Institute of Medical Research
Melbourne, Australia

#The California Institute of Technology
Pasadena, California, USA

INTRODUCTION

Acute myeloid leukemia is obviously a heterogeneous disease, different entities being distinguishable by differences in morphology, karyotype, disturbances in the expression of proto-oncogenes, clinical course etc. Some five years ago we made observations that indicated that one mechanism of leukemogenesis involved the malignant cell beginning to produce a hemopoietic growth factor that stimulated its own growth (Schrader and Crapper, 1983). Experiments in which cell-lines have been rendered leukemogenic by infection with retroviruses encoding hemopoietic growth factor genes, have formally confirmed the hypothesis that the inappropriate production of an autostimulatory growth factor by a hemopoietic cell that was immortal but not leukemogenic, could convert the cell to a transplantable leukemia (Lang, et al., 1985; Campbell, et al., 1987 and Nienhuis et al., this volume). Here, we summarize experiments that suggest that perturbations of growth factor production may be a relatively frequent feature of the leukemogenic process and that genetic techniques may allow pinpointing of diseases that involve this mechanism.

Coincident Onset of Leukemogenesis And Activation of a Growth Factor Gene

The experiments that demonstrated that the onset of leukemogenic transformation could be due to the activation of a hemopoietic growth factor involved a murine cell-line (R6-X). This line was absolutely dependent for its growth upon the presence of one particular hemopoietic growth factor, interleukin-3 (Schrader, 1986). When cultured in vitro in the absence of IL-3, these cells not only ceased growing but rapidly died. Other experiments showed that IL-3-dependent cells failed to survive when injected into the skin of mice (Crapper et al., 1984). IL-3 is undetectable in the serum of normal mice (Crapper and Schrader, 1986) suggesting that, at least in the skin, these IL-3-dependent cells required IL-3 to survive in vivo as well as in vitro. In keeping with this interpretation of these experiments, injected IL-3-dependent cells persisted in the skin of mice that had been inoculated with an IL-3 secreting tumor and in which IL-3 was present in the serum (Crapper et al., 1984).

The parental IL-3-dependent cell line that we studied in our initial experiments on leukemogenesis was not a normal cell in that it had undergone karyotypic changes and was immortal. Nevertheless it remained dependent on IL-3 for growth and survival and did not give rise to disease when large numbers of cells were injected into syngeneic mice. In the key experiment, 50 million parental cells were cultured in agar in the absence of IL-3. The bulk of the cells behaved as expected and died. However, a very small number of cells survived and over the course of the next month, grew into colonies. These colonies were plucked, the cells recloned, and examined for their growth characteristics. Single cells derived from these colonies were able to survive in the absolute absence of an exogenous source of IL-3 and slowly increased in number, their rate of growth increasing with increasing cell-density. This density-dependent pattern of growth is characteristic of an autocrine mechanism in which a cell secretes a factor stimulating its own growth.

We tested supernatants of these cultures and found a factor with all the biological, biochemical and antigenic properties of murine interleukin-3. In experiments using IL-3 cDNA probes, we demonstrated that whereas the parental line contained no IL-3 messenger RNA, the autonomous colonies contained messenger RNA of apparently normal size. Southern blot analysis using a variety of restriction enzymes did not reveal any rearrangements of IL-3 genes, and the mechanism responsible for activation of the IL-3 gene in this clone has not been determined.

The critical experiment of injecting the variant, IL-3-producing cells into animals to determine whether there had also been a change in the growth in vivo, showed unequivocally that the variant cells had become leukemogenic. Thus syngeneic mice injected with as few as 10^5 cells developed a disseminated leukemia. Leukemic cells recovered from these mice possessed a karotypic abnormality characteristic of the parental cells, demonstrating that the parental cells were indeed derived from the injected clone (Schrader & Crapper, 1983).

There was good evidence that the leukemic cells were releasing IL-3 in vivo. The serum of these leukemic animals contained IL-3 and there were high levels of IL-3 in the medium of cultures in which cell suspensions from the enlarged spleens of leukemic animals had been incubated for 24 hours (unpublished data).

These experiments clearly demonstrated that the onset of production of an auto-stimulatory factor, in this case IL-3, by an immortalized hemopoietic clone, could transform the cells into transplantable leukemia. These experiments demonstrated that aberrant factor production could be a leukemogenic step, but did not address the question of whether this mechanism occurred in leukemias that had arisen in vivo.

Analysis of a Monocytic Leukemia That Arose in Vivo

In most cases of human myeloid leukemia, the leukemic cells clearly depend on the presence of hemopoietic growth factors for their growth and survival in vitro, at least at relatively low cell densities (discussed in Schrader, 1984). It was commonly considered that murine myeloid leukemias were different in this respect and did not require factors for growth. However we reasoned that the apparent factor-independence of cell-lines derived from murine myeloid leukemias could have resulted from artefactual selection for factor-independence in in vitro conditions, and might not truly reflect the biological properties of the original tumor cells. Therefore we examined a monocytic leukemia, WEHI-274, that had arisen in a mouse in vivo and had not been selected in vitro for growth at low cell-density or in the absence of growth factors.

We investigated the growth of leukemic cells taken from an animal carrying the WEHI-274 tumor by culturing them in vitro, in much the same way as have been leukemic cells from human patients i.e., at relatively low cell-densities in the presence or absence of a hemopoietic growth factor. These experiments indicated that the leukemic WEHI-274 cells taken freshly from the mouse, clearly responded to the presence of an exogenous hemopoietic growth factor, in this case IL-3. Thus at one week, at cell-densities below 10^5 cells/ml, there were no large colonies of leukemic cells unless an external source of a growth factor, which could be IL-3, was provided.

We plucked more than a hundred of these colonies and proceeded to analyze them in terms of their responsiveness to growth factors, their leukemogenicity and the possibility that they produced hemopoietic growth factors. These experiments indicated that the clones of WEHI-274 that we had isolated from this single animal fell into three classes. All were leukemogenic, however, their growth properties in vitro differed considerably. Clones of Class I and Class II showed an autocrine pattern of growth in vitro. This was expecially pronounced in the case of Class I colonies, which grew very poorly at low cell-densities unless a hemopoietic growth factor, which could be interleukin-3, granulocyte-macrophage colony-stimulating factor (GM-CSF) or macrophage colony-stimulating factor (CSF-1), was added. Cells of the third class were even more dependent than the Class I and Class II on exogenous growth factor for their growth in vitro, and even at high cell-densities, grew very poorly.

Class I Clones of WEHI-274

In the case of Class I clones, medium that had been conditioned by high density cultures contained an activity that had all the properties of IL-3. Thus it had similar molecular weight, eluted from reverse phase HPLC columns at a similar aceto-nitrile concentration, and reacted with anti-IL-3 antibody in the same way as did T cell-derived IL-3. Moreover it had all of the biological properties of T cell-derived IL-3)

Analysis of a Class I clone, WEHI-274.14, demonstrated that one copy of the IL-3 gene had been rearranged (Leslie et al., in preparation). Southern blot analysis with various restriction enzymes showed that this rearrangement had occurred about 1kb upstream of the normal IL-3 promotor. Experiments in progress suggest that this rearrangement has resulted from a chromosomal translocation, the nature of which is being clarified. Northern blot analysis of whole-cell polyadenylated RNA from the Class I clone WEHI-274.14 revealed a 10kb nuclear transcript, a number of splicing intermediates and a cytoplasmic transcript that was significantly larger than the normal, T cell-derived IL-3 mRNA. Thus in this Class I clone, there was unequivocal evidence of a pathological rearrangement-of one IL-3 gene that had resulted in abnormal transcripts and the production of IL-3 protein that was able to stimulate tahe leukemic cell's own growth.

Class II Clones of WEHI-274

A second class of clones derived from the WEHI-274-bearing mouse, again showed an autocrine pattern of growth, although they grew better at low cell-densities in the absence of exogenous growth factors than did the Class I clones. In this case, the factor present in medium conditioned by high density cultures had all the properties of a second T cell-derived hemopoietic growth factor, granulocyte-macrophage colony-stimulating factor.

Southern blot analysis of these clones again revealed a rearrangement of one GM-CSF gene (Leslie et al., in preparation). This rearrangement appeared to have occurred in considerable distance upstream of the GM-CSF gene (at least 3.5kb away), and the nature of this is still under investigation. Analysis of whole-cell polyadenylated RNA from clones of this second class, revealed abnormal GM-CSF gene transcripts. The largest transcript was a 10kb nuclear species, whereas the cytoplasm contained a smaller transcript, that was however, significantly larger than that present in activated T-cells. Thus in this second class of WEHI-274 clones, there had occurred a second, independent genetic change involving a second hemopoietic growth factor gene, encoding GM-CSF.

Class III Clones of WEHI-274

Clones of the third class differed strikingly from those of the first or second classes in that they showed no sign of producing a factor which stimulated their own growth. Thus supernatants of high density cultures of Class III WEHI-274 clones failed to stimulate their own growth, that of a variety of factor-dependent murine hemopoietic cell-lines, or that of normal bone marrow cells. Despite this, the Class III cells were exquisitely responsive to IL-3, GM-CSF or CSF-1 in vitro, growing very poorly in the absence of exogenous growth factors. When injected into syngeneic mice however, the Class III cells gave rise to a disseminated leukemia.

One possible explanation for their rapid growth in vivo but not in vitro, was that in vivo they stimulated the production of a hemopoietic growth factor or factors by the host. Experiments in which it was shown that serum from an animal bearing a Class III clone stimulated the growth of WEHI-274-3 cells, supported this notion.

More recent in vitro experiments have clarified the nature of the interaction between WEHI-274 Class III clones and host cells. These experiments demonstrated that when supernatants of WEHI-274 Class III clones were added to cultures of murine fibroblasts, they stimulated the release of activities which supported the in vitro growth of both WEHI-274.3 cells and of the factor-dependent cell line FDCP-1 (Leslie et al., in preparation). Thus the Class III WEHI-274 cells were also utilizing growth factor- related mechanisms of growth. In this case, the autostimulation was not direct but involved a paracrine mechanism in which the leukemic cell was releasing a factor which stimulated the relase of a hemopoietic growth factor.

The nature of the factor released by the Class III clones of WEHI-274 is still under investigation; obvious possibilities are substances such as interleukin-1 or tumor necrosis factor, known to stimulate the relase of hemopoietic growth factors from fibroblasts. The nature of the factor released by the fibroblasts is also under investigation; GM-CSF is one possible candidate although a novel factor could be involved.

Heterogeneity of WEHI-274

These experiments suggested that the original WEHI-274 leukemia was a heterogeneous disease, but in themselves gave no information on the relationship between the various classes of clones and the question of whether the clones shared a common ancestry. Clear-cut evidence on this point came from Southern blot analysis in which the three classes of clones were probed with a c-myb genomic probe. Southern blot analysis demonstrated clearly that all three clones possessed an identical rearrangement of one copy of the c-myb gene. Experiments by Gonda et al., have demonstrated that this rearrangement has resulted from an insertion of a defective Moloney Leukemia Virus proviral sequence into the first intron of one c-myb gene resulting in production of an altered c-myb message.

Clearly all the three classes of WEHI-274 cells shared a common ancestor in which one c-myb gene had been rearranged. It is reasonable to postulate that the clone in which the c-myb gene has been rearranged, arose in an environment in which significant amounts of GM-CSF and IL-3 or other growth factors were being produced. Such an environment would be expected to have been present in the original animal which had been infected with the Abelson murine leukemia virus and a Moloney helper virus, since it is known that infection of mice with murine leukemia viruses leads to chronic stimulation of the immune system and the release of T cell lymphokines such as IL-3 (Lee & Ihle, 1981). It is not clear whether the clone in which the c-myb gene had been rearranged would have been a transplantable leukemia or whether it was immortalized but still factor-dependent and analogous to the IL-3-dependent clone discussed in our first, in vitro experiments above.

Certainly it is evident that the genetic changes that led to leukemogenesis or progression of the disease, involved at least three growth factor-related routes. Two of these involved direct auto-stimulation mediated by IL-3 or GM-CSF, and the third indirect or paracrine mechanisms of auto--stimulation, the precise details of which have yet to be elucidated. There was no evidence that any of the clones had integrated v-abl sequences.

The experiments with WEHI-274 provide an example of heterogeneity in a tumor population, a not unexpected finding. What was striking, however, was that in two instances, heterogeneity involved independent activation of different hemopoietic growth factor genes. The fact that autostimulatory mechanisms were involved in two independent lines of evolution that had occurred in the one disease, suggests that this mechanism is likely to be a relatively frequent one in myeloid leukemogenesis. Moreover, the experiments in which GM-CSF and IL-3 rearrangements were identified using genetic techniques demonstrate that these should be a particularly useful way of identifying diseases involving this mechanism. It is important to note that the presence of messenger RNA for hemopoietic growth factors in hemopoietic cells may not necessarily reflect a pathological event. Thus there is now good evidence that cells such as macrophages can produce hemopoietic growth factor such as GM-CSF and G-CSF and CSF-1 under physiological circumstances (Thorens et al., 1987). On the other hand, evidence of pathology of the gene in the form of gene rearrangements of abnormal transcripts, clearly suggests an involvement of the corresponding growth factor in the disease process.

Mechanisms of Auto-Stimulations and Therapeutic Implications

The simplest mechanism of auto-stimulation is that described by Sporn and Todaro (1980) as autocrine. In this situation stimulation of the tumor cell occurs when growth factors that have been secreted by the cell interact with the specific receptor on the surface of that cell. This mechanism is, in theory, susceptible to interruption by antibodies to the growth factor or the receptor or by antogonists of the growth factor.

A second more complex situation arises if it is postulated that interaction of the growth factor and its receptor could occur internally during production of these molecules in the endoplasmic reticulum. In this instance, extracellular measures directed against a growth factor or its receptor, would be unlikely to be effective and other strategies aimed at interfering with transcription of the gene e.g. involving antisense nucleic acids would be indicated.

A third scenario would be that the growth factor and receptor would ineract in the endoplasmic reticulum but would not deliver their growth-promoting signal until the factor-receptor complex had been inserted into the cell-membrane and had had the opportunity to interact with receptor--associated proteins involved in transmembrane signalling. In this instance it is possible that antibodies to the factor or receptor or factor-antagonists, could influence the autostimulatory process. It will be important to determine the relative importance of these mechanisms and to design strategies to determine which are operating in given cells.

Structure-Function Studies on Hemopoietic Growth Factors

We have been investigating the use of automated peptide synthesis to produce both structural analogs of hemopoietic growth factors and antibodies specific for defined regions of the molecules. The aim of these studies is to define the regions of IL-3 and GM-CSF that interact with the receptors. Ultimately this information will aid in the production of neutralizaing antibodies or of antagonists that can block the interaction of these growth factors with their receptors. Clark-Lewis and colleagues (1986) have demonstrated that biologically active analogs of IL-3 and GM-CSF can be produced by total chemical synthesis using an automated peptide synthesizer. Total chemical protein synthesis enables one to rapidly produce a large number of analogs. Experiments to date have demonstrated that large portions of the C-terminal parts of the IL-3 molecule, can be eliminated without totally abrogating the ability of analogs to interact with the membrane receptor (Clarke-Lewis et al., 1986). Similar experiments with analogs of GM-CSF, have shown that the first 14 residues of the molecule can be deleted without greatly affecting the biological activity of the molecule (Clark-Lewis et al., 1987) but the integrity of residues 16-18 are critical for activity.

Antibodies specific for defined peptides corresponding to portions of the IL-3 amino acid sequence, have been shown to bind native IL-3 and have shed some light on the parts of the molecule that are important in interaction with cells (Ziltener et al., 1987). Extension of these studies on analogs, together with direct crystallographic analysis of hemopoietins and ultimately their receptors, should in the future, lead to the design of new therapeutics, including antagonists, some of which may find use in the treatment of leukemia and other disease.

ACKNOWLEDGEMENTS

We thank Ms. Joanne Ringham, Mr. Gary Coe and Ms. Francis Lee for technical assistance. The work was supported by the NHMRC, Canberra, Australia, The Biomedical Research Centre and R01CA38684-01 from the N.I.H.

REFERENCES

Campbell, H.D., Fung, M.C., Hapel, A.J. and Young, I.G. Cloning expression of the Murine Interleukin-3 Gene. Lymphokines 13: 239-260.

Clark-Lewis, I.,Aebersold, R.A., Ziltener, H., Schrader, J.W., Hood, L.E. and Kent, S.B.H. (1986). Automated chemical synthesis of a protein growth factor for hemopoietic cells, interleukin 3. Science 231: 134-139.

Clark-Lewis, I., Lopez, A.F., Vadas, M., Schrader, J.W., Hood, L., Kent, S.B.H. (1987). Chemical synthesis of hemopoietic growth factors: an approach to protein design. In: Protein Structure and Design; UCLA Symposia on Molecular and Cellular Biology, New Series, Volume 69. ed. D. Oxender. A.R. Liss, New York, NY, In Press.

Crapper, R.M. and Schrader, J.W. (1986). Role of T-cell lymphokine, persisting-cell stimulating factor (PSF), in graft-versus-host reactions. Immunology 57: 553-558.

Crapper, R.M., Thomas, W.R. and Schrader, J.W. (1984). In vivo transfer of persisting (P) cells; further evidence for their identity with T-dependent mast cells. J. Immunol. 133: 2174.

Lang, R.A., Metcalf, D., Gough, N.M., Dunn, A.R., and Gonda, T.J. (1985). Expression of a hemopoietic growth factor cDNA in a factor-dependent cell line results in autonomous growth and tumorigenicity. Cell 43: 531-542.

Lee, J.C., Ihle, J.N. (1981). Increased responses to lymphokines are correlated with preleukemia in mice inoculated with Moloney leukemia virus. Proc. Natl. Acad. Sci. USA 78: 7712-1816.

Schrader, J.W., Ziltener, H.J., Leslie, K.B. (1985). Structural homologies among the hemopoietins. Proc. Natl. Acad. Sci. USA, Vol. 83, pp. 2458-2462.

Schrader, J.W. (1984). Hypothesis: Role of a single hematopoietic growth factor in multiple proliferative disorders of hematopoietic and related cells. The Lancet No. 8395, Vol 11, 133.

Schrader, J.W. (1986). The panspecific hemopoietin of activated T-lymphocytes interleukin 3. Ann. Rev. Immunol. 4: 205.

Schrader, J.W., Crapper, R.M. (1980). Autogenous production of a hemopoietic growth factor. Proc. Natl. Acad. Sci. USA 80: 6892.

Sporn, M.B., and Todaro, G.J. (1980). Autocrine secretion and malignant transformation of cells. New. Eng. J. Med. 303: 878-880.

Thorens, B., Mermod, J. and Vassalli, P. (1987). Phagocytosis and inflammatory stimuli induce GM-CSF mRNA in macrophages through post-transcriptional regulation. Cell 48: 671-679.

Ziltener, H.J., Clark-Lewis, I., Hood, L.E., Kent, S.B.H. and Schrader, J.W. (1987). Antipeptide antibodies of predetermined specificity recognize and neutralize the bioactivity of the pan-specific hemopoietin interleukin-3. J. Immunol. 138: 1105-1108.

STIMULATION OF HUMAN EARLY AND LATE ERYTHROPOIETIC PROGENITOR CELLS BY INSULIN: EVIDENCE FOR DIFFERENT MECHANISMS

Gunther Konwalinka, Christian J. Wiedermann, Andreas Petzer, Kurt Grunewald, Christoph Breier, Josef Patsch and Dietmar Geissler

Department of Internal Medicine
University of Innsbruck
A-6020 Innsbruck, Austria

ABSTRACT

In order to investigate cellular mechanisms involved in insulin stimulation of erythropoiesis, we have studied the response of early (BFU-e) and late (CFU-e) erythroid progenitor cells in a serum-free agar culture system. In this assay system, CFU-e proliferation occurred in media containing low-density lipoproteins, bovine serum albumin, transferrin and recombinant erythropoietin (rEPO). Insulin in physiological concentrations as low as 10^{-12}M, added directly to cultures, augmented CFU-e colony formation. This stimulatory effect was also seen when monocyte- and T lymphocyte-depleted cells from normal donors were cultured. In contrast, BFU-e was not stimulated by media devoid of insulin. Occurrence of BFU-e colonies required the presence of insulin in concentrations higher than 10^{-8}M. This insulin effect was not dependent on the presence of monocytes and T lymphocytes. Delayed addition studies of rEPO to insulin containing cultures revealed a slight but significant survival rate of CFU-e. A similar survival rate was found for BFU-e. From this, we conclude that insulin stimulates CFU-e by an EPO-like activity. For BFU-e, however, the decline in the number of bursts caused by EPO deprivation implies that insulin does not act directly as a burst-promoting activity but that it probably induces the release of this activity from non-adherent and T lymphocyte-depleted bone marrow cells.

INTRODUCTION

In both mouse and human models, proliferation and differentiation along the erythroid pathway require two distinct glycoprotein growth factors: erythroid burst-promoting activity (BPA)[1] at early stages, which was recently identified as one of the interleukin 3 activities[1], and erythropoietin (EPO) at later stages[2,3].

1. Abbreviations used in this paper: BFU-e, burst-forming unit, erythroid; BPA, burst-promoting activity; CFU-e, colony-forming unit, erythroid; rEPO, recombinant erythropoietin; BSA, bovine serum albumin; TF, transferrin; LDL, low-density lipoproteins; ECS, fetal calf serum; MNC, mononuclear cells.

In addition to the specific humoral factors of the erythropoietic lineage, the growth of erythropoietic progenitor cells is also affected by numerous endocrine hormones[4]. These hormones include androgens, glucocorticoids, thyroid hormones, growth hormones, platelet-derived growth factor, somatomedins and insulin[4,5].

It has already been shown that insulin not only stimulates the proliferation of late erythropoietic progenitor cells (CFU-e) but also of early ones (BFU-e) in the presence of EPO[6-9]. Direct effect of insulin on erythropoietic progenitor cells have been postulated[6,9,10]. The fact that pharmacological quantities of insulin were found to be active was explained by insulin functioning as a mitogen after binding to the insulin-like growth factor I receptor and stimulation of DNA synthesis and self-replication[10]. Additive effects of insulin and insulin-like growth factor I led Dainiak and Kreczko[8] to hypothesize that their activities were mediated by a similar receptor or postreceptor system.

Since insulin receptors are not only exposed on null cells but also on monocytes and lymphocytes[11], it seems possible that insulin induces indirect effects by releasing BPA-like molecules. The purpose of the present study was to investigate the possible cellular mechanisms of insulin for stimulation of erythropoiesis.

MATERIALS AND METHODS

Media

The medium used was McCoy's 5A medium, pH 8 (raised by addition of sodium bicarbonate)[12]. The medium contained bovine serum albumin (BSA), transferrin (TF), low-density lipoproteins (LDL) and recombinant (r) EPO (lot no. 6250; Kirin-Amgen, obtained from Amersham, Buckinghamshire, UK). Purified BSA (Behring Werke, Marburg, FRG) was treated with dextran/charcoal and deionized as described by Iscove et al.[13]. Fully iron-saturated human TF (Behring Werke, Marburg, FRG) was used. Recombinant human insulin (Eli Lilly GmbH, Vienna, Austria) was obtained from stock solution of 40 IU/ml.

Preparation of Lipoproteins

LDL were isolated from a plasma pool of normolipidemic individuals by sequential ultracentrifugation at a density of 1.019 - 1.063, as previously described[14]. On the basis of previous experiments, an LDL concentration of 0.06 mg/ml was used for CFU-e and BFU-e proliferation[12].

Bone Marrow Samples

Normal human bone marrow was obtained from adult volunteers by aspiration. Approximately 10 ml to 20 ml from one to two aspirations was collected into syringes containing preservative-free heparin.

Cell Separation Studies

The marrow suspensions were separated over Ficoll-Hypaque (1.077 c/cu cm; Nyegaard, Oslo, Norway) at 400 g for 40 min at 20°C, and the interface mononuclear cells collected, washed three times and resuspended in McCoy's 5A medium without fetal calf serum (FCS). When non-adherent and T lymphocyte-depleted bone marrow cells were prepared, the interface mononuclear cells (MNC) were resuspended in McCoy's medium with 20% FCS[15].

Non-adherent cells were obtained by two consecutive 90-min adherence procedures in 100-ml tissue culture dishes (Falcon, 4-1029-0). Only the cells obtained by gentle aspiration of the supernatant cell suspension into a pipette were used. The monocyte-depleted cells were then depleted of T lymphocytes by the erythrocyte-rosetting technique, using neuraminidase-treated sheep erythrocytes[15]. In brief, 5 x 10^6 cells were incubated with equal volumes of 1% sheep red blood cells at room temperature for 15 min and centrifuged at 250 g for 20 min. The pellet was kept at 4°C overnight. After resuspension, the cell mixture was gently layered over Lympho-prep and centrifuged at 400 g for 30 min. Interface cells were washed three times in Hank's balanced salt solution without CA and Mg ions and were finally resuspended in McCoy's medium without FCS. This cell suspension contained less than 1% monocytes and 3% T lymphocytes.

Serum-free CFU-e and BFU-e Assay

A miniaturized agar system previously described in detail[16] was used as the culture method. Briefly, MNC or monocyte- and T lymphocyte-depleted cells were suspended in McCoy's 5A medium containing 0.3% agar. Aliquots of 50 µl containing 3 x 10^4 MNC or 1.5 x 10^4 non-adherent and T lymphocyte-depleted cells were pipetted into four wells of flat-bottomed microtiter plates (Nunc, Denmark). After equilibration at room temperature, 50 µl of a mixture of McCoy's medium containing 2% deionized and delipidated BSA (final BSA concentration 2 mg/ml), 0.08% TF, LDL and rEPO were placed over the agar layer. A rEPO concentration of 0.6 U/ml was used for CFU-e and BFU-e proliferation. Insulin dilutions were prepared that gave final concentrations of 10^{-6}M, 10^{-7}M, 10^{-8}M, 10^{-9}M, 10^{-10}M, 10^{-11}M and 10^{-12}M (mU/l x 7,241 = pmol/l; 17). The effect of delayed rEPO or insulin addition was studied in the following manner. Non-adherent and T lymphocyte-depleted bone marrow cells were plated in the standard serum-free agar system with insulin at a concentration of 10^{-8}M and 10^{-6}M for CFU-e and BFU-e proliferation respectively. After varying periods of incubation, a 5-µl volume of rEPO was added to the liquid overlayer to achieve a final concentration of 0.6 U/ml. Incubation was then continued for seven and 14 days respectively. In the converse experiments, cultures with compositions identical to those of the previous experiments were set up, except that they contained 0.6 U/ml of rEPO and no insulin. A 4-µl volume of insulin was added after varying periods to the liquid overlayer, giving a final concentration of 10^{-8}M and 10^{-6}M for evaluation of CFU-e and BFU-e proliferation respectively. After seven days of incubation for CFU-e and 14 days for BFU-e colony number, the agar layers were processed as described previously[16]. The permanently preserved slides were scored under light microscope. CFU-e-derived colonies contained 8-64 hemoglobinized cells. Erythroid bursts were defined as aggregates containing more than 64 hemoglobinized cells after 14 days of incubation. The bursts often consisted of several large aggregates.

RESULTS

Effects of Insulin in CFU-e and BFU-e Colony Formation

Our previously described serum-free agar system, in which no erythroid colony formation occurred in the absence of rEPO, was used for the study of insulin effects. When bone marrow cells from four different helthy donors were tested for CFU-e proliferation in the presence of rEPO, the mean colony formation was 35/well with SE ± 5. In contrast, no BFU-e colony formation was found under these conditions. Fig. 1 shows dose-response curves for insulin in molar concentrations. As little as 10^{-12}M produced detectable augmentation in the CFU-e growth assay; peak activity was seen at 10^{-8}M, resulting in an approximately twoto threefold increase in CFU-e colony formation. Slightly less stimulating activity of insulin was seen at 10^{-6}M, and a further decrease was observed 10^{-5}M. As Fig. 1 also shows, BFU-e colony formation was only detectable when insulin in concentrations

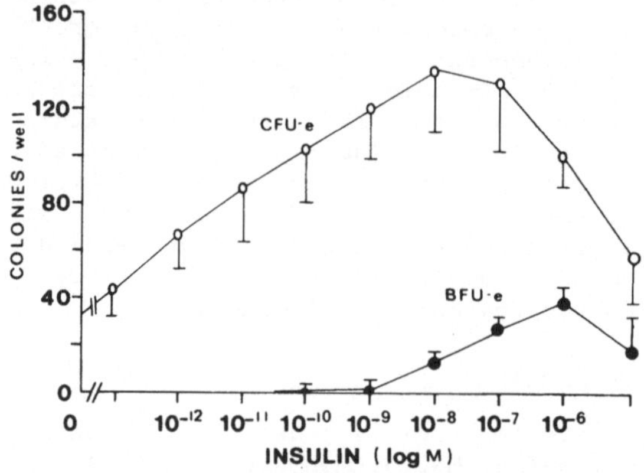

Fig. 1. Effect of insulin on growth of CFU-e (o) and BFU-e (•) in serum-free cultures with rEPO concentration of 0.6 U/ml. MNC were seeded in a concentration of 3×10^4 cells/well. Mean ± SE; n = 4.

higher than 10^{-8}M was added to serum-free cultures; peak activity was seen at 10^{-6}M with substantially less stimulating activity at 10^{-5}M.

Effects of Insulin on CFU-e and BFU-e Growth After Monocyte and T Lymphocyte Depletion

In order to reduce the endogenous production of erythropoietic stimulating activity by monocytes and T lymphocytes, these cells were removed by adherence and sheep rosette techniques (see Materials and Methods). Unseparated MNC and the non-adherent and T lymphocyte-depleted fraction of bone marrow cells were cultured in the presence of rEPO with and without insulin. As Fig. 2 shows, the stimulatory effects of insulin for CFU-e and BFU-e proliferation were comparable in the two cell preparations incubated for seven and 14 days for CFU-e and BFU-e respectively.

Effects of Delayed rEPO or Insulin Addition on CFU-e and BFU-e Colony Formation

The capacity of CFU-e and BFU-e to survive a period of culture in the absence of rEPO or insulin was tested with 1.5×10^4 monocyteand T lymphocyte-depleted bone marrow cells. Recombinant EPO or insulin was added delayed up to day 5 for CFU-e and up to day 7 for BFU-e proliferation. In the presence of insulin, both CFU-e and BFU-e colony formation decreased about 40% after one day and 80% after two days of rEPO deprivation. All erythroid colony formation was abolished by delaying the addition to day five (Fig. 3). In the presence of rEPO, the delayed addition of insulin resulted in only a small decrease in CFU-e colony formation. In contrast to the CFU-e data, BFU-e survival was found to be 75% after two days and up to 25% after five days of insulin deprivation. In control experiments where insulin was absent, rEPO did not give rise to any burst formation.

DISCUSSION

It has previously been shown that similar supraphysiological concentrations of insulin stimulate murine and human bone marrow CFU-e and BFU-e colony formation in serum-supplemented with serum-free cultures[5-9]. In

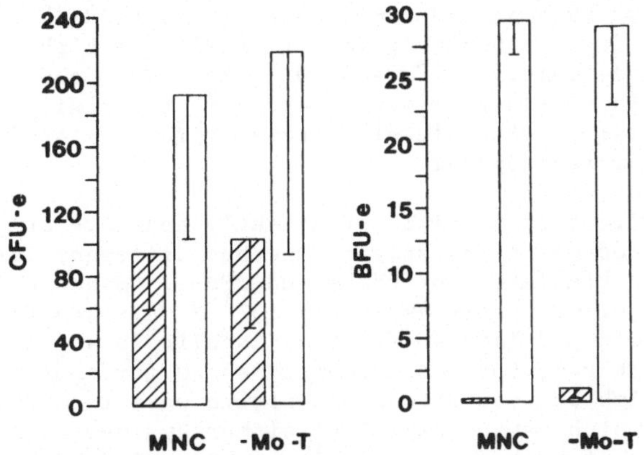

Fig. 2. Effect of insulin on growth of CFU-e (left panel) from BFU-e (right panel) colony formation obtained from MNC (3×10^4 cells/well) and monocyte- and T lymphocyte-depleted cells (1.5×10^4 cells/well) in serum-free medium. A concentration of 0.6 U/ml rEPO was used. Hatched bars without insulin, unhatched bars with 10^{-8}M and 10^{-6}M insulin for CFU-e and BFU-e proliferation respectively. Mean ± SE; n = 3.

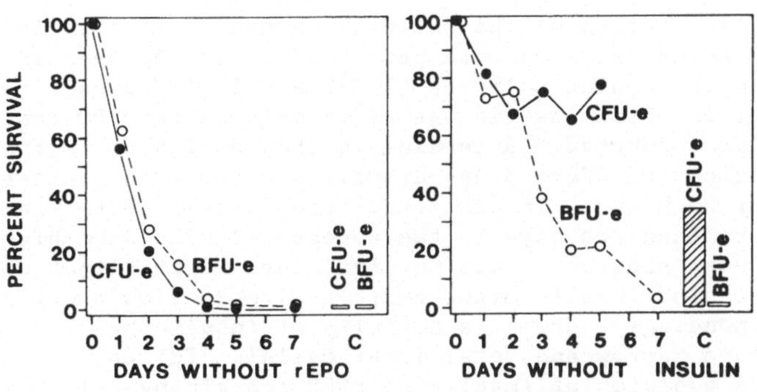

Fig. 3. Survival of CFU-e (●) and BFU-e (o) in the absence of added rEPO (left panel) and insulin (right panel) in serum-ree culture of monocyte- and T lymphocyte-depleted human bone marrow cells (1.5×10^4 cells/well). For details of delayed rEPO and insulin, see Materials and Methods. C = controls: The mean erythroid colony number of three experiments with rEPO and insulin present from initiation of cultures was taken as 100%. Mean; n = 3. Bars indicate CFU-e and BFU-e survival without rEPO (left panel) and without insulin (right panel).

this communication, we show in a serum-free agar system[15,16] that in the presence of rEPO the dose-response curve for insulin is biphasic with different optimal concentrations, namely 10^{-8}M and 10^{-6}M for CFU-e and BFU-e proliferation respectively. In contrast to other reports[8], the addition of high concentrations of insulin is absolutely necessary for erythroid bursts formation in our serum-free system.

The obviously different sensitivity of early and late erythropoietic progenitor cells forced us to investigate whether different cellular mechanisms are involved in stimulation of BFU-e and CFU-e. Since physiological concentrations of insulin, namely as low as 10^{-12}M, already augmented CFU-e proliferation, a direct modulatory effect of insulin is suggested, possibly by interaction with TF receptors, as recently shown for somatomedins[18]. On the other hand, high concentrations of insulin were required for initiation of burst formation, which can be explained either by insulin activation of insulin-like growth factor I receptors[10] or by an indirect effect of insulin via release of BPA-like molecules from accessory cells. The latter mechanism of action is substantiated by a recent investigation by Dainiak et al[19], showing that thyroid hormones augment erythroid burst formation indirectly by releasing BPA-like molecules. A similar mechanism has been suggested for enhanced erythroid burst formation by insulin-like growth factor I and platelet-derived growth factor[8,19].

To prove our hypothesis that, rather than acting directly on BFU-e progenitors, insulin stimulates normal human erythroid burst formation by enhancing the release of paracrine hormones from accessory cells, we tested the effects of insulin at a concentration of 10^{-6}M on target cells depleted of monocytes and T lymphocytes. These depletion procedures, however, had no apparent effect on the capacity of insulin to stimulate BFU-e colony formation. This might have been expected since several investigators showed that T lymphocytes have only a limited role in modulating the growth of normal human bone marrow BFU-e progenitors[19-21]. It has been suggested that B lymphocytes may be an important source of BPA[19,22].

For further elucidation of the cellular mechanism of insulin in CFU-e and BFU-e proliferation, studies were performed on the delayed addition of rEPO to insulin containing cultures (10^{-8}M and 10^{-6}M respectively). Converse experiments, in which insulin was added delayed to rEPO containing cultures, were also conducted. According to the results of Iscove and Guilbert[23], the number of CFU-e drops dramatically from the outset when EPO is added after 24 hours. In our serum-free system, CFU-e was able to survive for more than two days in the absence of rEPO when high concentrations of insulin were present. This survival data suggest that not only erythropoietin but also insulin might exert a direct stimulatory action on CFU-e. Correspondingly, EPO-like activity of insulin has been proposed in murine adult bone marrow and fetal liver cells[7]. In the converse experiments with delayed addition of insulin to rEPO containing cultures, it was found that delayed addition of insulin caused only a minor reduction in CFU-e survival. Even after a delay up to day 5, the number of CFU-e counted on day 7 was higher than in control cultures containing only rEPO, which suggests a direct role of insulin.

Studies of delayed addition of rEPO to insulin containing cultures and converse experiments with delayed addition of insulin to rEPO containing cultures revealed a rapid decline in the number of BFU-e, suggesting that insulin and rEPO possess no intrinsic BPA activity. The BFU-e survival rate of 75% after two days and up to 25% after five days of rEPO deprivation, however, supports the hypothesis that insulin may release BPA molecules from accessory cells such as B lymphocytes or null cells. The possibility also remains that contaminating monocytes and T lymphocytes remaining after depletion contribute to the relase of BPA-like molecules. Whether this

BPA-releasing effect is mediated by insulin receptors or other growth promotion receptors with affinity for insulin, is presently under investigation.

We conclude that our data indicate different cellular mechanisms of insulin for CFU-e and BFU-e proliferation in vitro.

ACKNOWLEDGEMENTS

Our grateful thanks to Monika Niedermuhlbichler and Gabriele Trobinger for their expert technical assistance.

REFERENCES

1. Yokota, T., F. Lee, D. Rennick, C. Hall, N. Arai, T. Mosmann, G. Nabel, H. Cantor, and K. Arai. 1984. Isolation and characterization of a mouse cDNA clone that expresses mast-cell growth-factor activity in monkey cells. Proc. Natl. Acad. Sci. USA 81: 1070-1074.
2. Jacobs, K., C. Shoemaker, R. Rudersdorf, S.D. Neill, R.J. Kaufmann, A. Mufson, J. Seehra, S.S. Jones, R. Hewick, E.F. Fritsch, M. Kawakita, T. Shimizu, and T. Miyake. 1985. Isolation and characterization of genomic and cDNA clones of human erythropoietin. Nature 313: 806-810.
3. Lin, F., S. Suggs, C. Lin, J.K. Browne, R. Smallins, J.C. Egrie, K.K. Chen, G.M. Fox, F. Martin, Z. Stabinsky, S.M. Badrawi, P. Lai and E. Goldwasser. 1985. Cloning and expression of the human erythropoietin gene. Proc. Natl. Acad. Sci. USA 82: 7580-7584.
4. Golde, D.W. 1978. Hormonal modulation of erythropoiesis in vitro. In: In Vitro Aspects of Erythropoiesis. M.J. Murphy Jr., C. Peschle, A.S. Gordon, and E.A. Mirand, editors. Springer Verlag, New York. 81-85.
5. Dainiak, N. 1985. Role of defined and undefined serum additives to hemopoietic stem cell culture. In Hemopoietic Stem Cell Physiology. E.P. Cronkite, N. Dainiak, R.P. McCaffery, J. Palek, P.J. Quesenbery, editors. Liss, New York. 59-68.
6. Bersch, N., J.E. Groopman and D. W. Golde. 1982. Natural and biosynthetic insulin stimulates the growth of human erythroid progenitors in vitro. J. Clin. Endocrinol. Metab. 55: 1209-1211.
7. Kurtz, A., W. Jelkmann and C. Bauer. 1983. Insulin stimulates erythroid colony formation independently of erythropoietin. Br. J. Haematol. 53: 311-316.
8. Dainiak, N., and S. Kreczko. 1985. Interactions of insulin, insulinlike growth factor II, and platelet-derived growth factor in erythropoietic culture. J. Clin. Invest. 76: 1237-1242.
9. Geffner, M.E., S.A. Kaplan, N. Bersch, B.M. Lippe, W.G. Smith, R.A. Nagel, T.V. Santulli Jr., C.H. Li, and D.W. Golde. 1987. Leprechaunism: In vitro insulin action despite genetic insulin resistance. Ped. Res. 22: 286-291.
10. Flier, J.S., P. Usher, and A.C. Moses. 1986. Monoclonal antibody to the type I insulin-like growth factor (IgF-I) receptor blocks IgF-I receptor-mediated DNA synthesis: Clarification of the mitogenic mechanisms of IgF-I and insulin in human skin fibroblasts. Proc. Natl. Acad. Sci. USA 83: 664-668.
11. Chen, P., S. Kwan, T. Hwang, B.J. Chiang and C. Chou. 1983. Insulin receptors on leukemia and lymphoma cells. Blood 62: 251-255.
12. Konwalinka, G., C. Breier, D. Geissler, C. Peschel, C.J. Wiedermann, J. Patsch and H. Braunsteiner. 1988. Proliferation and Differentiation of Human Erythropoiesis in Vitro: Effect of Different Human Lipoprotein Species. Exp. Hematol. 16: 125-130.

13. Iscove, N.N., L.J. Guilbert, and C. Weymann. 1980. Complete replacement of serum in primary culture of erythropoietin-dependent red cell precursors (CFU-e) by albumin, transferrin, iron, unsaturated fatty acid, lecithin and cholesterol. Exp. Cell Res. 126: 121-131.
14. Havel, R.J., H.A. Eder, and J.H. Bragdon. 1955. Distribution and chemical composition of ultra-centrifugally separated lipoproteins in human serum. J. Clin. Invest. 34: 1345-1351.
15. Konwalinka, G., D. Geissler, C. Peschel, C. Breier, K. Grunewald, R. Odavic and H. Braunsteiner. 1986. Human erythropoiesis in vitro and the source of burst-promoting activity in a serum-free system. Exp. Hematol. 14: 899-903.
16. Konwalinka, G., C. Peschel, J. Boyd, D. Geissler, M. Ogriseg, R. Odavic and H. Braunsteiner. 1984. A miniaturized agar culture system for cloning human erythropoietic progenitor cells. Exp. Hematol. 12: 75-79.
17. Schroder, K.S., S. Raptis, W. Beischer. 1976. Radio-immunologische Bestimmunung von STH, Insulin und C-Peptid. Med. Welt. 27: 1720-1726.
18. Davis, R.J., M. Faucher, L.K. Racaniello, A. Carruthers and M.P. Czech. 1987. Insulin-like growth factor I and epidermal growth factor regulate the expression of transferrin receptors at the cell surface by distinct mechanisms. J. Biol. Chem. 262: 13126-13135.
19. Dainiak, N., D. Sutter, and S. Kreczko. 1986. L-tri-iodothyronine augments erythropoietic growth factor release from peripheral blood and bone marrow leukocytes. Blood 68: 1289-1297.
20. Zuckerman, K.S. 1981. Human Erythroid burst-forming units. Growth in vitro is dependent on monocytes but not T lymphocytes. J. Clin. Invest. 67: 702-710.
21. Kirschner, J., H. Ozer, M. O'Leary, D. Highyand, M. Marinello, and H. Preisler. 1985. T cell depletion of bone marrow does not inhibit in vitro hematopoiesis. Blut. 50: 201-206.
22. Dainiak, N., A. Najman, V. Kulkarni, D. Sutter, S. Kreczko, L. Feldman, C. Baillou, and G. Leblanc. 1984. Regulation of erythroid burst proliferation by B lymphocytes from healthy donors and patients with chronic lymphocytic leukemia (CLL). Blood 64: 113-118. (Suppl.)
23. Iscove, N.N. and L.J. Guilbert. 1978. Erythropoietin-independence of early erythropoiesis and a two-regulator model of proliferative control in the hemopoietic system. In In Vitro Aspects of Erythropoiesis. M. J. Murphy. Springer Verlag, New York. 3-7.

SERINE PROTEASES PROMOTE HUMAN CFU-GM IN METHYLCELLULOSE CULTURE SYSTEMS

Samuel Gross, Diana A. Worthington-White and Cheryl M. Smith

Department of Pediatrics
University of Florida College of Medicine
Gainesville, Florida

INTRODUCTION

Amino acid sequences and three dimensional structural analysis of serine proteases show sequence homology in a group of compounds, which include thrombin, Cl' esterase, urokinase, trypsin, plasminogen, snake venom, kallikrein, and the γ subunit of mouse submandibular gland nerve growth factor. Common to the structure of these enzymes are active sites consisting of aspartic acid, histidine and serine. The similarities between the various serine proteases are striking in a number of details, including cellular location, chemical structure, molecular weight and function.

Essentially all of the serine proteases have a well-defined function in normal hemostasis and in reproduction physiology. There is strong evidence, as well, for a major role in promoting cell proliferation and growth. More than twenty years ago it was shown that kallikrein significantly increased the mitotic rate of thymus lymphocytes in rats[1]. Later expansion of this work revealed kallikrein-induced increases in mitotic activity in rat bone marrow as well as thymocytes[2]. Even more recently, kallikrein has been found to be chemotactic for polymorphonuclear leukocytes[3] and to be capable of promoting wound healing. The latter effect is strikingly illustrated in studies on X-irradiated rats, wherein increased survival follows closely upon the efficacy of kallikrein-induced (accelerated) wound healing[4].

In the last fifteen years, the biological significance of serine proteases has come into more precise focus. It is now clear that they have roles far greater than initially expected. The similarities in chemical structure, amino acid sequence homology, antigenicity and function[5-8], including responses to naturally-occurring and synthetic inhibitors within defined biological systems, strongly support the view that they provide signals in a bioregulatory system underscoring conservation and specificity.

Within such framework, it is reasonable to assume that a serine protease, capable of activating a specific functional aspect, as for example, granulocyte chemotaxis, should have the prior capability of inducing granulocyte growth and differentiation. In 1972, well before we demonstrated their CFU-GM effects, and well before their respective chemotactic effects were described, Levi-Montalcini[9] and others[10-12] proposed a relationship between γ-NGF and kallikrein based on the then available knowledge of high concentrations of kallikrein in mouse and human salivary secretions and snake venom. Additional support has also been gained in the recent discovery of rich sources of NGF in the guinea pig prostate[13], long known to be rich in kallikreins.

The role of the serine proteases as growth factors is unclear but their presence within the system appears to be more than adventitious. In brief, there are a number of hematopoietic growth factors, apart from erythropoietin and IL-2 (or T-cell growth factor)[14], included under the generic term "colony stimulating factor": G-CSF for granulocytes[15], M-CSF for macrophages[16], GM-CSF for both granulocytes and macrophages[17], MC-CSF for mast cells (derived from a con A-activated mouse T-cell clone)[18], and multi--CSF (also known as IL-3[19], burst cell factor[20], or hematopoietic growth factor[21]). Several of the CSF's have already been cloned, and include a human GM-CSF from a complimentary DNA isolated with the aid of the monkey COS cell system to produce GM-CSF[22]. This protein appears to have significant homology with mouse GM-CSF, cloned from mouse lung mRNA, but not with serine proteases. Other isolated and characterized CSF's include a mouse cDNA clone for a mouse GM-CSF, which bears no homology to its functionally related IL-3[19], a cloning for mouse IL-3 (in complete agreement with IL-3 purified from the mouse monomyelocytic line, WEHI-3B)[23], which is a GM-CSF and seemingly activated by proteolytic cleavage catalyzed by serine proteases of the 'kallikrein family'[19], and a cDNA encoding human macrophage-specific colony-stimulating factor (CSF-1), which apparently requires proteolytic cleavage in order to release the CSF[23].

METHODOLOGY

Bone Marrow CFU-GM Assay System

The CFU-GM assay in these experiments is a modification of the method of Iscove, et al[24]. Normal human marrow (3-5 ml) is processed using Ficoll--hypaque density centrifugation in order to obtain a mononuclear cell fraction. Once collected, the cells are washed in α-media (GIBCO, Grand Island, NY) and resuspended to a concentration of 2×10^6 cells/ml. The suspension is plated on 35 x 10 mm tissue culture dishes (LUX, Newberry Park, CA) in a total volume of one ml of α-media containing 0.9% methylcellulose (DOW Chemical, Midland, MI), 0.6% l-glutamine (GIBCO), 10% cells, and 10% protease. Control cultures contained 20% media conditioned by normal human peripheral leukocytes, which serves as a source of colony-stimulating factor (CSF). The cultures are incubated in a humidified 5% CO_2 atmosphere up to 14 days, at which time colonies containing greater than 40 cells are enumerated using an inverted microscope. Cytological distinction is carried out by micropipetting individual colonies from representative plates onto glass slides, following which they are air-dried, fixed in methanol and Wright-Giemsa stained.

Purity of all serine proteases is obtained by subjecting the compounds to electrophoresis on polyacrylamide gels (SDS-PAGE) and by Fast Protein Liquid Chromatography (Pharmacia's FPLC).

Protease Inactivation

D-phe-pro-arg chloromethyl ketone (ACMK), irreversibly binds to the protease enzyme active site histidine. 6.4×10^{-5}M ACMK is reacted with 3.7×10^{-6}M protease for three hours at 25°C in 0.1M Tris-HCl + 0.02M EDTA, pH 7.5. The mixture is then dialyzed against 0.1M Tris-HCl, pH 7.5. This method allows for less than 0.1% residual enzymatic activity compared to the native proteases, using the chromogenic substrate benzoyl DL-arginine p-nitroanilide HCl (BAPNA), and measured spectrophotometrically at 410 nm.

The second method of protease inactivation involves utilizing the naturally-occurring inhibitor, alpha-1-antitrypsin. Alpha-1-antitrypsin binds to the enzyme catalytic site rendering the enzymatic portion of the protease inactive. 1 µg of alpha-1-antitrypsin is reacted with varying doses of the purified trypsin fractions at 25°C for one hour. The resultant mixture is then added to the culture system just prior to plating in methylcellulose.

RESULTS

Purification and Dose Response

All serine proteases were purified on FPLC as described in the methodology. Growth studies were carried out on the pre and post purification specimens. It is notable, in terms of trypsin, wherein two peaks were identified via the FPLC analysis. Peak #1 was found to be trypsin and Peak #2, chromotrypsin (Figure 1).

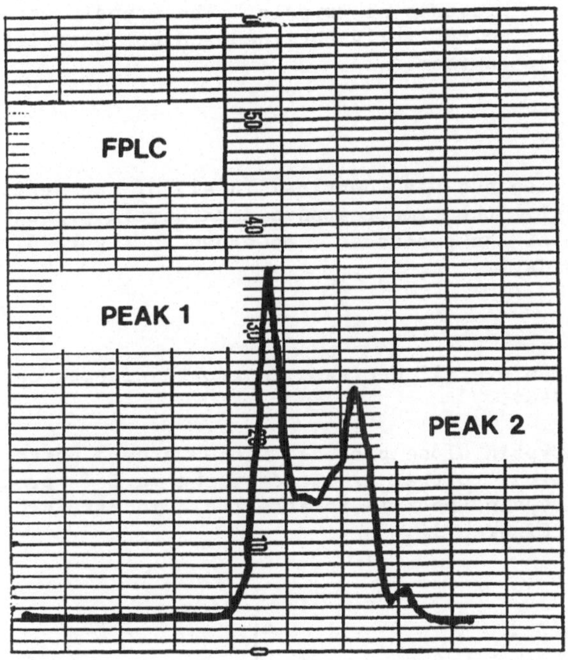

Fig. 1. Trypsin FPLC fractions.

Trypsin, prior to purification, resulted in an erratic dose response (Figure 2). Following purification and elutriation of the two "trypsin" peaks, two distinct CFU-GM dose response curves were generated, as shown in Figure 3. Indeed, following superimposition of the curves in Figure 3, one can reproduce the curve of the "unpurified" trypsin in Figure 1. PAGE chromatography, as well as FPLC, established the chymotrypsin nature of peak #2, which was further verified by co-chromatography with purified chymotrypsin. Purified chymotrypsin produced identical CFU-GM results (Figure 4). Plasminogen and kallikrein also affected CFU-GM curves similar to trypsin, the only difference being the peak concentrations (Figure 5).

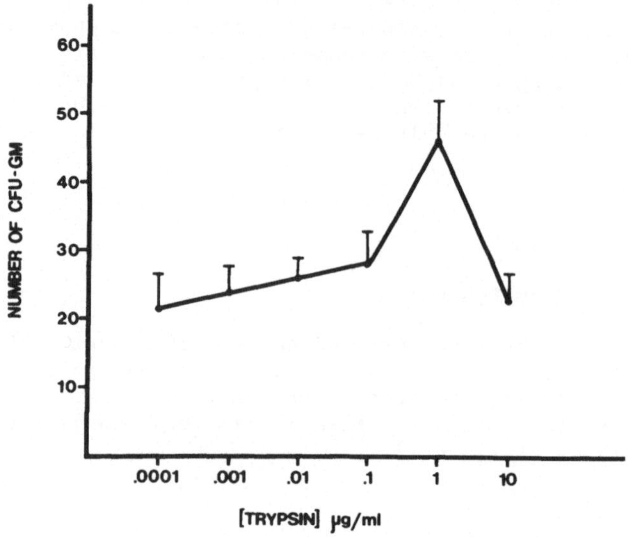

Fig. 2. The growth of bone marrow CFU-GM in the presence of trypsin. The trypsin was commercially obtained and not purified further.

Time Response Curves

Time to peak response was identical in all of the serine proteases studied. Accordingly, the time response data herein presented denotes the average of all of the proteases. As shown in Figure 6, time to peak growth averaged 10 days.

Inhibition Studies

ACMK affected an overall decrease in growth without changing the growth patterns (Figures 2-5).

Alpha-1-antitrypsin alone produced CFU-GM growth patterns, save for chymotrypsin, identical to that produced by the various proteases. When co-cultured with trypsin (peak 1) it affected a modest, albeit insignificant CFU-GM inhibition (Figure 7).

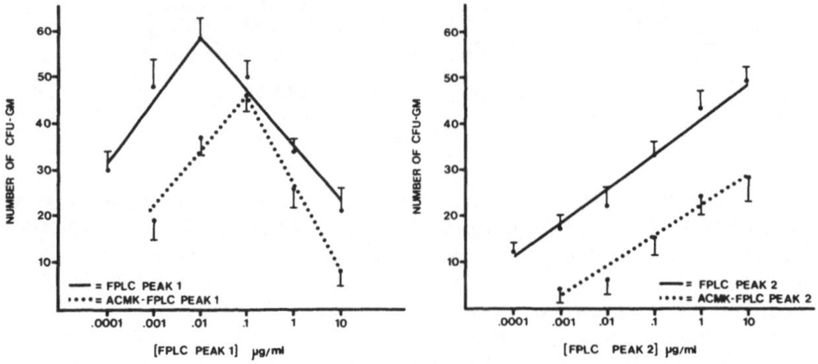

Fig. 3. The growth of CFU-GM colonies when cultured with the FPLC trypsin peaks (±SEM).

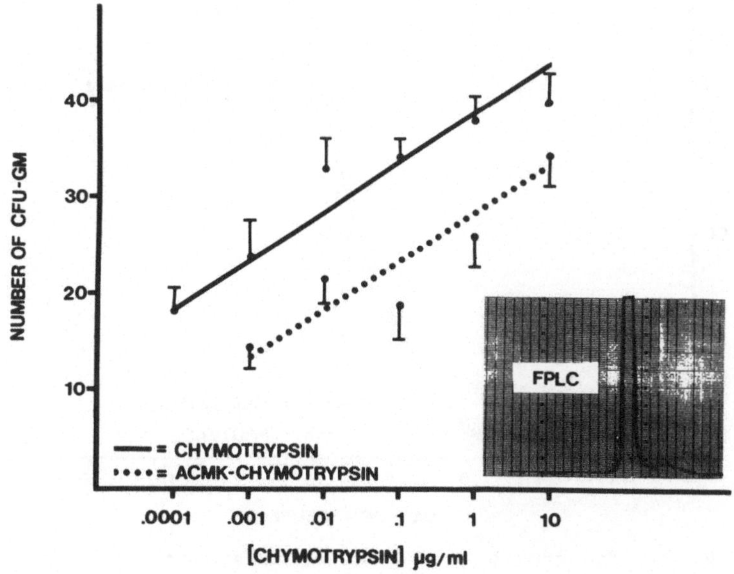

Fig. 4. The growth of bone marrow CFU-GM in the presence of chymotrypsin, purified via FPLC (±SEM).

DISCUSSION

The ability of a variety of serine proteases to effect CFU-GM growth in methylcellulose culture systems underscores the presumption that control and modulation of cell growth and differentiation are guided by a number of signals, all of which function in a highly refined and as yet incompletely understood fashion. Moreover, as we gain additional insight into the mechanism(s) by which these highly specialized "protein cutters" interact, the evidence will, in all likelihood, suggest a system capable of responding to clusters of messages, the responses designed to conserve energy whilst maintaining specificity. Thus, it is not surprising that the kallikreins can induce growth and differentiation of precursor GM's and, although not necessarily temporally, support migration and chemotaxis of the same cell lines.

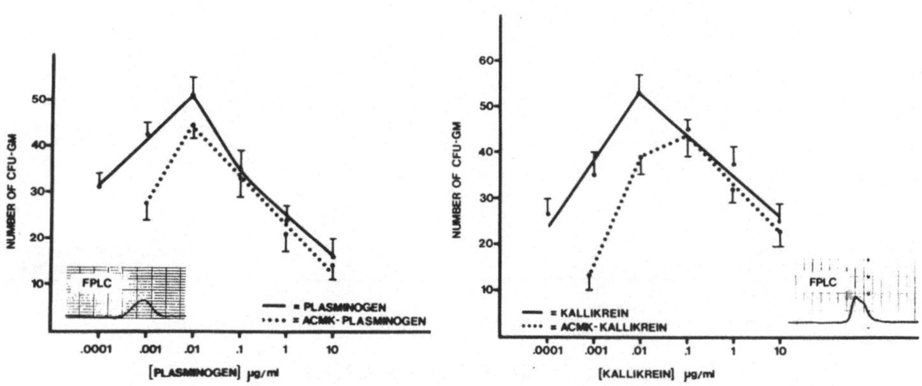

Fig. 5. The growth of bone marrow CFU-GM colonies in the presence of plasminogen (left graph) and kallikrein (right graph) (±SEM).

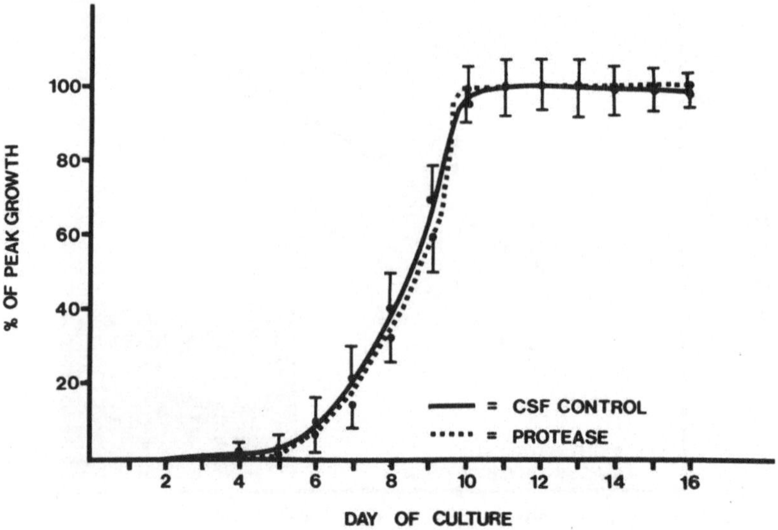

Fig. 6. The growth of CFU-GM colonies over time. The graph is represented as a percentage of peak growth over time.

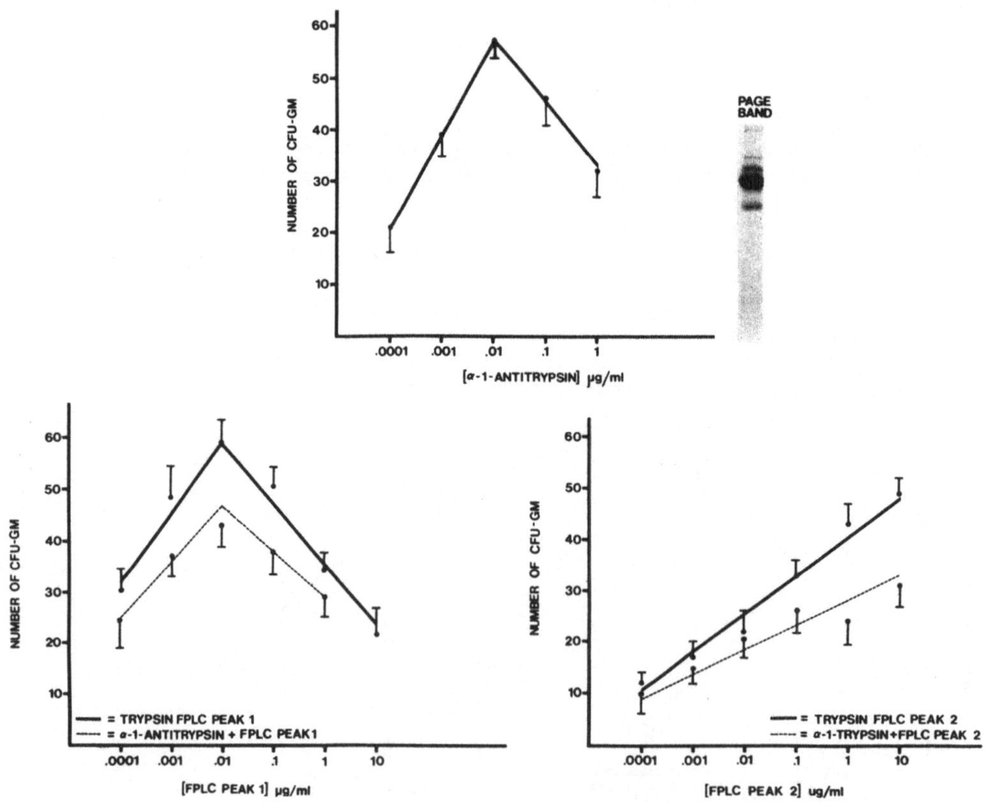

Fig. 7 The growth of bone marrow CFU-GM in the presence of alpha-1-antitrypsin, either alone, or in combination with FPLC trypsin fractions.

Whether the serine proteases act directly on sets of committed cells or function as intermediates remains a moot point. The fact that many of the newly cloned (non-erythroid) hematopoietic growth factors fail to show significant homology with the serine proteases is strong evidence for an interacting role. As an example of the latter, the recently cloned GM-CSF from the mouse monomyelocytic line, WEHI-3B-D+, appears to be activated by proteolytic cleavage catalyzed by the serine proteases of the kallikrein type, and it has also been shown that the macrophage-specific colony stimulating factor, CSF-1, requires a serine protease in order to expedite its release. Further evidence in support of an interacting effect is the constant finding that the degree of response to the various serine proteases is approximately 40-50% of our lymphocyte derived controls. Further support for this hypothesis is derived from embryologic evidence wherein the first hematopoietic colonies, the so-called "blood islands" are situated in the circulation rather than in a static tissue and that recent embryologic data strongly support the likelihood that the first growth factor is clearly dependent on certain of these circulating proteases. Indeed, the data from these studies, wherein (1) ACMK failed to inactivate the enzyme active site and (2) the trypsin specific inhibitor, alpha-1-antitrypsin, did likewise (and, in fact, functioned as a growth promoter akin to the proteases) provides additional evidence in support of an interactivating effect, i.e., multiple signals by preformed, activated compounds capable of performing a number of different but related functions.

REFERENCES

1. Perris, A.D., and J.F. Whitfield. 1969. The mitogenic action of bradykinin on thymic lymphocytes and its dependence on calcium. Proc Soc. Exp. Biol. Med. 130: 1198-1201.
2. Rixon, R.H., J.F. Whitfield, and J. Bayliss. 1971. The stimulation of mitotic activity in the thymus and bone marrow of rats by kallikrein. Horm. Metab. Res. 3: 279-284.
3. Schapira, M., C.F. Scott, L.A. Boxer, and R.W. Colman. 1983. Activation of human polymorphonuclear leukocytes by human plasma kallikrein. Adv. Exp. Med. Biol. 156: 747-753.
4. Mandel, P., J. Rodesch, and J.M. Mantz. 1973. The treatment of experimental radiation lesions by kallikrein. In: Kininogenases 1, ed. by G.L. Haberland and J.W. Rohen, pp. 171-188, Schattauer, Stuttgart, New York.
5. Kraut, J. 1977. Serine proteases: structure and mechanism of catalysis. Ann. Rev. Biochem. 46: 331-358.
6. Stroud, R.M. 1974. A family of protein-cutting proteins. Sci Am 231: 74-88.
7. Pisano, J.J. 1975. Chemistry and biology of the kallikrein-kinin system. In: Proteases and Biological Control, ed. by E. Reich, D.B. Rifkin and E. Shaw, pp. 199-222, Cold Spring Harbor Laboratory.
8. Neurath, H., and K.A. Walsh. 1976. Role of proteolytic enzymes in biological regulation (a review). Proc. Natl. Acad. Sci. USA 73: 3825-3832.
9. Levi-Montalcini, R., R.H. Angeletti, and P.U. Angelietti. 1972. The nerve growth factor. In: The Structure and Function of Nervous Tissue, ed. by G.H. Bourne, pp. 1-38, Academic Press, New York, London.
10. Bothwell, M.A., W.H. Wilson, and E.M. Shooter. 1979. The relationship between glandular kallikrein and growth factor-processing proteases of mouse submaxillary gland. J. Biol. Chem. 254: 7287-7294.
11. Habener, J.F., H.T. Change, and J.T. Potts. 1977. Enzymic processing of proparathyroid hormone by cell-free extracts of parathyroid glands. Biochem. 16: 3910-3917.

12. Schacter, M., B. Maranda, and C. Moriwaki. 1978. Localization of kallikrein in the coagulation and submandibular glands of the guinea-pig. J. Histochem Cytochem 26: 318-321.
13. Metcalf, D. 1981. In: Tissue Growth Factors, ed. by R. Baserga, pp. 343-384, Springer, New York.
14. Nicola, N.A., D. Metcalf, M. Matsumoto, and G.R. Hohnson. 1983. Purification of a factor inducing differentiation in murine myelomonocytic leukemia cells. Identification as granulocyte colony--stimulating factors. J. Biol. Chem. 258: 9017-9021.
15. Stanley, E.R., and P.M. Heard. 1977. Factors regulating macrophage production and growth. Purification and some properties of colony stimulating factors from medium conditioned by mouse L Cells. J. Biol. Chem. 252: 4305-4312.
16. Burgess, A.W., J. Camakaris, and D. Metcalf. 1977. Purification and properties of colony stimulating factors from mouse lung-conditioned medium. J. Biol. Chem. 252: 1998-2003.
17. Wokota, T., F. Lee, D. Rennich, et al. 1984. Isolation and characterization of a mouse cDNA clone that expresses mast-cell growth factor activity in monkey cells. Proc. Natl. Acad. Sci. USA. 81: 1070-1074.
18. Fung, M.C., A.J. Hapel, S. Ymer, et al. 1984. Molecular cloning of CDNA for murine interleukin-3. Nature 307: 233-237.
19. Iscove, N.N., C.A. Roitsch, N. Williams, and L.J. Guilbert. 1982. Molecules stimulating early red cell, granulocyte, macrophage and megakaryocyte precursors in culture: Similarity in size, hydrophobicity, and charge. J. Cell Physiol. Suppl 1: 65-78.
20. Bazill, C.W., M. Haynes, J. Garland, and T.M. Dexter. 1983. Characterization and partial purification of a hematopoietic growth factor in WEHI-3 cell conditioned medium. J. Biochem. 210: 747--759.
21. Wong, G.G., J.S. Witek, P.A. Temple, et al. 1985. Molecular cloning of the complimentary DNA and purification of the natural and recombinant proteins. Science 228: 810-815.
22. Gough, N.M., J. Gough, D. Metcalf, et al. 1984. Molecular cloning of cDNA encoding a murine hematopoietic growth regulator, granulocyte-macrophage colony stimulating factor. Nature 303: 763-767.
23. Kawasaki, E.S., M.B. Ladner, A.M. Wang, et al. 1985. Molecular cloning of a complimentary DNA encoding human macrophage-specific colony-stimulating factor (CSF-1). Science 230: 291-296.
24. Iscove, N.N., J.S. Senn, J.E. Till, and E.A. McCulloch. 1971. Colony formation by normal and leukemic marrow cells in culture: Effects of conditioned medium from normal leukocytes. Blood 37: 1-5.

MOLECULARLY CHARACTERIZED FACTORS GOVERNING THE GROWTH OF MURINE MULTIPOTENT STEM CELLS IN SERUM-DEPLETED MARROW CULTURES

Francis C. Monette and George Siqounas

Department of Biology
Boston University
Boston, MA 02215

ABSTRACT

The growth requirements of normal murine marrow-derived multipotent stem cells (CFU-GEMM) in a simple clonal cell culture system substantially devoid of exogenous serum proteins was assessed. The ability of murine interleukin-3 (Il-3), recombinant human erythropoietin (rEPO), and a crystalline preparation of the protoporphyrin hemin to support colony growth in "serum-free" cultures was examined by titration. The results suggest that both Il-3 and hemin are limiting for multipotential colony growth in "serum-free" cultures, but that EPO is not. In addition, the 'sensitivity' of CFU-GEMM to each growth factor appeared to increase in the "serum-free" environment as evidenced by a "shift-to-the-left" in all the titration curves. Nearly half of the GEMM colonies grew to full maturity in the absence of exogenous EPO. Given the optimal concentration of each growth factor, high colony growth was consistently observed in the "serum-free" cultures, with a range from 65% to 119% of the serum control level. It is therefore concluded that supplementation of murine marrow cultures with Il-3 and hemin alone may provide the necessary setting for studying the factors which modulate the growth of multipotent stem cells in a serum-free environment.

INTRODUCTION

The availability of a simple serum-free culture system for the routine growth of mammalian hematopoietic stem cells would provide a significant technical advance in our ability to decipher the mechanisms governing the proliferation and differentiation of these cells upon which hemopoiesis relies. Thus far, however, attempts to grow adult hematopoietic cell progenitors under totally serum-free conditions have met with limited success owing to a lack of definition of the principal factors responsible for hematopoietic colony formation and/or the unavailability of homogeneous preparations of candidate regulators. Recently, however, many hematopoietic growth factors (e.g., GM-CSF, Interleukin-3) as well as hormones (e.g., erythropoietin) have been cloned and/or purified to homogeneity[1-6]. The general availability of defined growth factors, therefore, has been a prerequisite for the development of highly-defined culture systems whose development can now proceed.

Some success in growing hematopoietic cell progenitors (e.g., CFU-e) in "serum-free" culture systems has been reported by several groups[8-14]. These culture systems usually require a battery of additives including phospholipids, cholesterol, serum albumin and the like, all of which are very difficult to purify to homogeneity[15]. In most cases these "serum-free" culture systems would more accurately be described as 'serum-depleted' owing to the requirement for the addition of substantial amounts of serum albumin. To date, there have been few, if any, reports defining a successful serum-free culture system for the more-primitive multipotent stem cells (i.e., CFU-GEMM/CFU-Mix) which form large, multicentric colonies of granulocytic, erythroid, monocytic/macrophagic, and megakaryocytic cells in semi-solid cultures[16-18]. It is not yet clear at the present time which factors are absolutely essential for hematopoietic colony formation in such cultures, let alone the identity of factors and hormones which are capable of modulating such growth. However, the work with interleukin-3 (Il-3) appears to strongly implicate this lymphokine as a principal regulator of multipotent cell growth, at least *in vitro*[4-6,10,13,19-22]. A similar, although probably reduced role has also been suggested for the hematopoietic growth factor GM-CSF[4,13,22-24]. Whether the hormone erythropoietin (EPO) plays an essential role in the early development of colonies derived from multipotent stem cells is not yet clear. For example, several groups[9,12] have reported an "EPO-independent" colony formation in "serum-free" cultures which stand in contrast to the work of others[6,8,10]. A third family of modulators of hematopoietic cell growth *in vitro* is suggested by the work of several groups[25,29] with the protoporphyrin hemin. This relatively simple metalloporphyrin has been shown to greatly augment the growth of erythroid cell progenitors from the multipotential cell level of development forward towards the CFU-e stage[25,26,28,29]. Although its effects on non-erythroid cell stages may be significant[30,31], hemin's ability to augment erythroid differentiation is most striking[25,26].

Hemin (ferric chloride protoporphyrin IX) is a most interesting molecule in many ways. Although its relatively simple molecular structure has long been appreciated (Figure 1), as has its ability to act as an inducer of erythroid differentiation in transformed cell lines[32,33] as well as its ability to control its own synthesis in erythrocyte lysates[34], the full dimension of hemin's biological activity is just becoming apparent (Table 1). Since some of hemin's activities are exerted on non-hematopoietic cells as well, the possibility remains that it may have an important and general role in the regulation of cellular activities, including the expression of differentiated properties and thus the control of cellular proliferation and/or differentiation at both the translational and transcriptional levels.

It would therefore appear that a number of factors are capable of modulating the growth of primitive hematopoietic cells under *in vitro* conditions. However, it is still unclear how these factors (i.e., Il-3, EPO, and hemin) interact to promote colony growth. The availability of purified and highly-defined growth factors is an important step towards this over-all goal. The development of a serum-free (or serum-depleted) bioassay would provide the impetus to more-critically assess the role of these hematopoietic growth factors at the cellular level.

We describe here the development of a relatively simple culture system which is capable of supporting the growth of murine CFU-GEMM without the addition of exogenous serum, which is supplemented essentially with Il-3, EPO, hemin, and serum albumin.

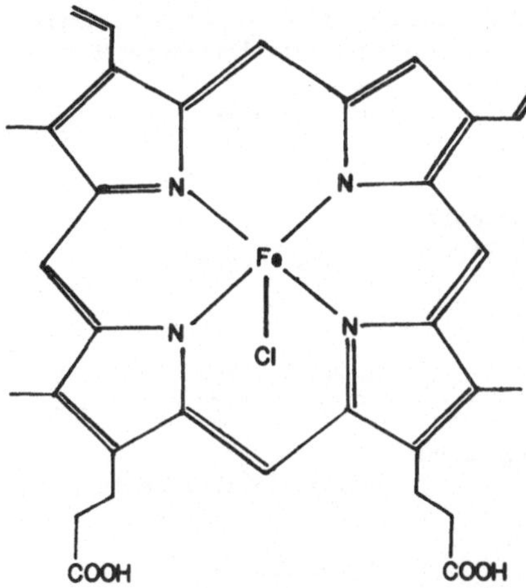

Fig. 1. The molecular structure of hemin (ferric chloride protoporphyrin IX (Reproduced with permission of Experimental Hematology, from reference 57).

MATERIALS AND METHODS

Detailed methods have been published elsewhere[26,45,46]. CD_1 strain female bone marrow was utilized as the cell source throughout these investigations. Alpha medium (Flow Laboratories, Rockville, MD) was suplemented with amino acids, pyruvate, L-glutamine, penicillin & streptomycin as described[47]. Crystalline bovine hemin (Type I, Sigma Chemical Co., St. Louis, MO) was prepared[25] fresh prior to use at a final concentration of 0.2 uM unless otherwise noted. One-half unit of a highly-purified human recombinant EPO (r-EPO) preparation (AMGen Biologicals, Thousand Oaks, CA) (Sp. Act. > 70,000 units/mg protein) was added to all cultures unless otherwise noted. A partially-purified preparation of murine IL-3 (Genzyme, Boston, MA) (Sp. Act. > 5000 units/mg) was employed at a final concentration of 20 units/ml unless otherwise noted. For serum titration studies, several different lots of fetal bovine serum (FBS) were utilized (Reheis Chemical Co., Phoenix, AZ). The methylcellulose assay[48] was utilized to grow CFU-GEMM with the following additions: 10 mg/ml BSA, 0.2 mg/ml L-asparagine, and 10^{-4} M 2-mercaptoethanol. The usual final concentrations of r-EPO, Il-3, and hemin were: 0.5 unit/ml, 20 units/ml, and 0.2 mM, respectively, unless otherwise noted. The usual cell plating density ranged from 75 to 150 x 10^3 cells/ml. Cultures were grown for a period of 11 days at 5% CO_2 in room air at 37°C. CFU-GEMM colonies were defined as large aggregates of > 200 cells consisting of erythroid, granulocytic, monocytic/macrophagic, and occasionally, megakaryocytic, eosinophilic, and/or mast cells. The student's t-test on paired independent variables was used to compare control groups to experimental groups.

RESULTS

The optimal serum requirement of murine CFU-GEMM grown in methylcellulose was observed to be at 5% FBS (Figure 2). Higher amounts of FBS added to cultures supplemented with Il-3, r-EPO, and hemin proved inhibitory ($\leq$ 0.02 at 25% FBS). However, in cultures receiving no exogenous serum the colony growth was 64% of the optimum level and did not differ signifi-

Table 1. Some biological Activities of Ferric Chloride Protoporphyrin IX (Hemin) in Erythroid and Non-erythroid Cells.

Activity	Cell System	Reference
ERYTHROID CELLS:		
Activation of globin synthesis via inhibition of the 'hemin-controlled translational repressor'	reticulocyte	34
Enhances synthesis of α - and β - globin m-RNA in:	human K-562 cells	36, 37
	mouse erythroleukemia cells	32
Induces transferrin-independent hemoglobin synthesis	human K-562(h) cells	39
Binds to specific surface membrane receptors	MEL cells	40
Enhancement of membrane protein synthesis	reticulocyte	35
Enhances colony development *in vitro*	mouse marrow CFU-e	28, 29
	mouse marrow BFU-e	25
	human marrow BFU-e	27
	mouse marrow CFU-GEMM	26
Enhances *in vitro* growth	Mouse erythroleukemia cells	41
Enhancement *in vivo*	Mouse marrow BFU-e	56
Accelerated morphological cell maturation	Erythroblast	38
NON-ERYTHROID CELLS:		
Mitogenic	Human T-cells	43
Involved in receptor-mediated aggregation	blood platelet	44
Enhances adipocyte differentiation	3T3 cells	42
Enhances *in vitro* growth	mouse marrow GM-CFU	41

cantly ($p \geq 0.05$) from the number observed at 5% FBS. These studies clearly establish the possibility that multipotent stem cells can be grown *in vitro* without the addition of exogenous serum. The Il-3 requirement of normal murine marrow CFU-GEMM grown without serum but in the presence of 'optimal' levels of r-EPO and hemin, is depicted in Figure 3. The growth of CFU-GEMM clones was dependent upon the addition of Il-3 since no colonies were observed when this lymphokine was omitted from the culture medium. Optimal colony growth occured at an Il-3 concentration of approximately 5-10 units per ml beyond which no further increase in colony number was observed. The EPO requirement of CFU-GEMM grown under "serum-free" conditions was assessed in cultures supplemented with 'optimal' levels of Il-3 (20 units/ml), hemin (0.2 mM) but lacking FBS (Figure 4). Although the addition of up to 0.5 unit/ml r-EPO increased colony levels nearly two-fold, nearly half of the optimal colony level was observed in cultures without serum (11.4 ± 1.1 per 10^5 cells). These studies suggest the possibility that either very low levels of EPO are capable of supporting colony growth, or alternatively, that nearly half of the CFU-GEMM can mature fully in the absence of EPO when cultured in the presence of optimal concen-

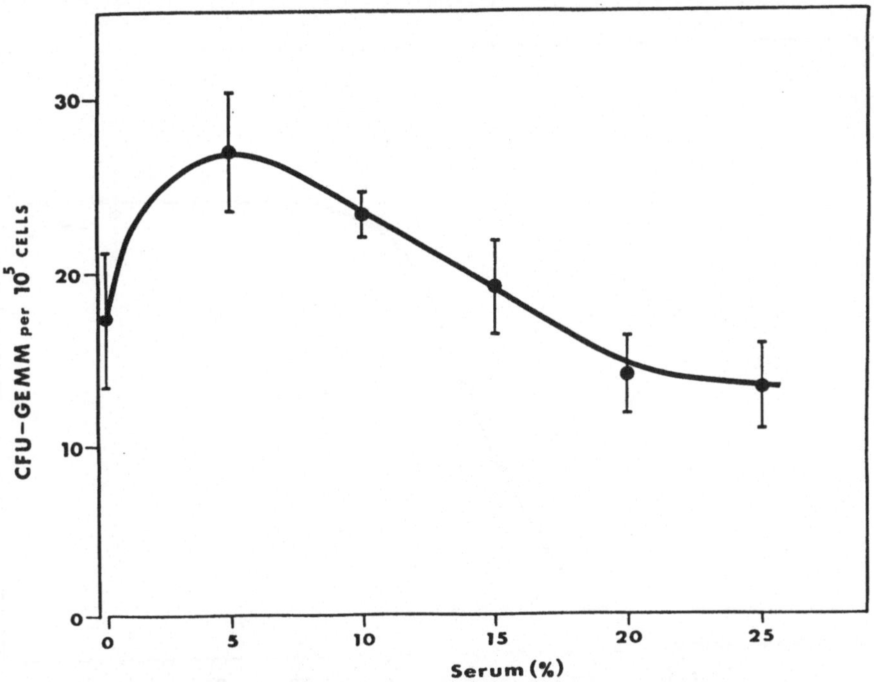

Fig. 2. Fetal bovine serum titration of murine marrow CFU-GEMM. Results were averaged from four separate studies employing two different serum lots. The average colony number observed at 5% serum (26.9 ± 3.4) was not significantly ($p \leq 0.05$) higher than that noted in the no serum group (17.3 ± 3.9). However, colony number was significantly ($p \leq 0.02$) reduced at the higher serum levels. Cultures contained 30 units/ml Il-3, 2.0 units/ml r-EPO, 200 uM hemin, and 1% BSA. (Reproduced with permission of Experimental Hematology, from reference 46).

trations of both Il-3 and hemin. This latter possibility is currently being tested with the use of EPO antiserum.

Figure 5 compares the results of five separate hemin titration studies of CFU-GEMM growth in cultures deprived of FBS but supplemented with 'optimal' levels of Il-3 (20 units/ml) and r-EPO (1.0 unit/ml). In the present methylcellulose culture system, the addition of exogenous hemin was absolutely necessary for colony growth as no colonies were observed in the absence of hemin or even at very low concentrations of the metalloporphyrin ($\leq$ 0.025 mM) In contrast, CFU-GEMM grown in the presence of serum exhibited some growth in the absence of hemin. In both the 'serum-free' as well as the serum containing culture system, the half-maximal hemin requirement was observed at 0.05 mM hemin. Likewise, both culture systems exhibited a substantial reduction in colony growth with the addition of 3 to 6 times the optimal hemin level.

It would appear from the forgoing that CFU-GEMM can be grown without exogenous serum when cultured with 'optimal' concentrations of growth factors as follows: 10 to 20 units Il-3/ml; 0.5 to 1.0 unit r-EPO/ml; and 0.1 to 0.2 mM hemin. Figure 6 summarizes the results of three separate cell dose titration studies where CFU-GEMM were grown without FBS at optimal concentrations of each growth factor. Linear regression analysis indicated that the relationship between colony frequency and the number of cells

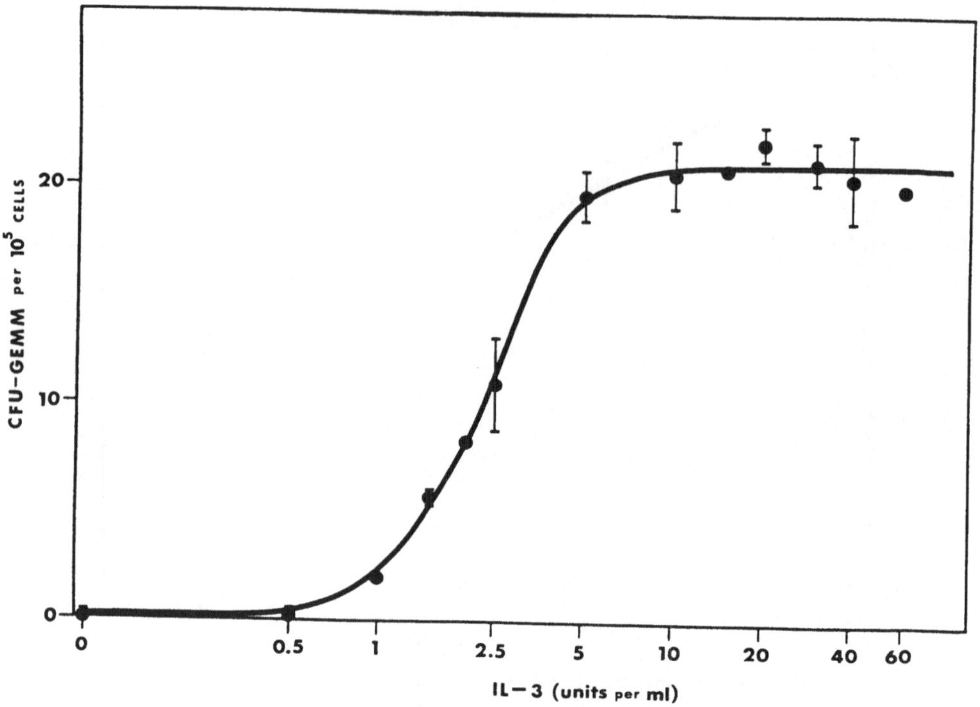

Fig. 3. Growth of murine CFU-GEMM in "serum-free" methylcellulose cultures titered against murine interleukin-3 (Il-3) and expressed as ± 1 standard deviation. Results are from two (no error bars) to eight separate studies. Cultures contained 2.0 units/ml r-EPO and 200 uM hemin. (Reproduced with permission of Experimental Hematology, from reference 46).

plated was linear and the titration curve extrapolated to the origin, suggesting that over the cell dose range assessed, the marrow cells themselves were limiting for colony formation.

These studies strongly suggest that the optimal growth of multipotential colonies in the absence of serum factors requires the addition of Il-3, EPO, as well as iron protoporphyrin IX. Since the titration studies demonstrated that only Il-3 and hemin were essential for colony growth under 'serum-free' conditions, it was of general interest to determine the nature of the interaction, if any, between Il-3 and hemin. Towards this goal, colony growth was compared in 'serum-free' or serum-containing (10% FBS) cultures when the lymphokine was added separately or in combination with hemin. Table 2 summarizes the results of from 8 to 16 separate studies. These data clearly show a synergistic effect on the part of hemin when combined with Il-3. No colonies were observed in the 'serum-free' system when Il-3 was omitted, even with optimal hemin supplementation, when 3.6 ± 1.3 colonies were observed in cultures containing FBS + Il-3. However, in both culture systems, the addition of both Il-3 and hemin greatly augmented the observed colony level suggesting some sort of interaction between these two growth factors in promoting the growth of these colonies *in vitro*.

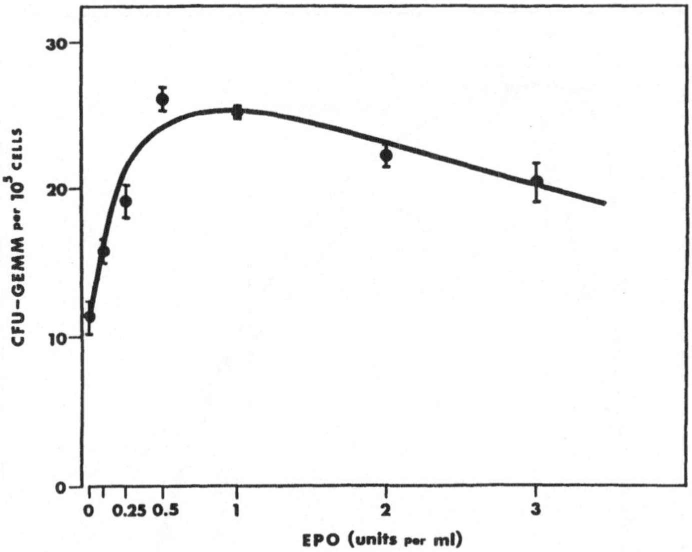

Fig. 4. Growth of murine CFU-GEMM in "serum-free" methylcellulose cultures titered against human recombinant EPO (r-EPO). Results are from three separate studies and are expressed as the mean ± 1 standard deviation. The observed colony number at 3 units r-EPO (20.2 ± 1.5) was significantly ($p \leq 0.05$) lower than that observed at 0.5 unit per ml (26.1 ± 0.7). Cultures contained 20 to 30 units/ml Il-3 and 200 uM hemin. (Reproduced with permission of Experimental Hematology, from reference 46).

DISCUSSION

It would now appear that murine marrow-derived multipotent stem cells can be reproducibly grown in methylcellulose clonal cell cultures deprived of exogenous serum. Two factors appear to be limiting for CFU-GEMM growth in such cultures: the lymphokine interleukin-3 and the iron protoporphyrin hemin. No colonies were observed in the absence of either factor in this study, in spite of high concentrations of the second factor. The "shift-to-the-left" in the dose response curves for Il-3, r-EPO, and hemin, when compared to those obtained with serum-containing cultures[46,49], suggests that much less growth factor is necessary for optimal colony development under serum-depleted conditions. In this set of experiments, CFU-GEMM were consistently observed in 'serum-free' cultures and averaged 65% of the optimal serum-containing level. In addition, colony growth was highly reproducible as can be seen from the cell dose titration studies (Figure 6) and was independent of the cell plating level. In another report representing a larger group of studies[46] we have demonstrated that colony growth in 'serum-free' cultures can actually approximate the level observed with serum supplementation and suggest that it may not be necessary to further-supplement serum-free cultures except with Il-3, EPO, and hemin. However, it should be noted that these cultures contained up to 10 mg/ml of bovine serum albumin. This protein is not only difficult to purify, it may be contaminated with significant levels of transferrin[8], lipids[11], lipoproteins[15], and other proteins[15] although not with Il-3[14] nor EPO [6,9]. It will therefore be essential to either obtain homogeneous preparations of albumin or, alter]natively, other proteins for the establishment of truly defined serum-free cultures.

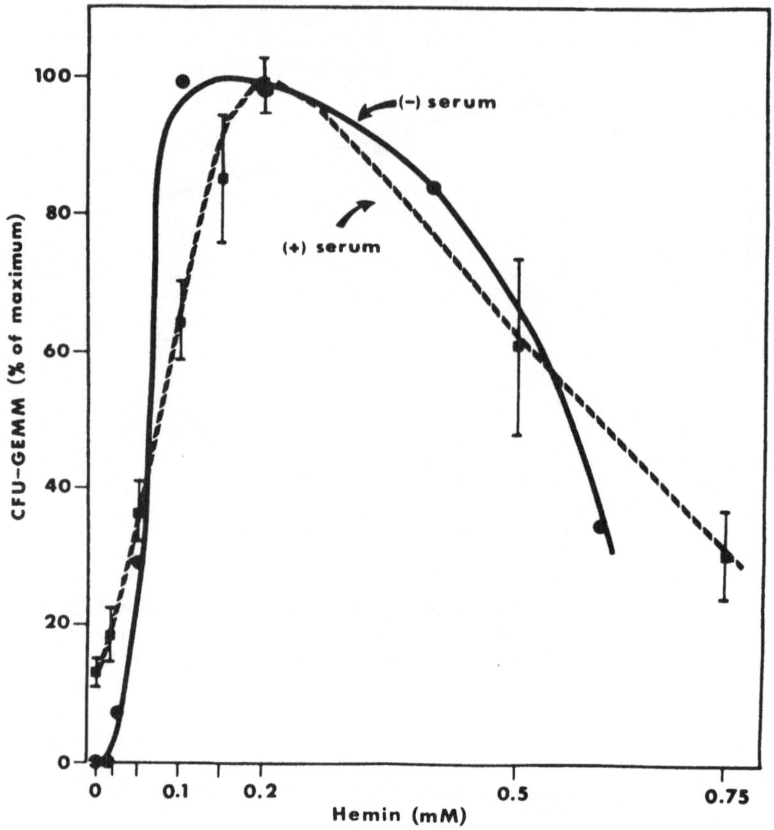

Fig. 5. Growth of murine CFU-GEMM in "serum-free" methylcellulose cultures titered against crystalline bovine hemin (ferric chloride protoporphyrin IX). The average of a total of 5 separate studies is shown. Data are depicted as the mean ± 1 standard deviation. The colony frequency is expressed relative to the maximal colony number since colony growth was higher in cultures supplemented with serum. Cultures contained 20 units/ml Il-3 and 2.0 units/ml r-EPO. (Reproduced with permission of Experimental Hematology, from reference 46).

The virtual absolute dependency of CFU-GEMM growth for Il-3 and hemin in this culture system raises the possibility that both factors play a pivotal role in colony growth *in vitro.* Although the colony requirement for Il-3 has been firmly established[1,5,22], the exact role of hemin in supporting colony growth is less well understood. From the data presented (Table 2) it seems clear that the protoporphyrin is capable of interacting in a synergistic manner with the lymphokine to promote colony development *in vitro*, but its mechanism of action remains elusive. It is becoming increasingly apparent that numerous other factors, including those hematopoietic growth factors which have been well-characterized, are capable of interacting with Il-3 in a synergistic fashion[50-52]. Recent studies have suggested that the manner by which hematopoietic growth factors interact with their target that the manner by which hematopoietic growth factors interact with their target cells may soon be revealed. Interleukin-3, for example, has been shown to interact with its marrow target cells by direct and specific binding through membrane receptors[53]. In contrast, hemin appears to interact with its target cells via both direct and indirect

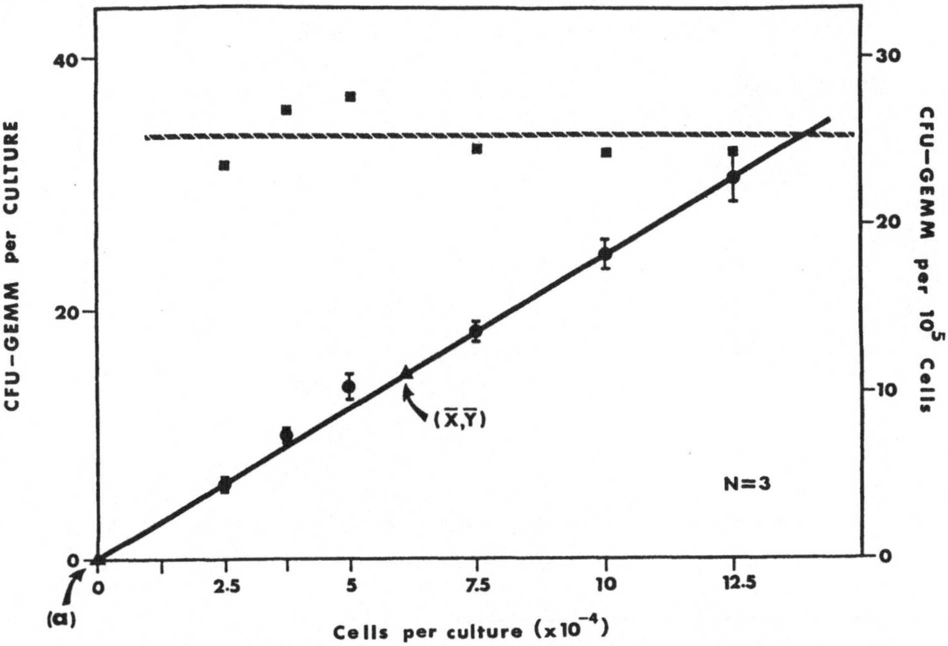

Fig. 6. Cell dose titration of murine marrow CFU-GEMM in "serum-free" cultures. Average of 3 separate studies. Line of best fit was drawn by linear regression analysis (a = 0.05; $\overline{y}$ = 14.9; $\overline{x}$ = 6.1). The slope of the line is 2.44. The dashed line represents colony efficiency expressed relative to 10^5 marrow cells and was drawn by eye. Data points shown are the mean ± 1 standard error. Cultures contained optimal concentrations of Il-3, hemin and r-EPO. (Reproduced with permission of Experimental Hematology, from reference 46).

mechanisms. For example, several groups have demonstrated the existence of heme receptors on the surface of murine[40] and human[54] erythroleukemia cells whereas others[55] have shown clearly that heme interacts with specific membrane receptors on its target cells by first binding to the serum glycoprotein hemopexin.

The "serum-free" culture system reported here, even though not truly defined in terms of all of its components, should help provide the framework for establishing a much more highly-defined culture system for the growth of multi-potent stem cells in vitro. With the aid of such systems, it should soon be possible to precisely define the factors modulating early events in mammalian hemopoietic cell development if not their mechanism of action at the molecular level.

ACKNOWLEDGMENTS

The work described herein was made possible by PHS grants AM 35325 and DK 37366 from the N.I.H. The care of all animals used in these studies was in accordance with P.H.S. and Boston University guidelines.

Table 2. Synergistic Action of Hemin

Factors Added§		Number CFU-GEMM per 10^5 cells			
Il-3	Hemin	(n)	+ Serum	(n)	"Serum-Free"
-	-	(3)	0.0	(3)	0.0
-	+	(4)	0.7 ± 0.5	(4)	0.0
+	-	(4)	3.6 ± 1.3	(5)	0.0
+	+	(8)	27.3 ± 1.4	(16)	23.6 ± 1.0

§, erythropoietin (r-EPO) was added to all cultures at a concentration of 0.5 unit per ml.

n, number of separate studies.

REFERENCES

1. Ihle, J.N., J. Keller, S. Oroszlan, L.E. Henderson, T.D. Copeland, F. Fitch, M.B. Prystowsky, E. Goldwasser, J.W. Schrader, E. Palaszynski, M. Dy, and B. Lebel. 1983. Biologic properties of homogeneous interleukin 3. I. Demonstration of WEHI-3 growth factor activity, mast cell growth factor activity, P. Cell stimulating factor activity, colony-stimulating factor activity, and histamine-producing cell-stimulating factor activity. J. Immunol. 131: 282-287.
2. Clark-Lewis, I., S.B.H. Kent, and J.W. Schrader. 1984. Purification to apparent homogeneity of a factor stimulating the growth of multiple lineages of hemopoietic cells. J. Biol. Chem. 259: 7488-7494.
3. Jacobs, K., C. Shoemaker, R. Rudersdorf, S.D. Neill, R.J. Kaufman, A. Mufson, J. Seehra, S.S. Jones, R. Hewick, E.F. Rritsch, M. Kawakita, T. Shimizu, and T. Miyake. 1985. Isolation and characterization of genomic and cDNA clones of human erythropoietin. Nature 313: 806-810.
4. Cutler, R.L., D. Metcalf, N.A. Nicola, and G.R. Johnson. 1985. Purification of a multipotential colony-stimulating factor from pokeweed mitogen-stimulated mouse speen cell conditioned medium. J. Biol. Chem. 260: 6579-6587.
5. Hapel, A.J., M.C. Fung, R.M. Johnson, I.G. Young, G. Johnson, and D. Metcalf. 1985. Biologic properties of molecularly cloned and expressed murine interleukin-3. Blood. 65: 1453-1459.
6. Suda, J., T. Suda, K. Kubota, J.N. Ihle, M. Saito, and Y. Miura. 1986. Purified interleukin-3 and erythropoietin support the terminal differentiation of hemopoietic progenitors in serum-free culture. Blood 67: 1002-1006.
7. Iscove, N.N., L.J. Guilbert, and C. Weyman. 1980. Complete replacement of serum in primary cultures of erythropoietin-dependent red cell precursors (CFU-E) by albumin, transferrin,

iron, unsaturated fatty acid, lecithin and cholesterol. Exptl. Cell Res. 126: 121-126.
8. Stewart, S., B. Zhu, and A. Axelrad. 1984. A "serum-free" medium for the production of erythropoietic bursts by murine bone marrow cells. Exptl. Hematol. 12: 309-318.
9. Goodman, J.W., E.A. Hall, K.L. Miller, and S.G. Shinpock. 1985. Interleukin 3 promotes erythroid burst formation in "serum-free" cultures without detectable erythropoietin. Nat. Acad. Sci. Proced. USA 82: 3291-3295.
10. Cormier, F., P. Baines, N. Lucien, and G.A. Boffa. 1985. Complete replacement of serum in cultures of murine primitive erythroid and multipotential progenitor cells: absolute requirement for spleen conditioned medium. Cell Differen. 17: 261-269.
11. Dainiak, N., S. Kreczko, A. Cohen, R. Pannel, and J. Lawler. 1985. Primary human marrow cultures for erythroid bursts in a serum-substituted system. Exptl. Hematol. 13: 1073-1079.
12. Eliason, J.F., and N. Odartchenko. 1985. Colony formation by primitive hemopoietic progenitor cells in serum-free medium. Natl. Acad. Sci. Proced.USA 82: 775-779.
13. Koike, K., E.R. Stanley, J.N. Ihle, and M. Ogawa. 1986. Macrophage colony formation supported by purified CSF-1 and/or interleukin 3 in serum-free culture: Evidence for hierarchical difference in macrophage colony-forming cells. Blood. 67: 859-864.
14. Konwalinka, G., D. Geissler, C. Peschel, C. Breier, K. Grunewald, R. Odavic, and H. Braunsteiner. 1986. Human erythropoiesis in vitro and the source of burst-promoting activity in a serum--free system. Exptl. Hematol. 14: 899-903.
15. Dainiak, N. 1985. Role of defined and undefined serum additives to hematopoietic stem cell culture. In: Cronkite, E.P., N. Dainiak, R.P. McCaffery, P.J. Quesenberry (eds.), Alan R. Liss Inc., Hematopoietic Stem Cell Physiology, New York, p. 59-76.
16. Metcalf, D., G.R. Johnson, and T.E. Mandel. 1979. Colony formation in agar by multipotential hemopoietic cells. J. Cell. Physiol. 98: 401-420.
17. Fauser, A.A., and H.A. Messner. 1979. Identification of megakaryocytes, macrophages, and eosinophils in colonies of human bone marrow containing neutrophilic granulocytes and erythroblasts. Blood. 53: 1023-1027.
18. Hara, H., and K. Noguchi. 1981. Clonal nature of pluripotent hemopoietic precursors in vitro. Stem. Cells. 1: 53-60.
19. Spivak, J.L., R.R.L. Smith, and J.N. Ihle. 1985. Interleukin 3 promotes the in vitro proliferation of murine pluripotent hematopoietic stem cells. J. Clin. Invest. 76: 1613-1621.
20. Migliaccio, A.R., and J.W.M. Visser. 1986. Proliferation of purified murine hemopoietic stem cells in serum-free cultures stimulated with purified stem cell-activating factor. Exptl. Hematol. 14: 1043-1048.
21. Suda, T., J. Suda, M. Ogawa, and J.N. Ihle. 1985. Permissive role of interleukin 3 (IL-3) in proliferation and differentiation of multipotential hemopoietic progenitors in culture. J. Cell. Physiol. 124: 182-190.
22. Iscove, N.N., C.A. Roitsch, N. Williams, and L.J. Guilbert. 1982. Molecules stimulating early red cell, granulocyte, macrophage, and megakaryocyte precursors in culture: Similarity in size, hydrophobicity and charge. J. Cell. Physiol. (Suppl. 1): 65-78.
23. Metcalf, D. 1985. The granulocyte-macrophage colony-stimulating factors. Science. 229: 16-22.
24. Broxmeyer, H.E., D.E. Williams, G. Hangoc, S. Cooper, S. Gillis, R.K. Shadduck, and D.C. Bicknell. 1987. Synergistic myelopoietic actions in vivo after administration to mice of combinations

of purified natural murine colony-stimulating factor 1, recombinant murine interleukin 3, and recombinant murine granulocyte/macrophage colony-stimulating factor. Proc. Natl. Acad. Sci. USA 84: 3871-3875.

25. Monette, F.C., and S.A. Holden. 1982. Hemin enhances the in vitro growth of primitive erythroid progenitor cells. Blood 60: 527-530.
26. Monette, F.C. and G. Sigounas. 1984. Factors affecting the proliferation and differentiation of clonogenic hematopoietic stem cells in vitro. Blood Cells 10: 261-286.
27. Lu, L., H.E. Broxmeyer, P.A. Meyers, M.A.S. Moore, and H.T. Thaler. 1983. Association of cell cycle expression of Ia-like antigenic determinants on normal human multipotential (CFU-GEMM) and erythroid (BFU-E) progenitor cells with regulation in vitro by acidic isoferritins. Blood. 61: 250-256.
28. Porter, P.N., R.H. Meintz, and K. Mesner. 1979. Enhancement of erythroid colony growth in culture by hemin. Exptl. Hematol. 7: 11-16.
29. Ibrahim, N.G., J.D. Lutton, and R.D. Levere. 1982. The role of haem biosynthetic and degradative enzymes in erythroid colony development: The effect of haemin. Brit. J. Haematol. 50: 17-28.
30. Rothmann, J., C.F. Hertogs, Z. Malik, and D.H. Pluznik. 1983. Hemin stimulating effect on colony formation of leukemic and bone marrow cells. Exptl. Hematol. 11: 147-153.
31. Holden, S.A., H.N. Steinberg, E.A. Matzinger, and F.C. Monette. 1983. Further characterization of the hemin-induced enhancement of primitive erythroid progenitor cell growth in vitro. Exptl. Hematol. 11: 953-960.
32. Ross, J., and D. Sautner. 1978. Induction of globin mRNA accumulation by hemin in extracts of Friend erythroleukemia cells. J. Biol. Chem. 253: 7124-7126.
33. Mager, D., and A. Bernstein. 1979. The role of heme in the regulation of the late program of Friend cell erythroid differentiation. J. Cell. Physiol. 100: 467-479.
34. Gross, M. 1979. Control of protein synthesis by hemin: Evidence that the hemin-controlled translational repressor inhibits formation of 80S initiation complexes from 48S intermediate initiation complexes. J. Biol. Chem. 254: 2370-2377.
35. Deutsch, J., and O.O. Blumenfield. 1977. Effect of hemin, sodium periodate and concanavalin A on in vitro biosynthesis of rabbit reticulocyte membrane proteins. Biochem. Biophys. Res. Commun. 79: 1224-1230.
36. Fuhr, J.E., E.G. Bamberger, C.B. Lozzio, and B.B. Lozzio. 1981. Induction of hemoglobin synthesis in original K562 cell line. Blood Cells. 7: 389-395.
37. Charnay, P., and T. Maniatis. 1983. Transcriptional regulation of globin gene expression in the human erythroid cell line K562. Science 220: 1281-1283.
38. Bonanou-Tzedaki, S.A., M. Sohi, and H.R.V. Arnstein. 1981. Regulation of erythroid cell differentiation by haemin. Cell Differen. 10: 267-279.
39. Viola, L., R. Biagini, R. Barbieri, and R. Gambari. 1987. Inhibition of hemoglobin accumulation by monoclonal antibodies to the human transferrin receptor is reversed by hemin. Expt. Hematol. 15: 1145-1152.
40. Galbraith, R.A., S. Sassa, and A. Kappas. 1985. Heme binding to murine erythroleukemia cells. J. Biol. Chem. 260: 12198-12202.
41. Rothmann, J., C. F. Hertogs, and D.A. Pluznik. 1977. Replacement of serum by hemolysate as growth promoter for murine leukemic and normal hemopoietic progenitor cells in culture. Exptl. Hematol. 5: 117-124.

42. Chen, J.J. and I.M. London. 1981. Hemin enhances the differentiation of mouse 3T3 cells to adipocytes. Cell. 26: 117-122.
43. Stenzel, K.H., A.L. Rubin, and A. Novogradsky. 1981. Mitogenic and comitogenic properties of hemin. J. Immunol. 127: 2469-2473.
44. Mslik, Z., D. Creter, A. Cohen, and M. Djaldetti. 1983. Haemin affects platelet aggregation and lymphocyte mitogenicity in whole blood incubations. Cytobios. 38: 33-38.
45. Monette, F.C., and G. Sigounas. 1987. Sensitivity of murine multipotential stem cell colony (CFU-GEMM) growth to interleukin-3, erythropoietin, and hemin. Exptl. Hematol. 15: 729-734.
46. Monette, F.C., and G. Sigounas. 1988. Growth of murine multipotent stem cells in a simple "serum-free" culture system: Role of interleukin-3, erythropoietin, and hemin. Exptl. Hematol. 16: (In Press).
47. McLeod, D.L., M.M. Shreeve, and A.A. Axelrad. 1974. Improved plasma culture system for production of erythrocytic colonies in vitro: quantitative assay method for CFUe. Blood. 44: 517-534.
48. Iscove, N.N., and F. Sieber. 1975. Erythroid progenitors in mouse bone marrow detected by macroscopic colony formation in culture. Exptl. Hematol. 3: 32-43.
49. Monette, F.C., and G. Sigounas. Hemin Acts synergistically with interleukin-3 to promote the growth of multipotent stem cells (CFU-GEMM) in "serum-free" cultures of normal murine bone marrow. (Submitted)
50. Broxmeyer, H.E., D.E. Williams, G. Hangoc, S. Cooper, S. Gillis, R.K. Shadduck, and D.C. Bicknell. 1987. Synergistic myelopoietic action in vivo after administration to mice of combinations of purified natural murine colony-stimulating factor 1, recombinant murine interleukin 3, and recombinant murine granulocyte/macrophage colony-stimulating factor. Proc. Natl. Acad. Sci. (USA). 84: 3871-3875.
51. Chen, B.E.-M., and C.R. Clark. 1986. Interleukin 3 (Il-3) regulates the in vitro proliferation of both blood monocytes and peritoneal exudate macrophages: synergism between a macrophage lineage-specific colony-stimulating factor (CSF-1) and Il 3. J. Immunol. 137: 563-570.
52. Schmitt, E., B. Fassbender, K. Beyreuther, E. Spaeth, R. Schwarzkopf, and E. Rude. 1987. Characterization of a T-cell-derived lymphokine that acts synergistically with IL 3 on the growth of murine mast cells and is identical with IL 4. Immunobiol. 174: 406-419.
53. Nicola, N.A., and D. Metcalf. 1986. Binding of iodinated multipotential colony-stimulating factor (Interleukin-3) to murine bone marrow cells. J. Cell Physiol. 128: 180-188.
54. Majuri, R., and R. Grasbeck. 1987. A rosette receptor assay with haem-microbeads. Demonstration of a haem receptor on K562 cells. Eur. J. Haematol. 38: 21-25.
55. Taketani, S., H. Kohno, and R. Tokunaga. 1987. Cell surface receptor for hemopexin in human leukemia HL60 cells. Specific binding, affinity labeling, and fate of heme. J. Biol. Chem. 262: 4639-4643.
56. Monette, F.C., S.A. Holden, M.J. Sheehy, and E.A. Matzinger. 1984. Specificity of hemin action in vivo at early stages of hematopoietic cell differentiation. Exptl. Hematol. 12: 782-787.
57. Monette, F.C., and S.A. Holden. 1982. Hemin enhancement of primitive erythroid progenitors in vitro: Relationship to burst--promoting activity (BPA). Exptl. Hematol. 10: (Suppl. 12): 281-294.

MARROW RETICULO-FIBROBLASTOID COLONIES (CFU-RF DERIVED) SPONTANEOUSLY RELEASE AN ERYTHROID COLONY (BFU-E) ENHANCING FACTOR

Carlos A. Izaguirre[1], William M. Ross [2,4], and
Elizabeth Y. Hsu[3]

Departments of Medicine[1], Physiology[2] and Pediatrics[3]
University of Ottawa
Ottawa, Canada, K1N 8M5

Defense Research Establishment[4]
Ottawa, Canada

ABSTRACT

The marrow microenvironment is composed of an extracellular matrix as well as a heterogeneous population of cells. Isolation of the various cell types and analysis of their function is necessary for a better understanding of their roles in hemopoiesis. We have recently reported a colony assay for a cellular component of the marrow microenvironment. The assay consists of a plasma clot-methylcellulose marrow culture. The stimulator is PHA-stimulated leukocyte conditioned medium (PHA-LCM) and hydrocortisone (5×10^{-5}M). The fibrin strands appear to act as a substrate for the growth of Reticulo-Fibroblastoid colonies derived from the CFU-RF precursor. RF colonies can be subcultured forming adherent layers when transferred to liquid cultures. Confluent adherent layers can be maintained for long periods of time by changing medium every 3 to 5 days. Supernatants derived from unstimulated RF cultures (RF-CM) were tested for growth promotion of hemopoietic precursors. We found: (1) RF-CM by itself does not induce colony formation. (2) In the presence of erythropoietin, RF-CM enhances the growth of BFU-E. (3) Recombinant IL 4 also enhances BFU-E formation, but in our assays IL 4 induced fewer colonies than RF-CM and the colonies were smaller. (4) Because neither IL 4 nor RF-CM, by themselves, can stimulate colony formation, we compared the effect of RF-CM on assays that are known to show other IL 4 functions. RF-CM did not induce proliferation of PHA induced blast T cells, a known property of IL 4. In conclusion, RF cells in culture appear to secrete, spontaneously, a factor that enhances BFU-E formation, and, unlike GM-CSF and 1L3, RF-CM has no effect on other hemopoietic colony forming cells. IL 4 has similar properties, but RF-CM appears not to have the B- and T-cell growth factor activities of Interleukin 4.

INTRODUCTION

The growth and development of hemopoietic cells is regulated by a complex interaction of hemopoietic stem cells and various populations of auxiliary cells. These include T-lymphocytes[1,2], monocytes[3,4] and cells described as marrow stromal cells. Considerable progress in the analysis of these stromal components was made after introduction of the long term liquid bone marrow culture method by Dexter et al[5]. These cultures demonstrated heterogeneity of cell types within the adherent layer of stromal cells[6]. They included fibroblasts, macrophages, endothelial cells, adipocytes, and a group of reticulo-epithelioid cells. This complex nature of the LTBMC monolayer has made difficult detailed analyses of the functions of individual subpopulations. Recently we have reported a colony assay[7,8] that identifies a precursor able to give rise to colonies composed of reticulo-fibroblastoid cells. Such colonies can be observed either in cultures of fresh marrow or in cultures derived from cells of the LTBMC monolayer, but not in peripheral blood cell cultures. Immunofluorescence and biosynthetic studies revealed that cells within the colonies contained and produced collagen types I, III, IV and V, laminin, and fibronectin and are negative for factor VIII, collagen type II (Table 1) and for markers usually associated with myelo and lymphopoiesis. The ability to synthesize laminin and collagen type IV as well as culture requirements, distinguish CFU-RF derived cells from previously characterized[9] marrow fibroblasts (CFU-F). These findings indicate that CFU-RF and CFU-F are different components of the marrow stromal population. The relationship of CFU-RF to hemopoietic precursors was investigated using patients with Chronic myeloid leukemia and bone marrow transplant recipients. Cells within CFU-RF derived colonies were uniformly negative for the Philadelphia chromosome, thus making it unlikely that they belonged to the malignant hemopoietic clone. CFU-RF derived colonies in bone marrow transplant recipients were found to be exclusively of host origin. Both observations support the view that CFU-RF is not part of the repertoire of hemopoietic stem cells and does not belong to the subset of stromal cells that have been reported to be transplantable[10] or that belong to the leukemic clone in Chronic myeloid leukemia[11].

The ability to clone a subset of marrow stromal cells has permitted us to examine their potential functional capabilities. In this paper we present a preliminary study of conditioned medium (RF-CM) collected from subcultured CFU-RF derived colonies (RF-CM). We showed that RF-CM contains a spontaneously released growth factor(s) capable of inducing BFU-E colony formation in the presence of erythropoietin. RF-CM does not support growth of any other hemopoietic stem cell in the absence of erythropoietin and does not have T cell growth activity, thus we presume that the factor is not IL 3, GM-CSF[12] or IL 4[13].

MATERIALS AND METHODS

Samples

Bone marrow cells were collected from pediatric patients undergoing bone marrow aspiration for diagnostic or follow-up purposes. Mononuclear cells of density less than 1.077 gms/ml were obtained by centrifugation in Ficoll-Paque (Pharmacia, Uppsala, Sweden). Nonadherent cells were prepared as described previously[14].

CFU-RF Assay and Subculture

Mononuclear bone marrow cells were cultured at a concentration of $2x10^5$/ml in methylcellulos (0.9%) supplemented with Iscove's modified medium IMDM, (Gibco, Grand Island, NY.), 10% phytohemagglutinin (Wellcome H15) stimulated leukocyte conditioned medium (PHA-LCM)[15], calcium chloride $4x10^{-5}$M, 2-mercaptoethanol $5x10^{-5}$M, (Sigma, St. Louis, MO), 1 unit of sheep erythropoietin (Connaught Laboratories, Willowdale, Canada), 10^{-6}M

Table 1. Phenotypic characteristics of CFU-RF derived cells, skin fibroblasts and umbilical vein endothelial cells(1).

	Collagen Type				Factor
	III	IV	V	Laminin	VIII:Rag
CFU-RF derived cells	+	+	+	+	-
Skin fibroblast	+	-	-		-
Endothelial cells	+	+	+	+	+

(1) Modified from reference 8.

hydrocortisone (HC), and 30% citrated normal human plasma. One milliliter of the mixture was placed into 35 mm. Petri dishes (Lux, Lab-Tek Division, Miles Laboratories, Naperville, IL) and incubated at 37°C in a humidified incubator supplemented with 5% CO_2. After 14 days of incubation, cultures were scored for reticulo-fibroblastoid colonies.

Colonies were removed by Pasteur pipette, transferred into phosphate buffered saline (PBS) and disaggregated mechanically by repeated aspiration through a 0.2 ml pipette tip. The resulting cell suspensions derived from pooled colonies were washed and placed into 25 cm^2 tissue culture flasks (Corning #25100) containing IMDM, supplemented with 30% fetal calf serum (Flow Lab, Rockville, MD). The cells settled down and were allowed to adhere overnight to the bottom of the dishes. Non-adherent cells were removed and fresh growth medium added. The medium was completely changed every three to four days. Supernatants were collected every 5 to 7 days after the cells reached confluent growth.

BFU-E Assay

Non adherent bone marrow cells were cultured at 5×10^4 cells/ml in 0.9% methylcellulose, 30% FCS, 5% PHA-LCM, 1% Bovine serum albumin, 5×10^{-5}M 2-mercaptoethanol and IMDM. After 2 weeks in culture at 37°C, 5% CO_2, BFU-E colonies were counted.

T Cell Growth Factor Activity Assay

Blood mononuclear cells were stimulated with 0.03% PHA-P (DIFCO) and cultured at 37°C, 5% CO_2. After 3 days the cells were harvested, washed extensively and used as targets (PHA-blasts) for growth factors[16]. PHA--blasts were resuspended at 10^6 cells/ml and 0.1 ml of cells were mixed with 0.1 ml of a dilution of various growth factors. The mixtures were done in a microtiter plate. After 3 days in culture, ^{3}H-thymidine incorporation was measured.

B Cell Colony Assay

B cell colonies were grown as previously described[17].

Growth Factors

RF-CM and PHA-LCM were prepared as described earlier. Recombinant human Interleukin 3 (IL 3), GM-CSF and Interleukin 4 (IL 4) were obtained from Genzyme (Boston, MA) and used at a concentration that gave optimal growth of BFU-E. Interleukin 2 (IL 2) was purchased from Amgen (Thousand Oaks, CA).

RESULTS

CFU-RF Colonies and Subculture of RF Cells

Human bone marrow cells cultured in methylcellulose plasma clot gave rise to small colonies by day 12-14 which were composed of a network of tightly linked reticulo-fibroblastoid cells. Occasionally, cells within these colonies contained oil red O positive fat droplets. When both PHA-LCM and HC were added to the culture, plating efficiency was at 20-40 colonies per $2x10^5$ mononuclear cells. For all experiments described in this paper, stromal cells were obtained from cultures containing both HC and PHA-LCM and were subcultured as described above. Single cells obtained from individual or pooled colonies attached best to tissue culture plastic at a FCS concentration of 30%. Cells derived from the majority of colonies continued to proliferate and formed confluent layers after 7 to 10 days. Supernatants were then collected every 5 to 7 days, filter sterilized and stored at 4°C until used.

BFU-E Stimulating Activity in RF-CM

As shown in Table 2, four of six RF-CM were able to enhance BFU-E colony formation in the presence of erythropoietin. In the absence of erythropoietin no colony formation by hemopoietic precursors was observed. Such activity was similar to that of IL 3, GM-CSF and IL 4. However, with IL 4 the BFU-E derived colonies were small. RF-CM batches #5 and #6 were obtained from subcultures that had not reached confluence at the time of harvesting, studies of the kinetics of release of the factor during RF cell growth are in progress.

T Cell Growth Activity

As shown in Table 3, three different batches of RF-CM (batches #2, 3, and 4 from exp. 2, Table 2) did not support T cell proliferation when compared to IL 2 and IL 4. Proliferation above background levels was noticed but such enhancement could not be quantitated and may have been due to nonspeciric factors. IL 4 induced maximal T cell proliferation

Table 2. Effect of RF-CM on BFU-E colony formation. Comparison with PHA-LCM, Interleukin 4, Interleukin 3 and GM-CSF[1].

		No. of BFU-E colonies per 10^5 cells	
		Exp. 1	Exp. 2
Erythropoietin 2 U/ml		4	10
PHA-LCM	5%	24	14
RF-CM[2] #1	5%	12	ND[3]
	10%	15	ND
	15%	9	ND
RF-CM #2	10%	ND	45
RF-CM #3	10%	ND	23
RF-CM #4	10%	ND	5
RF-CM #5[3]	10%	ND	0
RF-CM #6[3]	10%	ND	0
Interleukin 4	500 U/ml	9	ND
	100 U/ml	9	13
Interleukin 3	50 U/ml	12	25
GM-CSF	50 U/ml	14	15

[1] All cell cultures contained erytropoietin.
[2] In the absence of erytropoietin, RF-CM did no induce colony formation of any class of hemopoietic progenitor.
[3] RF-CM #5 and #6 were collected from non-confluent cultures.

Table 3. Effect of RF-CM, recombinants IL 2 and IL 4 on T cell blasts proliferation.

		^{3}H-Thymidine incorporation c.p.m.
Experiment 1		
medium		110
rIL 2	100 U/ml	55,642
rIL 4	300 U/ml	26,644
RF-CM #1	50%	922
#2	50%	1,027
#3	50%	720
Experiment 2		
medium		3,374
rIL 2	100 U/ml	39,242
RF-CM #1[1]	20%	9,304
#2	20%	7,158
#3	20%	8,727

[1] RF-CM #1 was concentrated 5-fold before use.

at 300 u/ml a concentration that was within the range of optimal BFU-E enhancing activity.

B Cell Colony Formation

Batches 2, 3, and 4 of RF-CM lacked B cell colony formation activity (data not shown).

DISCUSSION

Little is known about mechanisms that regulate hemopoiesis in vivo. Our understanding is based largely on studies performed in vitro suggesting a co-operative interaction between hemopoietic precursors and various other cell populations. The long term bone marrow culture method[5,18] has facilitated the examination of hemopoietic stem cells maintained by the stromal monolayer generated by such cultures. Several studies revealed cellular heterogeneity within the adherent layer of stromal cells in vitro[6,19,20] and in vivo[21,22]. They include fibroblasts macrophages, endothelial cells, adipocytes, and a group of reticulo-epithelioid cells. While these studies have contributed to the appreciation of heterogeneity among stromal cells, the role of each component in hemopoiesis is currently not understood.

Insight into the role of each subcomponent of the stromal layer may be obtained by isolating the various cell types and examining their function independently of each other. Friedenstein et al[23,9] and Castro-Malaspina et al[9] have developed direct culture methods for bone marrow fibroblasts. The fibroblastic nature of these cells was documented by their reaction with specific anticollagen antibodies. Progenitor cells for these colonies were termed CFU-F. Recently, we developed a semisolid culture method for human marrow cells that promoted growth of colonies of cells with a reticulo-fibroblastic appearance[7,8].. The development of techniques permitting subculture and clonal expansion of cells derived from such colonies have allowed studies of their phenotype and origin[8]. In this paper we report that cell derived from CFU-RF progenitors spontaneously release a factor(s) that enhances BFU-E colony formation. Such a factor does not have the ability to induce CFU-GM colonies, does not stimulate T cell proliferation or B cell colony formation. Thus, it appears to be different from GM-CSF, IL3, IL4, and IL2. Aye[24] has reported a similar finding. CFU-RF derived colonies, subcultured as adherent cells were able to release constitutively a BFU-E promotion activity without granulocytic colony stimulating activity. However, upon stimulation with Interleukin 1, RF cells produced granulocytic CSF. These observations suggest strongly that CFU-RF derived colonies may play a crucial role in the physiology of hemopoiesis. Such cells can now be cloned and be easily identified by their ability to produce collagen IV and laminin and therefore the CFU-RF cell lineage can be studied independently from other cells of the microenvironment. Several questions remain to be answered, such as which specific growth factors are needed for CFU-RF colony formation, identification of the growth factor(s) released by RF cells spontaneously or after stimulation, their role in maintaining hemopoiesis in long-term marrow culture, cellular interaction with stem cells, etc. We expect that the assay described in this paper will be of help in elucidating the physiological role of CFU-RF derived cells in hemopoiesis.

REFERENCES

1. Nathan, D.G., L. Chess, and D.G. Hillman. 1978. Human erythroid burst-forming unit: T-cell requirement for proliforation in vitro. J. Exp. Med. 147: 324-329.
2. Kelso, A., and D. Metcalf. 1985. Characteristics of Colony-stimulating factor production by murine T-lymphocyte clones. Exp. Hematol. 13: 7-15.
3. Zuckerman, K.S., V. A. Pastel, and Goodrum. 1983. Production of human erythroid burst promoting activity by monocytes stimulated with bacterial lipopolysaccharide. Exp. Hematol. 11: 475-480.
4. Schrader, J.W. 1983. Bone marrow differentiation in vitro. CRC Crit. Rev. Immunol. 4(3): 197-277.
5. Dexter, T.M., T.D. Allen, and L.G. Lajtha. 1977. Conditions controlling the proliferation of haemopoietic stem cells in vitro. J. Cell Physiol. 91: 391-400.
6. Allen, T.D., and T.M. Dexter. 1984. The essential cells of the hemopoietic microenvironment. Exp. Hematol. 12: 517-521.
7. Izaguirre, C.A., F.E. Katz, and M.F. Greaves. 1982. Origin of marrow microenvironmental stromal cells in hemopoietic pluripotent stem cells. Blood 60, Suppl. 1:323a.
8. Lim, B., C. A. Izaguirre, M.T. Aye, L. Huebsch, J. Drouin, C. Richardson, M.D. Minden, and H.A. Messner. 1986. Characterization of reticulo--fibroblastoid colonies (CFU-RF) derived from gene marrow and long-term marrow culture monolayers. J. Cell Physiol. 127: 45-54.
9. Castro-Malaspina, H., R.E. Gay, G. Resnick, N. Napoor, P. Meyers, D. Chiarieri, S. McKenzie, H.E. Broxmeyer, and M.A.S. Moore. 1980. Characterization of human bone marrow fibroblast colony forming cells (CFU-F) and their progeny. Blood. 56: 289-301.
10. Keating, A., J.W. Singer, P.D. Killen, G.E. Striker, A.C. Salo, J. Sanders, E.D. Thomas, D. Thorning, and P.J. Fialkow. 1982. Donor origin of the in vitro hematopoietic microenvironment following marrow transplantation in man. Nature. 298: 280-283.
11. Singer, J.W., A. Keating, J. Cuttner, A.M. Gown, R. Jacobson, P.D. Killen, J.W. Moohr, V. Najfeld, J. Powell, J. Sanders, G.E. Striker, and P.J. Fialkow. 1984. Evidence for a stem cell common to hematopoiesis and its in vitro microenvironment: Studies of patients with clonal hematopoietic neoplasia. Leuk. Res. 8: 535-545.

12. Sieff, C.A. 1987. Hemopoietic growth factors. J. Clin Invest. 79: 1549-1557.
13. Peschel, C., W. E. Paul, J. Chara, and I. Green. 1987. Effects of B cell stimulatory factor -1/Interleukin 4 on hematopoietic progenitor cells. Blood 70: 254-263.
14. Messner, H.A., J.E. Till, and E.A. McCulloch. 1973. Interacting cell populations affecting granulopoietic colony formation by normal and leukemic human marrow cells. Blood 42: 707-710.
15. Aye, M.T., Y. Niho, J.E. Till, and E.A. McCulloch. 1974. Studies of leukemic cell populations in culture. Blood. 44: 205-219.
16. Spits, H., H. Yssel, Y. Takebe, N. Arai, T. Yokota, F. Lee, K.I. Arai, J. Banchereau, and J.E. deVries. 1987. Recombinant interleukin 4 promotes the growth of human T cells. J. Immunol. 139: 1142-1147.
17. Izaguirre, C.A., M.D. Minden, A.F. Howatson, and E.A. McCulloch. 1980. Colony formation by normal and malignant human B-lymphocytes. Brit. J. Cancer. 42: 430-437.
18. Gardner, S., and H.S. Kaplan. 1980. Long term culture of human bone marrow cells. Proc. Natl. Acad. Sci. USA 77: 4756-4759.
19. Bentley, S.A. 1984. Studies of bone marrow stromal cells in vitro. Prog. Clin. Biol. Res. 154: 179-207.
20. Zuckerman, K.S., and M.S. Wicha. 1983. Extracellular matrix production by the adherent cells of long-term murine bone marrow cultures. Blood 61: 540-547.
21. Lichtman, M.A. 1981. The ultrastructure of the haemopoietic environment of the marrow. A review. Exp. Hematol. 9: 391-410.
22. Weiss, L. 1980. The haemopoietic microenvironment of bone marrow: An ultra structural study of the interactions of blood cells, stroma and blood vessels. CIBA Found. Symp. 71: 3-19.
23. Friedenstein, A.J., R.K. Chailakhjan, and K.S. Lalykina. 1970. The development of fibroblast colonies in monolayer cultures of guinea pig bone marrow and spleen cells. Cell Tissue Kinet. 3: 393-403.
24. Aye, M.T. 1987. A novel growth factor produced by cells sub cultured from marrow stromal (CFU-RF) colonies. Blood Suppl. 1:166a.

INCREASED BPA PRODUCTION MODULATES EPO SENSITIVITY OF CIRCULATING BFU-E IN SICKLE CELL ANEMIA

Helena Croizat and Ronald L. Nagel

Albert Einstein College of Medicine
Division of Hematology, Department of Medicine
Bronx, New York

ABSTRACT

We have examined the possibility that permanent "stress" hemopoiesis in sickle cell anemia (SS) patients leads to the modification of the behavior of circulating 14 day erythroid progenitor cells (BFU-E). In these patients we find that peripheral blood BFU-E are increased in number and have high sensitivity to erythropoietin (Epo). Maximal number of BFU-E are generated from peripheral blood of SS patients at 0.3-0.75 Epo/ml of culture compared to 1.5-2.0 U Epo/ml of culture in normals. Peripheral blood adherent cells depletion leads to the shift of Epo dose response curve, so that the Epo sensitivity of BFU-E significantly decreases. This result suggests that apparent Epo hypersensitivity reflects, in fact, an increased production of a burst promoting activity (BPA) by SS peripheral blood light density adherent (PB-LDA) cells. Experiments with conditioned medium by SS PB-LDA cells confirmed this interpretation. When peripheral blood light density non-adherent (PB-LDNA) cells of SS patients or normal individuals were plated in the presence of various concentrations of SS PB-LD cells conditioned medium and constant amounts of Epo, a dose dependent increase of the number of BFU-E was observed. When the same target cells were plated in the presence of PB-LD cells conditioned medium from normal individuals, such effect does not occur. We conclude that increased BPA production may play a role in the erythropoietic regulation during constant hemopoietic stress in sickle cell anemia and might partially explain the lower than expected Epo levels in these patients.

INTRODUCTION

The ability of hemopoietic stem cells to proliferate and differentiate in semisolid cultures, when stimulated by appropriate hemopoietin, provides an assay for studying the regulation of hemopoietic progenitors from normal individuals or patients[1,2]. Using methylcellulose cultures we have examined the possibility that permanent "stress" hemopoiesis in sickle cell anemia (SS) patients could lead to the modification of the behavior of circulating 14 day erythroid progenitor cells (BFU-E).

Previous studies[3,4] suggested that bone marrow CFU-E of SS patients exhibit an increased sensitivity to regulatory factor erythropoietin. It was also reported[5] that sickle cell patients may have an increased number of circulating BFU-E which Epo requirements are higher than those of normals[6]. However, the number of patients included in these studies was too low to permit any final conclusion. The present report on larger populations of SS patients strongly supports that continuous stress of SS anemia results in the modification of the circulating 14 day erythroid progenitor cells regulation; they are increased in number in 50% of the patients and exhibit spurious hypersensitivity to erythropoietin due to an increased production of BPA like activity by light density mononuclear cells.

MATERIALS AND METHODS

Mononuclear cells were separated from 20-30 ml of heparinized peripheral blood by 30-min centrifugation at 400 g in Ficoll-Hypaque (Pharmacia Fine Chemical AB, Uppsala, Sweden). The cells from the interface were washed two times with alpha medium supplemented with 2% heat-inactivated fetal calf serum (FCS), enumerated, and then diluted with alpha medium to a 3 x 10^6 cells/ml for immediate plating or to 6 x 10^6 cells/ml for separation of light density non adherent cell population and preparation of conditioned medium.

The methylcellulose culture system used was modified from that described by Ogawa et al.[5]. The cells were plated in Multiwell Tissue Culture plates (0.5 ml/well) in a final concentration 3 x 10^5 cells/ml. The plates were incubated at 37°C in 5% CO_2 in air at 100% humidity. Colonies were counted at day 14 under dissecting or inverted microscope.

Light density non adherent cells (LDNA cells) conditioned medium (CM) were obtained as follows: 18 x 10^6 light density mononuclear cells were allowed to adhere to 50 x 15 mm tissue culture dishes in 3 ml of alpha medium with 2% FCS for 24 hrs. At the end of the incubation the non adherent cell suspension was removed, centrifuged and the conditioned medium was harvested and saved for further use. The non adherent cell pellets were resuspended in alpha medium with 2% FCS to a final concentration of 3 x 10^6 cells per ml for plating in methylcellulose culture.

RESULTS

When nucleated cells were plated in 0.5 ml of methylcellulose culture with 1.5 or 3U/ml of erythropoietin, we observed in approximately half of the patients that circulating BFU-E are increased in number up to 112 ± 8.2 of 14 day erythroid colonies were generated per well in the presence of 3U Epo/ml and up to 121 ± 3.9 in the presence of 1.5U Epo/ml (Table I). The remaining patients had a number of 14 day erythroid colonies within the normal range (9 ± 2 to 18.7 ±1.4 colonies per well). The colony size was also visibly larger in the group of patients with a high number of circulating BFU-E than in the second group and controls (data not shown).

SS patients' circulating erythroid progenitors gave rise to a maximum number of 14 day colonies in the presence of 0.3 - 0.75 U Epo/ml of culture; while those of normal individuals required 1.5 - 2.0 U Epo/ml of culture (Fig. 1). Using erythropoietin dose response curves, we determined that 50% stimulatory concentrations of Epo required by circulating BFU-E of SS patients is 0.12 U of Epo/ml, much lower than those of controls (approximately 0.3 U Epo/ml).

While SS circulating BFU-E exhibited increased sensitivity to Epo, they are not able to generate colonies in the absence of added Epo.

In order to test the hypothesis whether the Epo hypersensitivity of circulating BFU-E is real or spurious due to the increased production of BPA like factors we first compared the number of 14 day erythroid colonies generated by LD mononuclear cells with those produced by LD non adherent cells of the same patient. We found that adherent cells depletion leads

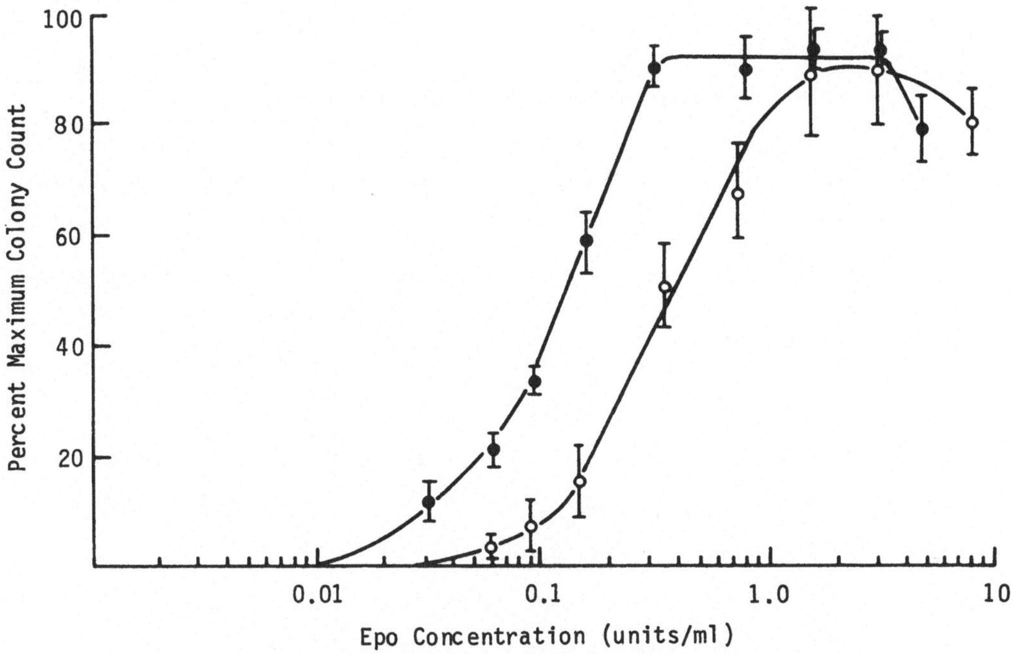

Fig. 1. Erythropoietin dose response curve of 14 days erythroid progenitor cells of SS patients and normal individuals. These results represent the mean ± S.E. of Percent Maximum count of 8 experiments (-●-) and 4 experiments (-○-) [3 x 10^5 light density mononuclear cells were plated in 0.5 ml of culture medium in the presence of various concentrations of erythropoietin. Maximum number of erythroid colonies ranged from 23.3 ± 2.4 to 138.7 ± 12.9 per well.]

to the decrease of the number of 14 day erythroid colonies, so that the Epo sensitivity of 14 day erythroid progenitors appeared similar to that of normal individuals (50% stimulatory concentration of Epo approximately 0.3 U/ml) (data not shown). These results suggested to us that the presence of light density adherent cells stimulate erythroid colony formation at any given Epo concentration and thus the increased Epo sensitivity of SS

Table I. Number of 14 Day Erythroid Colonies Per Well

SS Patients			Normal Individuals		
Exp#	1.5U Epo/ml	3U Epo/ml	Exp#	1.5U Epo/ml	3U Epo/ml
14	9.0 ± 2.0	14.6 ± 2.5	1	ND	11.5 ± 2.1
15	13.6 ± 2.0	17.6 ± 4.9	7	ND	13.5 ± 1.5
16	11.6 ± 1.1	14.0 ± 5.0	17	19.5 ± 2.9	20.0 ± 2.0
21	18.7 ± 1.4	16.3 ± 4.0	26	18.0 ± 1.5	18.0 ± 2.3
27	17.0 ± 2.7	19.0 ± 1.8	30	22.5 ± 9.4	13.2 ± 1.7
31	45.0 ± 4.0	51.2 ± 4.1	35	17.3 ± 4.3	30.0 ± 2.3
32	33.0 ± 2.1	ND	63	9.5 ± 3.6	8.0 ± 3.1
33	67.0 ± 1.5	72.0 ± 2.2			
40	121.0 ± 3.9	112.0 ± 8.2			
47	103.0 ± 3.9	85.0 ± 6.9			
48	70.5 ± 7.2	ND			
62	80.0 ± 14.1	76.0 ± 17.9			
65	40 ±	39.7 ± 4.8			

The number of erythroid colonies generated in experiments 31-65 are significantly higher ($p < 0.001$) than those in experiments 14-27 and controls.

circulating BFU-E is due to the modulation of Epo requirement by increased production of BPA by LDA cells.

To further analyze this observation, we studied the effect of LD mononuclear cells conditioned medium from SS patients or normal individuals on 14 day erythroid colonies formation.

Medium conditioned by LD mononuclear cells (SS or controls) prepared as described in Materials and Methods, was assayed in methylcellulose culture containing (0.75 U Epo/ml) and LD non adherent cells from SS patients or normals, at final concentrations of 5%, 10% and 20%. As shown in Table II, SS CM medium promotes in a dose dependent manner 14 day erythroid colonies formation up to 100% of control value. Stimulatory effect can be seen independently of the source of target cells (SS or controls). In contrast, as shown in Table III, no burst promoting activities could be domonstrated in the conditioned medium by mononuclear cells of normal individuals.

DISCUSSION

These studies present evidence that sickle cell anemia patients are heterogeneous with respect to the number of circulating erythroid progenitor cells. Approximately half of the patients studied had a high number of circulating BFU-E, as previously reported by Ogawa et al.[5], while the remainder appear to have a normal number of circulating BFU-E. Furthermore, we have also noted that the size of colonies generated by mononuclear cells from patients with a high concentration of circulating BFU-E, was larger than those produced by the second group.

The exact significance of circulating progenitor cells in blood is not known. It has been suggested that circulating progenitors represent a population distinct from that found in the bone marrow and may play a specific role in emergency situations as in the response to radiotherapy[7], antigenic stimulation[8], etc.

Table II. The number of 14 day erythroid colonies per well generated by LD non adherent cells of SS patients or normals in the presence of various concentrations of SS-CM and constant amount of Epo (0.75U Epo/ml)

Target cells	0% CM	5% CM	10% CM	20% CM
SS exp 40	82.6 ± 10.0	143.6 ± 9.0	139.0 ± 13.0	175.0 ± 4.0
SS exp 47	65.2 ± 5.0	104.2 ± 6.4	ND	111.2 ± 4.4
SS exp 60	15.6 ± 1.1	20.6 ± 2.0	24.6 ± 2.3	30.0 ± 5.6
SS exp 61	49.0 ± 9.8	58.8 ± 4.7	74.3 ± 2.1	104.3 ± 12.8
SS exp 62	69.0 ± 7.0	69.3 ± 2.5	88.0 ± 3.6	98.3 ± 7.6
Control exp 39	23.5*	23.3*	47.5*	35.3*
Control exp 60	11.3 ± 2.5	11.5 ± 2.2	12.5 ± 2.0	23.0 ± 4.2

*SE is not done because only 2 wells per experiment point.

The number of erythroid colonies generated in the presence of 20% SS CM is significantly different ($p < 0.01$) from those generated in the absence of SS CM.

Table III. The number of 14 day erythroid colonies per well generated by LD non adherent cells of SS patients or normals in the presence of various concentrations of CM from normal individuals and constant amount of EPO (0.75U Epo/ml)

Target cells	0% CM	5% CM	10% CM	20% CM
SS exp 40	82.6 ± 10.0	96.0 ± 6.4	107.0 ± 5.4	77.3 ± 7.8
SS exp 47	65.2 ± 5.0	68.3 ± 4.0	91.5 ± 8.4	88.0 ± 7.0
SS exp 60	15.6 ± 1.1	14.0 ± 2.0	13.0 ± 3.0	13.7 ± 1.7
SS exp 62	69.0 ± 7.0	64.2 ± 10.0	62.7 ± 12.0	64.2 ± 16.2
Control exp 39	23.5*	20.0*	29.0*	30.0*
Control exp 63	11.3 ± 2.5	9.0 ± 2.6	8.0 ± 0.8	9.5 ± 0.7

*SE is not done because only 2 wells per experiment point.

The number of erythroid colonies generated in the presence of different concentrations of CM from normal individuals are not significantly different from those generated in the absence of CM.

The increased concentration of circulating progenitors has also been described in myeloproliferative disorders[8,9] and a correlation between the clinical course and the number of circulating stem cells has been suggested. According to these observations it is possible that the number of circulating BFU-E detected in SS patients may reflect the severity of the disease: in response to the hemopoietic stress of SS anemia, amplification and subsequent release of BFU-E to the circulation may occur. Those patients with high numbers of circulating BFU-E may release progenitors to the circulation from earlier compartments (normally quiescent or slowly proliferative) after amplification. The fact that erythroid colonies generated by mononuclear cells from patients with increased numbers of BFU-E have high proliferative capacity, as demonstrated by the size of the colonies, support this hypothesis.

Epo dose response curves generated by LD mononuclear cells showed that SS 14 day erythroid progenitors exhibit an increased sensitivity to Epo, which seems to be in contradiction with data reported by Nathan and colleagues[10], suggesting that Epo requirement for SS BFU-E stimulation is higher than those for normals, but in agreement with recent observations[11] that BFU-E from β-thalassemic individuals were more sensitive to Epo than controls. The changes in Epo response described in this paper might reflect either the modification of BFU-E itself, differences between BFU-E subpopulations, and/or an increase in production of BPA in the culture. To clarify this point we compared Epo dose response curves of BFU-E generated by mononuclear cells with those produced by light density non adherent cells of the same patients. We observed that peripheral blood adherent cell depletion leads to the shift of the Epo dose response curve, so that the Epo sensitivity of BFU-E significantly decreases. This result suggests: first, that apparent Epo hypersensitivity reflects, in fact, an increased production of BPA-like factors by SS light density adherent cells and second, that some of the circulating SS progenitors need this activity in order to form 14 day colonies.

The results of exchange experiments confirmed that light density adherent cells of SS patients produced constitutively BPA-like activity. When LD non adherent cells from SS patients or normal individuals were plated in the presence of various concentrations of CM by SS LD cells and a constant amount of Epo, a dose dependent increase of the number of 14 day erythroid colonies was observed. When the same target cells were plated in the presence of CM by LD cells of normal individuals such effect does not occur. These data suggest that increased BPA production may play a role in the erythropoietic regulation during constant hemopoietic stress in sickle cell anemia and might partially explain the lower than expected Epo levels in these patients.

In conclusion the data presented here showed that sickle cell anemia affects the regulation of circulating erythroid progenitors and that the phenotypic changes observed may reflect a difference in the severity of SS disease.

ACKNOWLEDGEMENTS

We thank Liz Ezzone for excellent secretarial assistance in the preparation of the manuscript and Kathy Conroy for blood sample collection. This work was supported by Grant #HL21016 and training Grant #T32 HL07556.

REFERENCES:

1. Metcalf, D. 1977. In vitro cloning of normal and leukemic cells. Recent Results in Cancer Research, New York, Springer-Verlag, 1-227.
2. Seiff, C.A. 1987. Heamtopoietic growth factors. J. Clin. Invest. 79: 1549-1557.
3. Lutton, J.D., E.A. Schmalzer, A.N. Rao, S.P. Rao, and R.D. Levere. 1980. Erythroid colony studies on sickle cell anemia in hypoproliferative crisis. Am. J. Hematol. 8: 15-21.
4. Pennathus-Das, R., E. Alpen, E. Vichinsky, J. Carcia, and B. Lubin. 1984. Evidence for the presence of CFU-E with increased in vitro sensitivity to erythropoietin in sickle cell anemia. Blood 63: 1168-1171.
5. Ogawa, M., O.C. Grush, R.F. O'Dell, H.Hara, and M.D. MacEachern. 1977. Circulating erythropoietic precursors assessed in culture: Characterization in normal men and patients with hemoglobinopathies. Blood 50: 1081-1093.
6. Nathan, D.G., B.P. Alter. 1980. F-cell regulation. Ann. N.Y. Acad. Sci. 344: 219-232.
7. Koeffler, H.P., M.J. Cline, and D.W. Golde. 1979. Spenic irradiation in myelofibrosis: Effect on circulating myeloid progenitor cells. Br. J. Haematol. 43: 69.
8. Metcalf, D., M.A.S. Moore. 1971. Haemopoietic cells. Frontiers of biology, North-Holland Publishing Co., 24: 107.
9. Chervenick, P.A. 1973. Increase in circulating stem cells in patients with myelofibrosis. Blood 41: 67.
10. Croizat, H., D. Amato, D.L. McLeod, D. Eskinazi, and A.A. Axelrad. 1983. Differences among myeloproliferative disorders in the behavior of their restricted progenitor cells in culture. Blood 62: 578-584.
11. Johnson, G.R., C.G. Begley, and R.N. Matthews. 1987. Transfusion-dependent β-thalassemia: In vitro characterization of peripheral blood multipotential and committed progenitor cells. Exp. Hematol. 15: 394-405.

PRODUCTION OF ERYTHROPOIETIN BY AN ESTABLISHED HUMAN RENAL CARCINOMA CELL LINE: IN VITRO AND IN VIVO STUDIES

Daniel Shouval[1], and Judith B. Sherwood[2]

[1]The Liver Unit, Department of Medicine A
Hadassah University Hospital
Jerusalem, Israel 91120

[2]Division of Hematology, Department of Medicine
Albert Einstein College of Medicine
Bronx, NY 10461

The inverse relationship between tissue oxygenation and circulating erythropoietin levels has been well documented in clinical and in vivo experimental studies[1]. Although a considerable body of information exists on the correlation between erythropoietin production in vivo and such parameters involved in oxygen supply as hemoglobin concentration, red cell mass, ambient pO_2, and hemoglobin-oxygen affinity, little information exists on regulation of erythropoietin synthesis and secretion at the cellular level.

Cell culture systems have been invaluable in studying the regulation of hormone synthesis and secretion, and several investigators have attempted to develop in vitro model systems that produce erythropoietin on a continuous and reproducible basis. Erythropoietin production has been observed in cultures of normal kidney cells such as goat renal glomeruli[2], rat mesangial cells[3], and organ cultures of rat kidney[4], and transformed cells such as human renal carcinomas[5-8], a human testicular carcinoma[9], and a mouse erythroleukemia cell line[10]. The major problem with both normal and transformed cells in culture is the loss of normal regulatory mechanisms, usually within a few passages from the primary cultures.

As described by Sherwood and Shouval[6], we obtained tumor tissue from a female patient with renal cell carcinoma and erythrocytosis, in July 1981. A cell suspension prepared from the tissue was either placed immediately into tissue culture or injected into athymic mice that were further immunosuppressed by irradiation from a ^{137}Cs source or by injection with rabbit anti-mouse lymphocyte serum (ALS) plus anti-interferon globulin. The nude mouse was chosen as temporary host for the primary human renal tumor in order to provide the best available physiologic conditions for selection and survival of the desired cells. The rationale for the use of the immunosuppressed nude mouse is based on a previous observation[11] that suppression of natural killer (NK) cell activity in nude mice contributes to a significant increase in tumorigenicity of established human carcinoma cell lines.

The tumor cells that were directly established and maintained in tissue culture, without passage through the nude mouse, dedifferentiated after 3-6 weeks of culture. One tumor developed in the athymic mouse that was immunosuppressed by irradiation from a ^{137}Cs source or by injection with rabbit anti-mouse lymphocyte serum (ALS) plus anti-interferon globulin. The nude mouse was chosen as temporary host for the primary human renal tumor in order to provide the best available physiologic conditions for selection and survival of the desired cells. The rationale for the use of the immuno-suppressed nude mouse is based on a previous observation[11] that suppression of natural killer (NK) cell activity in nude mice contributes to a significant increase in tumorigenicity of established human carcinoma cell lines.

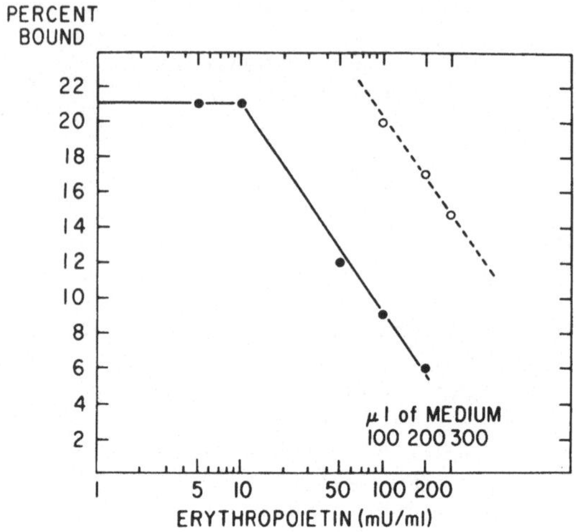

Fig. 1. Radioimmunassay dose-response curves of human urinary erythropoietin (●) and culture supernatant medium from renal carcinoma cell line RC-1(O)[6].

The tumor cells that were directly established and maintained in tissue culture, without passage through the nude mouse, dedifferentiated after 3-6 weeks of culture. One tumor developed in the athymic mouse that was immuno-suppressed by a combination of anti-interferon globulin and ALS. This nude mouse tumor and the human tumor from which it was derived were morphologically similar. Cells obtained from the nude mouse tumor were established in culture; they adapted easily to tissue culture conditions and gave rise to our cell line. Two sublines were cloned by limiting dilution and established in culture; the RC-1 clone produces erythropoietin while the RC-2 clone does not. The RC-1 cell line is a stable, transformed human renal clear cell carcinoma cell line that produces erythropoietin in vitro and that has maintained this function for several years and many passages in culture[6].

The human origin of these cells is supported by the human female karyotype of the RC-1 cloned cells and by the hybridization of DNA from RC-1 cells (passage 37) with a ^{32}P-labeled human genomic probe (second intervening sequence of the human gamma globin gene), under stringent hybridization conditions[6,12]. Electron microscopic analysis of the cloned cells revealed features characteristic of human renal clear cell carcinoma - i.e. lamellar inclusion bodies, elongated mitochondria, and scattered glycogen granules[6]. The pinocytotic vesicles observed in the RC-1 cells are characteristic of the proximal convoluted tubule, which is considered to be the cell type from which the renal carcinoma originates[13]. It is therefore possible that the erythropoietin-producing cell is a tubule cell.

The erythropoietin produced by this cell line cross-reacts with our rabbit anti-human erythropoietin serum. In the radioimmunoassay, displacement curves produced by increasing amounts of RC-1 culture medium were parallel to that produced by human urinary erythropoietin used as standard, suggesting that the tumor erythropoietin is structurally similar to the native human hormone (Fig 1). The tumor factor is biologically active *in vitro* and *in vivo*. In the *in vitro* mouse CFU-E assay[14], media from RC-1 cultures produced a dose-dependent increase in CFU-E number, in contrast to control samples (Table 1)[6].

To study the *in vivo* effect of the tumor erythropoietin, Shouval and Sherwood[15] injected RC-1 and RC-2 cells subcutaneously into nude mice. Both cell lines are highly tumorigenic, producing tumors at relatively low injection densities. Histologic analysis of these tumors showed large vacuolated cells, arranged in a glandular pattern, as described for the clear cell carcinoma in the human host[6,15].

The RC-1 tumor-bearing mice developed significant splenomegaly and some hepatomegaly which was roughly correlated with tumor weight, with a signif-

Table 1. Mouse CFU-E assay of assay of erythropoietin by RC-1 cells in culture.

Sample	Sample volume, μl	CFU-E per 5×10^5 cells*	Mean % of control
Control (saline)	—	68	100
		72	
EPO standard, units			
0.01	—	88	120
		80	
0.05	—	116	167
		120	
0.10	—	124	183
		132	
0.50	—	192	286
		208	
1.0	—	232	331
2.0	—	240	351
		252	
Passage 37 cells†	50	692	991
		696	
	100	1144	1634
		1143	
Human fibroblasts‡	50	64	91
(control cells)	100	76	103
		68	
Control medium	50	192	274
(10% fetal calf serum)	100	164	237
		168	

EPO, erythropoietin.
*Bone marrow cells from BALB/c male mouse in Iscove's medium with methyl-cellulose and 10% fetal calf serum.
†Maintained in DMEM/F-12 medium with 5% fetal calf serum.
‡Maintained in DMEM/F-12 medium with 10% fetal calf serum.

icant increase in spleen weight occurring when the tumors were greater than 400 mg (spleen, 2.7-25.1 fold greater than control mice, with the lower value representing the lowest tumor weight; liver, 1.2-2 fold greater than controls). The RC-2 tumor-bearing mice did not develop splenomegaly or hepatomegaly.

Hemoglobin and hematocrit values increased after RC-1 cell injection (Table 2). In a series of studies, mice were injected with 6 x 10^6 RC-1 cells, and red cell values determined for specific tumor volumes and days after cell implant. We observed a linear correlation between RC-1 tumor volume and weight versus hemoglobin (linear regression coefficient 0.823) and hematocrit (linear regression coefficient 0.850). Mean hemoglobin and hematocrit values (3-5 mice per point) rose from 16 g/dl and 49% (tumor volumes 0.7 cm^3). to 21 g/dl and 62% (tumor volumes 9 cm^3), respectively, within several weeks. Values for the 10 control mice were 16 ± 1 g/dl and 48 ± 3%, respectively. Both blood volumes and red cell mass increased significantly in RC-1 tumor bearing animals as compared to tumor-free controls (2.3-fold greater red cell mass in RC-1 tumor bearing mice as compared with controls and 2-fold greater blood volume in RC-1 tumor bearing mice as compared with PLC/PRF/5 human hepatoma tumor bearing control mice, at 76 days). The linear relationship between tumor mass versus hemoglobin, hematocrit, blood volume, and red cell mass was maintained only as long as tumor volume did not exceed approximately 900 mm^3 (approximately 80-90 days after tumor cell injection); when tumors became larger and necrotic, these values showed unpredictable fluctuations. In contrast, no changes in parameters of ery-

Table 2. Relationship Between Tumor Weight and Erythropoiesis

Mouse No.	Tumor Type	Tumor Weight (gm)	Hemoglobin (g/dl)	Hematocrit (%)
#1	RC-1	0.070	20.7	61
2	"	0.090	19.5	64
3	"	0.430	21.9	70
4	PLC/PRF/5	0.80	-	38
5	"	0.90	-	37.5
6	No Tumor		16.6	49
7	"		16.7	47
8	"		17.3	52
9	"		16.4	51

Athymic mice were injected subcutaneously with 10^7 RC-1 renal carcinoma cells or human hepatoma cells PLC/PRF/5. A control group of 4 mice received no tumor cell injection. 35 days after injection, hemoglobin and hematocrit values were determined, tumor bearing mice and control mice were sacrificed, and tumors excised and weighed.

throcytosis were observed in the control tumor-bearing mice (RC-2 cells and PLC/PRF/5 human hepatoma cells) and in the uninjected control mice, indicating that mechanical factors associated with tumor mass and other experimental manipulations did not affect erythropoietin levels. This correlation provides further evidence that the transplanted tumor cells are secreting erythropoietin and that this human tumor derived erythropoietin is biologically active in nude mice.

Large amounts of erythropoietin were extracted[15,16] from tumors obtained from 6 RC-1 tumor-bearing mice (Table 3). In comparison, similarly-prepared extracts of normal tissue contained significantly less erythropoietin (0.7 units/g of normal rabbit kidney, 0.3-0.6 units/g of normal dog kidney, 0.11 units/g of normal rat kidney), indicating synthesis of erythropoietin by RC-1 tumors.

In these studies[15], we have demonstrated that red blood cell production is progressive and directly proportional to tumor mass in the athymic mouse bearing tumors derived from the RC-1 human renal carcinoma cell line. This unique in vivo model system provides the opportunity to study ectopic erythropoietin production, as well as regulation of in vivo erythropoietin production[15]. Ectopic production of hormones by many different types of tumors has been described, but the mechanism is poorly understood[17].

The studies of Sherwood and Shouval indicate that certain physiologic regulatory processes with respect to erythropoietin production are present in the RC-1 cells. Therefore, we have been utilizing the RC-1 cell line as an in vitro model system to study regulation of erythropoietin synthesis and secretion at the cellular level. When the RC-1 cells from passage 72 were incubated under varying oxygen concentrations, an inverse correlation between oxygen concentration and erythropoietin levels in the culture medium was observed (3% O_2, 1240 ± 100 mU/ml; 18% O_2, 871 ± 134 mU/ml; 98% O_2, 355 ± 17 mU/ml)[6].

Table 3. Tumor Erythropoietin Levels, Hemoglobin, and Hematocrit in RC-1 Tumor-Bearing athymic Mice

Mouse No.	Hematocrit (%)	Hemoglobin (g/dl)	Erythropoietin (units/g tumor)
396	58	18.8	2.97
137	51	18.9	3.00
138	54	19.5	3.20
133	65	20.3	3.35
50	-	-	3.37
97	70	21.9	4.10

Athymic mice were injected subcutaneously with 4-9 x 10^6 RC-1 renal carcinoma cells. 45-60 days after injection, hemoglobin and hematocrit values were determined, the mice were exsanguinated, and an aliquot of tumor tissue was snap frozen at -180°C.

Cyclic 3',5'-adenosine monophosphate (cAMP) is known to regulate the release of many polypeptide hormones in vivo and in vitro[18]. A "first messenger" acts at the cell membrane, usually via a receptor, to stimulate membrane adenylyl cyclase-mediated cAMP production and activation of cAMP-dependent protein kinase[18,19]. Elevated cAMP stimulates the rapid release of hormone[18] within 30 minutes after exposure of cells to the nucleotide, as has been reported for parathyroid hormone[20-22], growth hormone[23,24], thyroid hormone[25], insulin[26], thyroid-stimulating hormone[27], and others.

Sherwood and Shouval have utilized the RC-1 cell line to demonstrate that cAMP mediates the release of erythropoietin in a manner similar to that observed for other hormones[28]. Erythropoietin production by our renal carcinoma cell line was stimulated by 8-bromo-cAMP and by 3-isobutyl-1-methylxanthine (MIX). MIX, a cyclic nucleotide phosphodiesterase inhibitor, prevents the degradation of intracellular cAMP, thus allowing its accumulation[21]. The primary mechanism of action of 8-bromo-cAMP is protein kinase activation[29].

As shown in Table 4[28], when cells were incubated in medium ± 10^{-4} M cAMP for periods of 1-4 hours, erythropoietin levels in the supernatant media from cAMP-treated cells were significantly higher ($p<0.02$) than in the media from control cultures without cAMP. Converseley, erythropoietin levels in the cell pellets - i.e. intracellular erythropoietin - were significantly lower ($p<0.02$) in the cAMP-treated cultures than in the controls. Likewise, 4-hour incubation of the RC-1 cells in the presence of 10^{-4} M MIX was associated with erythropoietin levels in the supernatant media that were significantly greater than in the control cultures. As shown in Figure 2[28], cAMP produced a rapid and enhanced release of erythropoietin within 15 minutes of exposure of the cells to the nucleotide. Erythropoietin levels continued to increase for approximately 30 minutes, after which the levels in the medium remained relatively constant, not significantly different from the 30 minute value. In contrast, no erythropoietin was detected in the control culture media until after 2 hours of incubation (this unstimulated hormone release suggests some constitutive production).

It is important to ask if regulation of erythropoietin production in the RC-1 cell line is similar to that in normal, untransformed kidney cells. Therefore, we performed similar cAMP studies on normal rabbit renal cortical cells[30]. Erythropoietin release from both the normal cells and the RC-1 cells was stimulated by the same concentration range of cAMP (10^{-2} - 10^{-4} M cAMP). The response to cAMP involved a rapid and enhanced release of erythropoietin that occurred within 30 minutes of exposure of the cells to the nucleotide. The control cells did not release erythropoietin within this time period[30].

The short-term effects of cAMP, increased erythropoietin in the medium within 15-30 minutes after cAMP stimulation associated with a concomitant depletion of intracellular erythropoietin, are consistent with secretion of preformed erythropoietin. The immediate release of an intracellular pool of preformed hormone in response to cAMP stimulation has been extensively reported in other endocrine systems.

These tissue culture studies, demonstrating that the RC-1 cells have normal physiologic regulatory processes with respect to hormone production, suggest the utility of this cell line as an in vitro model system to study the regulation of erythropoietin synthesis and secretion. In addition, the in vivo studies demonstrate that the RC-1 cell line can be utilized to generate true polycythemia in athymic mice in a progressive and reproducible manner. This therefore provides the first reproducible animal model system useful for studies on ectopic erythropoietin production by human renal carcinomas and for studies on the effect of chronic high circulating bioactive erythropoietin levels on host progenitor cells and erythropoietin production.

Table 4. Effect of 8-Bromo-cAMP on Intracellular and Extracellular Erythropoietin

Experiment No.	Cell Passage No.	Effector	Sample Assayed	Incubation (hours)	Erythropoietin (mU per 10^6 Cells)
1	37	cAMP	Supernatant	4	110 ± 1
			Cells		118 ± 39
		Control	Supernatant	4	ND*
			Cells		310 ± 38
2	86	cAMP (10^{-4} mol/L)	Supernatant	1	605 ± 123
			Cells		ND*
		Control	Supernatant	1	111 ± 17
			Cells		136 ± 40
3	87	cAMP (10^{-4} mol/L)	Supernatant	2	1143 ± 88
			Cells		37 ± 5
		Control	Supernatant	2	71 ± 18
			Cells		540 ± 40

Cells were incubated at confluence with 10^{-4} mol/L of 8-bromo cAMP in 2 mL of F12:Dulbecco's modified Eagle's medium. Erythropoietin values were determined in the radioimmunoassay. Each value represents the mean ± SEM for three plates, each measured in triplicate. All values are significantly different from values for control cells.

*ND, not detectable, <10 mU/mL. Control medium without cells had no detectable erythropoietin.

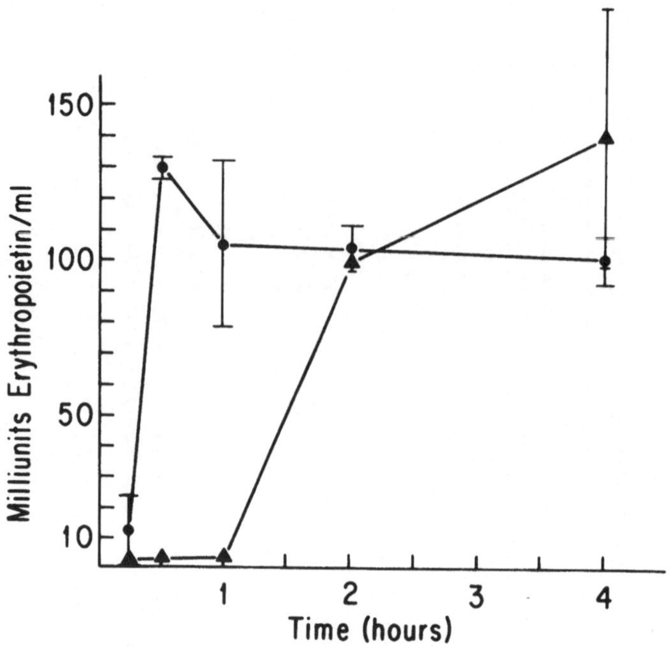

Fig. 2 Effect of cAMP on the time course of erythropoietin secreation by the human renal carcinoma cell line[28].

Cells from passage 105 were grown to confluence (approximately 6 x 10^6 cells/60 mm plate) and incubated in standard culture medium with (●-●) or without (▲-▲) 10^{14} M B-bromo cAMP. At the times shown, the supernatant medium was harvested from 3 replicate plates and the erythropoietin levels measured in the radioimmunoassay.

REFERENCES

1. Krantz, S.B., and L. Jacobson. 1970. Erythropoietin and the Regulation of Erythropoiesis. Univ. of Chicago P., Chicago.
2. Burlington, H., E.P. Cronkite, U. Reinecke, and E.D. Zanjani. 1972. Erythropoietin production in cultures of goat renal glomeruli. Proc Natl. Acad. Sci. USA. 69: 3547-3550.
3. Kurtz, A., W. Jelkmann, F. Sinowatz, and C. Bauer. 1983. Renal mesangial cell cultures as a model for study of erythropoietin production. Proc. Natl. Acad. Sci. USA. 80: 4008-4011.
4. Sherwood, J.B., S.H. Robinson, L.R. Bassan, S. Rosen, and A.S. Gordon. 1972. Production of erythrogenin by organ cultures of rat kidney. Blood 40: 189-197.
5. Sherwood, J.B., and E. Goldwasser. 1976. Erythropoietin production by human renal carcinoma cells in culture. Endocrinology 99: 504-510.
6. Sherwood, J.B., and D. Shouval. 1986. Continuous production of erythropoietin by an established human renal cell line: Development of the cell line. Proc. Natl. Acad. Sci. USA. 83: 165-169.
7. Hagiwara, M., I-Li Chen, R. McGonigle, B. Beckman, F.H. Kasten, and J. M. Fisher. 1984. Erythropoietin production in a primary cul-

ture of human renal carcinoma cells maintained in nude mice. Blood 63: 828-835.

8. Sytkowski, A.J., J. P. Richie, and K.A. Bicknell. 1983. New human renal carcinoma cell line established from a patient with erythrocytosis. Cancer Res. 43: 1415-1419.

9. Ascensao, J.L., F. Gaylis, D. Bronson, E.E. Fraley, and E.D. Zanjani. 1983. Erythropoietin production by a human testicular germ cell line. Blood. 62: 1132-1143.

10. Choppin, J., C. Lacombe, N. Casadevall, D. Muller, P. Tambourin, and B. Varet. 1984. Characterization of erythropoietin produced by IW32 murine erythroleukemia cells. Blood 64: 341-347.

11. Shouval, D., B. Rager-Zisman, P. Quan, D.A. Shafritz, B.R. Bloom, and L.M. Reid. 1983. Role in nude mice of interferon and NK cells in inhibiting tumorigenicity of human hepatocellular carcinoma cells infected with hepatitis B virus. J. Clin. Invest. 72: 707-717.

12. Southern, E.M. 1975. Detection of specific sequences among DNA fragments separated by gel electrophoresis. J. Mol. Biol. 98: 503-517.

13. Firminger, H.I. 1975. Atlas of Tumor Pathology. In: Armed Forces Institute of Pathology, Wash. D.C.

14. Iscove, N.N., F. Sieber, and K.H. Winterhalter. 1974. Erythroid colony formation in cultures of mouse and human bone marrow: analysis of the requirement for erythropoietin by gel filtration and affinity chromatography on agarose-concanavalin A. J. Cell Physiol. 83: 309-320.

15. Shouval, D., M. Anton, E. Galun, and J.B. Sherwood. 1988. Erythropoietin induced polycythemia in athymic mice following transplantation of human renal carcinoma cell line. Cancer Research 48: In Press.

16. Sherwood, J.B., and E. Goldwasser. 1978. Extraction of erythropoietin from normal kidneys. Endocrinology 103: 866-870.

17. Sherwood, L.M. 1984. Ectopic hormone syndromes. In: Contemporary Endocrinology, Vol. II. S.H. Ingbar, ed., Plenum Press, New York. 345-402.

18. Sutherland, E.W. 1970. On the biological role of cAMP. JAMA 214: 1281-1288.

19. Brown, E.M., and G.D. Aurbach. 1980. Role of cyclic nucleotides in secretory mechanisms and actions of parathyroid hormone and calcitonin. Vitamins Hormones 38: 206.

20. Abe, M., and L.M. Sherwood. 1972. Regulation of parathyroid hormone secretion by adenyl cyclase. Biochem. Biophys. Res. Commun. 48: 396-401.

21. Brown, E.M., S. Hurwitz, and G.D. Aurbach. 1977. Beta-adrenergic stimulation of cyclic AMP content and parathyroid hormone release from isolated bovine parathyroid cells. Endocrinology 100: 1696.

22. Williams, G.A., G.K. Hargis, E.N. Bowser, W.J. Henderson, and N.J. Martinez. 1973. Evidence for a role of adenosine 3',5'-monophosphate in parathyroid hormone release. Endocrinology 92: 687--691.

23. Kraicer, J., and A.E.H. Chow. 1982. Release of growth hormone from purified somatotrophs: Use of perifusion system to elucidate interrelations among calcium, adenosine 3',5'-monophosphate, and somatostatin. Endocrinology 111: 1173-1180.

24. Schofield, J.G. 1967. Role of cyclic 3',5'-adenosine monophosphate in the release of growth hormone in vitro. Nature 215: 1382-1383.

25. Tonoue, T., and J. Kitoh. 1978. Release of cyclic AMP from the chicken thyroid stimulated with TSH in vitro. Endocrinol. Jpn. 25: 105-109.

26. Robison, G.A., R.W. Butcher, and E.W. Sutherland. 1971. Cyclic AMP. In: Academic Press, Orlando, Florida.

27. Posternak, T., and G. Cehovic. 1971. Derivatives and analogues of cyclic nucleotides. Ann. N.Y. Acad. Sci. 185: 42-49.
28. Sherwood, J.B., E.R. Burns, and D. Shouval. 1987. Stimulation by cAMP of erythropoietin secretion by an established human renal carcinoma cell line. Blood 69: 1053-1057.
29. Free, C.A., M. Chasin, V.S. Paik, and S.M. Hess. 1971. Steroidogenic and lipolytic activities of 8-substituted derivatives of cyclic 3',5'-adenosine monophosphate. Biochem. 10: 3785-3789.
30. Sherwood, J.B. 1985. cAMP-stimulated release of erythropoietin by normal and neoplastic renal cells. Blood 66: 161a (Abstr.)

CONCLUDING REMARKS AND FUTURE DIRECTIONS

Alan S. Levine

Blood Diseases Branch
Division of Blood Diseases and Resources
National Heart, Lung, and Blood Institute
National Institutes of Health

The organizers and sponsors are to be commended for assembling this high quality, exciting, and enjoyable symposium: the third in a series of scientific interchanges spanning various aspects of hematopoiesis. Drs. Nader G. Abraham, Joao Ascensao, Mehdi Tavassoli, Esmail Zanjani, and especially Richard D. Levere and the Department of Medicine, New York Medical College deserve special recognition for this successful event to commemorate the 100th Anniversary of the National Institutes of Health.

The celebration of the National Institutes of Health Centennial Year began one year ago, and it is most appropriate that the symposium concludes a twelve-month period of festivities and scientific achievement. The symposium began with a perspective of the historical role that the National Institutes of Health has played in the partnership between academia and the Federal Government in biomedical research and improving the health of the people of the world. The NIH campus consists of dozens of buildings, most of which are situated on more than three-hundred acres in Bethesda, Maryland. Almost a half a century ago, President Franklin D. Roosevelt visited the campus on the occasion of the dedication of the buildings and grounds of the National Institutes of Health. His historic speech marked the beginning of this biomedical research partnership and of the NIH as we know it today. A recording of the full speech was played at the Opening Centennial Ceremonies at the NIH one year ago. Since this Symposium is in celebration of the NIH Centennial, it is appropriate to share an excerpt of this speech in this forum. Note the striking similarities of the policy issues and concerns of almost fifty years ago with those of contemporary times.

> ...Ladies and gentlemen, no where in the world, except in the Americas, is it possible for any nation to devote a great sector of its effort to life conservation rather than life destruction. All of us are grateful that we in the United States can still turn our thoughts and our attention to those institutions of our country that symbolize peace; institutions whose [mission] is to save life, and not to destroy it. It is for the dedication of these noble buildings to the service of man, that we are assembled here today. The National Institute of Health speaks the universal language of humanitarianism.

> It has been devoted throughout its long and distinguished history to furthering the health of all mankind; in which service it has recognized no limitations by international boundaries, and has recognized no distinctions of race, or creed or color. The total defense that we have heard so much about of late, that total defense which this Nation seeks, involves a great deal more than building airplanes, and ships, and guns, and bombs, for we cannot be a strong Nation unless we are a healthy Nation. So we must recruit not only men and materials, but also knowledge and science in service of national strength, and that is what we are doing here. . .
>
> President Franklin Delano Roosevelt
> October 31, 1940

Ten or fifteen years ago, if someone were to have asked you why you are so driven to undertake the kinds of laboratory investigations you enjoy, your response might have been, that some day, we may be able to treat debilitating diseases that are currently untreatable. Some day, we may be able to cure fatal genetic disorders. That "some day" is now, as evidenced by the truly remarkable advances reported at these proceedings and in the literature very recently. Dr. Zanjani presented data for *in utero* transplantation of hematopoietic stem cells, where fetal liver cells from a preimmune sheep fetus were transplanted into an unrelated preimmune sheep fetus. After birth, the hemoglobin isoelectric focusing pattern demonstrated chimerism. Even at nine months of age, the chimera contained a mixture of its pretransplant type B hemoglobin and the type A hemoglobin of the donor. This was accomplished without immunosuppression, without graft rejection, and without developing graft-versus-host disease. We saw a slide depicting a healthy and contented nine month old sheep. Along the same lines, the participants were shown a slide of a photograph of the first live birth of a rhesus monkey chimera that had been transplanted *in utero* with fetal liver cells from an unrelated preimmune fetus. Here again, no signs of graft-versus-host disease were seen at six months of age.

As early as 1960, the then National Heart Institute recognized the central role erythropoietin played in controlling red blood cell mass. However, at that time, little was known about the structure, function, and mechanism of action of this hormone. Knowledge of the physiology and biology of erythropoiesis was in its infancy. In an effort to encourage further research on erythropoiesis, the National Heart Institute took steps to make this precious substance more readily available to investigators. The Institute obtained extracts of human urinary erythropoietin from South America, arranged for its partial purification in the United States, and distributed it free to investigators worldwide, based on scientific merit as determined by peer review. This important program continued for almost twenty-five years, at which time, largely through the encouragement of the National Heart, Lung, and Blood Institute, the private sector assumed responsibility for providing erythropoietin and enhancing its supply through biotechnology.

This biomedical research "partnership" has certainly been fruitful. Recent reports in the literature of small scale clinical trials have suggested that recombinant human erythropoietin may be safe and effective in treating the anemia of end stage renal disease[1]. The results were remarkable in that all patients given erythropoietin responded and transfusion-dependent patients no longer required blood transfusions. Large multicenter trials are currently in progress.

At this meeting Dr. Golde provided an update of his exciting collaborative studies[2] indicating that recombinant GM-CSF can be safely administered to AIDS patients, resulting in dose-dependent increases in circulating leukocytes, neutrophils, eosinophils, and monocytes. More important, recombinant GM-CSF appears to correct defects in neutrophil function, which may provide the essential enhancement of host defense against infection. Interestingly enough, he reported beneficial results in aplastic anemia patients as well, but without any of the side effects that were observed in the AIDS patients.

Dr. Anderson presented remarkable data demonstrating successful gene therapy for adenosine deaminase deficiency in a primate model. His preliminary safety data are most encouraging and as soon as additional evidence of the safety of this protocol is obtained, plans will be made for applying this procedure in a baby with severe combined immunodeficiency caused by adenosine deaminase deficiency.

Although progress has been substantial and rapid in both basic and clinical aspects of hematopoiesis research, many major challenges remain to be addressed. It is important to determine the mechanism of action of the many hematopoietic growth factors on specific progenitor cell populations. This problem needs to be examined at the cellular level, at the molecular level, especially at the level of receptor binding, and at the genomic level. Is there second messenger signaling? Do hematopoietic growth factors "turn on" other hematopoietic growth factors or in some way influence gene expression? In order to address these questions in a systematic fashion, there is a need to improve techniques for enriching or even purifying progenitor cells. More widespread use of serum-free culture systems are essential for advancing our understanding of the interactions of growth factors and their target cells.

There is a need to determine the role of the hematopoietic microenvironment in blood cell production. Likewise, what is the role of the extracellular matrix, especially collagen, in hematopoiesis? Are endothelial cells important in controlling hematopoietic events?

Much work remains to be accomplished in the area of clinical applications of hematopoietic growth factors. Large scale studies need to be undertaken to determine the long-term efficacy and safety of recombinant human GM-CSF in AIDS patients. We saw data suggesting that GM-CSF is useful in treating aplastic anemia and this too should be followed-up. Long term studies are needed to determine if there is a decrease in morbidity and mortality due to infections. As pointed out by Dr. Golde, it is important to find out how GM-CSF works since it cannot be detected in the circulation. Many AIDS patients being treated with AZT have developed transfusion-dependent anemia. There is now a need to determine if recombinant erythropoietin will prevent anemia in these patients.

Are there other cytopenic conditions that would be responsive to treatment with GM-CSF, for example, Fanconi's anemia? Will GM-CSF be useful in speeding-up marrow reconstitution after bone marrow transplantation. If so, will this improve survival by shortening the period of immunoincompetence? It is important to determine whether other cloned growth factors, such as the multipotent interleukin-3, granulocyte colony-stimulating factor (G-CSF), and macrophage colony-stimulating factor (M-CSF), have a role in the clinical management of cytopenic conditions. If so, would "cocktails" consisting of two or more lineage-specific growth factors, be useful for stimulating the production of multiple cell types?

It is now well established that AIDS patients exhibit bone marrow dysplasia affecting one or more of the cell lines. Opportunities now exist for conducting basic research to determine the molecular and cellular mechanisms of HIV-induced bone marrow damage. There is a need to explore the effects of various recombinant growth factors on HIV-infected marrow.

Recombinant human erythropoietin appears to be efficacious in treating the anemia of end stage renal disease. However, it is important to determine if other anemias will be responsive to high levels of erythropoietin. Will it be useful prior to elective surgery, thereby enabling the collection of multiple units of blood in a short period of time for autologous transfusions?

In adults, fetal hemoglobin synthesis appears to increase under conditions of rapid erythroid regeneration. Therefore, it has long been postulated that high doses of erythropoietin may induce rapid erythroid production and a preferential synthesis of fetal hemoglobin. The availability of large quantities of pure recombinant erythropoietin has enabled this hypothesis to be tested and it was recently shown that fetal hemoglobin production in anemic or non-anemic baboons increases significantly upon administration of high doses of recombinant human erythropoietin[3]. It is now urgent to determine whether erythropoietin will induce similar increases in fetal hemoglobin in patients with sickle cell anemia or beta-thalassemia, and to determine whether these increases will favorably affect the clinical course of these hemoglobinopathies.

There are major hurdles to cross in the area of human gene therapy. We must improve vector infectivity. Efficient vectors developed for one species do not perform well in another species. Will vectors that function well in animal systems be useful in human protocols? There is a need to improve gene expression and ultimately to be able to regulate gene expression in order to correct hemoglobinopathies where precise balance between alpha and beta globin gene expression is essential. These two genes reside on different chromosomes and, therefore, extensive efforts will be required to solve the problem of regulation in this system. Our knowledge of the role of trans-acting factors is in its infancy. These proteins or peptides may hold the key to gene expression and regulation. At these proceedings, Dr. Bank discussed the intriguing concept of positive and negative trans-acting factors. For example, in K562 cells, where fetal and embryonic (but not adult) globin genes are active, there may be a lack of a positive trans-acting factor needed for beta-globin expression; or perhaps the presence of a negative trans-acting factor that inhibits beta-globin gene expression.

Finally, it is most important to demonstrate the safety of somatic cell gene therapy. Dr. Anderson's safety studies in primates, as reported in these proceedings, are indeed very encouraging. His data indicate it appears likely that even massive doses of his viral vector, injected directly into the blood stream of the animal, causes no ill effects. Studies such as this must continue to be pursued.

The Division of Blood Diseases and Resources of the National Heart, Lung, and Blood Institute is anxious to provide support for continued investigations addressing these, and other, important issues. The Division supports a broad spectrum of research activities in hematology and transfusion medicine, from basic research through clinical trials and technology transfer activities. In Fiscal Year 1986, the Division provided almost $125,000,000 in support of hematology research through a variety of grant and contract mechanisms. However, the primary emphasis continues to be on the investigator initiated research grant, where approximately one-third of approved grant applications receive support.

Equally important to the Division of Blood Diseases and Resources is the support of research training and research career development programs. The Division is proud of its record of accomplishment in this area, where traditionally at least fifty percent of approved applications receive funding. Support under these programs is available for investigators throughout various stages of their careers. The Division of Blood Diseases and Resources is especially interested in supporting scientists under the Research Career Development Award (RCDA or K04 program), the Clinical Investigator Award (CIA or K08 program), and the Physician Scientist Award (PSA or K11 program). An overview of these and other research training and career development programs, as well as the scientific programs of interest to the Division of Blood Diseases and Resources, is available[4].

REFERENCES

1. Eschbach, J.W., J.C. Egrie, M.R. Downing, J.K. Browne, and J.W. Adamson. 1987. Correction of the anemia of end-stage renal disease with recombinant human erythropoietin. N. Engl. J. Med. 316: 73-78.
2. Groopman, J.E., R.T. Mitsuyasu, M.J. DeLeo, D.H. Oette, and D.W. Golde. 1987. Effect of recombinant human granulocyte-macrophage colony-stimulating factor on myelopoiesis in the acquired immunodeficiency syndrome. N. Engl. J. Med. 317: 593-598.
3. Al-Khatti, A., R.W. Veith, T. Papayannopoulou, E.F. Fritsch, E. Goldwasser, and G. Stamatoyannopoulos. 1987. Stimulation of fetal hemoglobin synthesis by erythropoietin in baboons. N. Engl. J. Med. 317: 415-420.
4. Research and Academic Career Development Opportunities in Hematology. Division of Blood Diseases and Resources, National Heart, Lung, and Blood Institute: 1987.

INDEX